How Economics Shapes Science

HOW ECONOMICS SHAPES SCIENCE

PAULA STEPHAN

HARVARD UNIVERSITY PRESS
Cambridge, Massachusetts
London, England
2012

Library of Congress Cataloging-in-Publication Data

Stephan, Paula E.
How economics shapes science / Paula Stephan.
p. cm.
Includes bibliographical references and index.
ISBN 978-0-674-04971-0 (alk. paper)
1. Research—Economic aspects. 2. Science and state. I. Title.
HC79.R4S74 2012
500—dc23   2011013433

For Bill, always for Bill

Contents

Figures and Tables

Figures

Tables

Preface

THIS IS A BOOK that explores what economics has to do with science. The book also explores how science affects the economy, especially economic growth. Because much of public research occurs at universities and medical schools, especially in the United States, much of the book's focus is on how research is conducted and supported at universities. It is also about the consequences for universities of having the research enterprise—at least in the United States—so fully embedded in the university.

This is not to say that economics has a monopoly when it comes to factors that affect science or in providing a lens for examining science. Other disciplines—and their foci—contribute considerably to the study of science. Sociology, for example, contributes a great deal to the understanding of how science is organized and the reward structure of science. It is also not to say that science is the only factor that contributes to economic growth. Politics and values, for example, clearly play important roles.

Despite the title, the book draws on research and insights from several disciplines. Indeed, one of the factors that led me to study science was the opportunity to indulge my interest in and penchant for reading outside my—sometimes overly narrow—discipline of economics.

Some of the discussion in the book is highly descriptive, summarizing what is known about the various players and factors that influence research behavior and outcomes. This descriptive nature is by design. Throughout my thirty-plus years of studying science, I have been amazed at the number of people who venture to write about science and science policy without understanding the environment in which research takes place. One of my goals in writing this book is to lay out the scientific landscape in what I hope to be a somewhat engaging manner, so that those who wish to continue the study of the economics of science (and I am happy to say there are a growing number) can approach it with a more solid footing. I also hope to offer, from time to time, questions that warrant further research. I do not mean by this that I see myself as the first to examine these issues, and I certainly don't see myself as the most proficient. Far from it: my work—and that of other scholars in the field—owes an enormous debt to the luminaries who began the field a generation (or half a generation) before I began doing research in the area. They include Kenneth Arrow, Paul David, Zvi Griliches, Robert K. Merton, Richard Nelson, and Nathan Rosenberg.

But I did not only—or primarily—write the book for my peers or their students. I also wrote it for the considerable community that works at public research institutions, be they in the United States, China, Europe, or Japan. I also wrote it for policy makers, as well as for members of the general public who share an interest in the workings of public institutions and the study of science. It is my hope that a greater understanding of how economics shapes science can lead to more effective science policy and a better use of resources in the research enterprise.

Abbreviations

AAMC Associations of American Medical Colleges

AAU American Association of Universities

ANR L'Agence nationale de la recherche (France)

APS Advanced Photon Source, Argonne National Laboratory

ARRA American Recovery and Reinvestment Act

AUTM Association of University Technology Managers

BLS Bureau of Labor Statistics

CERN European Organization for Nuclear Research

CIS Community Innovation Surveys (Europe)

CMS Compact Muon Selanoid (at CERN)

CPS Current Population Survey

CNRS Centre national de la recherche scientifique (National Center for Scientific Research, France)

DARPA Defense Advanced Research Projects Agency (U.S.)

DGF Direct government funds

DOD Department of Defense (U.S.)

DOE Department of Energy (U.S.)

E-ELT European Extremely Large Telescope

ERC European Research Council

FIRB Fund for Investing in Fundamental Research (Italy)

GMT Giant Magellan Telescope

GRE Graduate Record Examination

GUF General university funds

H-1B visa A nonimmigrant visa that allows U.S. employers to hire noncitizens on a temporary basis in occupations requiring specialized knowledge hESC   The human embryonic stem cell policy implemented under President George W. Bush in 2001

HGP   Human Genome Project

HHMI   Howard Hughes Medical Institute

ITER   International Thermonuclear Experimental Reactor

LHC   Large Hadron Collider (at CERN)

MOU   Memorandum of Understanding

NASA   National Aeronautics and Space Administration

NIGMS   National Institutes of General Medical Science

NIH   National Institutes of Health

NIST   National Institute of Standards and Technology

NRSA   National Research Service Awards

NSCG   National Survey of College Graduates (Census administered; overseen by NSF)

NSF   National Science Foundation

OECD   Organization for Economic Co-operation and Development

OWL   Overwhelmingly Large Telescope

PSI   Protein Structure Initiative (NIH)

R&D   Research and development

RAE   Research Assessment Exercise (U.K.)

REF   Research Excellence Framework (to replace the Research Assessment Exercise, U.K.)

R01   Research project grant awarded by NIH, it is the agency's oldest grant mechanism used by the NIH to support research; generally investigator initiated.

S&E   Science and engineering

SDR   Survey of Doctorate Recipients (NSF-collected data)

SDSS   Sloan Digital Sky Survey

SED   Survey of Earned Doctorates (NSF-collected data)

SEPPS   National S&E PhD & Postdoc Survey

SER-CAT   Southeast Regional Collaborative Access Team

SKA   Square Kilometer Array

SMSA   Standard Metropolitan Statistical Area

Study Section   Scientific review groups at NIH, primarily made up of nongovernment experts

TMT   30-meter telescope

TTO   Technology Transfer Office

How Economics Shapes Science

CHAPTER ONE

What Does Economics Have To Do with Science?

THIS IS A BOOK about how economics shapes science as practiced at public research organizations. In the United States these are primarily universities and medical schools. But in Europe and Asia a considerable amount of public research is conducted at research institutes. The book's focus reflects the strong role that public research organizations play in creating knowledge. In the United States, for example, approximately 75 percent of all articles published in scientific journals are written by scientists and engineers working at universities and medical schools.[1] Of equal importance, almost 60 percent of basic research is conducted at universities and medical schools.[2]

What does economics have to do with science? Plenty, it turns out. Economics, after all, is the study of incentives and costs, of how scarce resources are allocated across competing wants and needs. Science costs money and incentives play a key role in science. At the extreme end of the cost spectrum is the Large Hadron Collider (LHC), which came on line (for the second time) in the fall of 2009 and cost approximately $8 billion (U.S.).[3] But there are numerous other examples. The personnel costs of a typical university lab with eight researchers is about $350,000 after fringe benefits but before taking into account the cost of the principal investigator's time or indirect costs.[4] Public research organizations routinely spend large sums of money building and maintaining research facilities and large sums of money on start-up packages for faculty hired to work in the new facilities. In recent years, these packages have become sufficiently large that a university routinely spends four to five times as much on the package as on the faculty member's annual salary.[5] Even mice, the ubiquitous research animal, can cost a substantial amount to buy and keep. Custom-made mice, designed with a predisposition to a specific disease or problem, such as diabetes, Alzheimer's disease, or obesity, can cost in the neighborhood of $3,500. The daily cost of keeping a mouse is around $0.18. Sounds cheap—until one realizes that some researchers keep a sufficient number of animals that the annual budget for mouse upkeep can be in excess of $200,000.[6]

The amount of money spent on scientific research in the public sector is substantial. The United States spends between 0.3 and 0.4 percent of its gross domestic product (GDP) on research and development at universities and medical schools. This represented almost $55 billion dollars in 2009 or approximately $170 per person.[7] While most other countries spend a smaller percent of GDP, several countries, including Sweden, Finland, Denmark, and Canada, spend a considerably higher percentage of their GDP on research and development at universities and medical schools.[8]

Costs

Costs affect the way research is conducted. Costs were a major factor in Europe's decision to settle for building the Exceedingly Large Telescope (E-ELT) rather than the Overwhelmingly Large Telescope (OWL)—with its much larger mirror—as originally planned.[9] Costs can derail large projects or at best delay them. Original plans called for the multi-billion-euro fusion reactor ITER to begin operation in 2016. Now the earliest that ITER can become operational is in 2018—and if it does become operational at that time, it will be a stripped-down version; additional components will be needed for power-producing plasmas.[10] Along the way, the costs of constructing ITER keep rising. New cost calculations made public in the spring of 2010, for example, suggest that Europe's contribution will be 2.7 times greater than the amount originally estimated; that of the United States will be about 2.2 times greater.[11]

Costs play a role in determining whether researchers work with male mice or female mice (females, it turns out, can be more expensive), whether principal investigators staff their labs with postdoctoral fellows (postdocs) or graduate students, and why faculty prefer to staff labs with "temporary" workers, be they graduate students, postdocs, or staff scientists, rather than with permanent staff. High electricity costs dictate that the LHC not run in the winter but rather during the rest of the year when electricity is considerably less expensive.[12] Costs are a major factor in determining what equipment at a university will be "core" and shared across labs rather than belonging to a specific lab. Costs—and the desire to minimize risk—have played a major role in the decision of universities to substitute non-tenure-track faculty for tenure-track faculty.

Costs affect the pace of discovery. When the human genome project began in 1990, it cost more than $10.00 to sequence a base pair. Sequencing costs fell rapidly, hitting less than a penny a base pair by 2007. That is now ancient history: since then, new generations of sequencing technology have been developed that have lowered the cost dramatically. Before this book sees the light of day, it is possible that the Archon X Prize for Genomics will be awarded to the first group to "build a device and use it to sequence 100 human genomes within 10 days or less . . . at a recurring cost of no more than $10,000 per genome."[13]

Incentives

Universities respond to incentives. In the early 2000s, universities went on an unprecedented building spree, developing new research facilities in the biomedical sciences. Within less than five years, construction and renovation costs for biomedical research facilities accelerated from $348 million annually to $1.1 billion annually at U.S. medical schools. (All figures are in 1990-adjusted dollars.)[14] The reason: the budget for the National Institutes of Health, the major funder of research in the biomedical sciences, doubled between 1998 and 2003, opening a panoply of what universities perceived to be new opportunities to expand their research efforts and, in the process, enhance their reputation. It was not the first time that U.S. medical schools responded to financial incentives. The substantial expansion of medical colleges over the past 40 years is widely attributed to the adoption of Medicare and Medicaid in 1965, which provided university medical schools with a new source of revenue.

Scientists and engineers respond to incentives as well. Money, despite statements to the contrary, is not unimportant. Actions speak louder than words. Scientists routinely move to take more lucrative-paying positions. A number of public universities have lost faculty in recent years because private universities, especially before the financial collapse of 2008, could often offer much more lucrative packages than their public sisters. Indeed, in the 2009–2010 academic year, only one public institution (UCLA) was among the top twenty research universities in terms of salaries paid to full professors—and it held the 20th position, paying $43,000 less than top-paying Harvard. Phones began to ring at Berkeley in 2009 soon after the California system imposed a substantial pay cut on its faculty. Full professors at Berkeley already earned about 25 percent less than their peers at Harvard and Columbia. Now they would earn even less.[15]

Scientists respond to incentives in choosing where to submit articles for publications. The number of articles submitted to the journal *Science,* for example, is significantly related to whether the scientist's home country offers a bonus or other monetary reward for publishing in the journal.[16] In some instances, the bonuses can be quite large—on the order of 20 to 30 percent of the scientist's base salary.

Financial incentives encourage university faculty to start new companies based on their research. In recent years, a number of scientists have made substantial sums of money by forming start-up companies or by receiving royalties from universities licensing patents on which they are an inventor. David Sinclair, a Harvard professor and founder of Sirtris Pharmaceuticals, received more than $3.4 million for the shares he held in Sirtris when Glaxo acquired the company in 2008. Robert Tjian received millions in 2004 when Tularik, the company he cofounded when he was a faculty member at the University of California–Berkeley, was sold to Amgen for $1.3 billion. Stephen Hsu, a professor of physics at the University of Oregon, received a substantial amount when Symantec paid $26 million in cash in 2003 for one of two software companies he had founded. László Z. Bitó, whose work led to the invention of the drug Xalantan for the treatment of glaucoma, has earned several million a year from the patent that Columbia University held on the drug. The patent is due to expire in 2011.[17] In 2005, three researchers at Emory University divided more than $200 million when Emory sold its royalty interest in emtricitabine, used in the treatment of human immunodeficiency virus (HIV), to Gilead Sciences and Royalty Pharma. Although rare, events such as these occur with sufficient frequency that, on the campus of almost every research university in the United States, two or three faculty members have become wealthy as a result of their research.

Neither do scientists, especially highly productive scientists, receive a pauper's pay. Full professors at the top of their game employed at private research universities in math earned an annualized salary of $180,000 a year in 2006 in the United States. Comparably ranked full professors at public universities earned $150,000. Those in the biological sciences earned $277,700 at private research universities; those at public universities earned $200,000.[18] It is no wonder that the United States has been a magnet for highly productive European scientists. Not only has there been a tradition of more support for investigator-initiated research in the United States, but salaries are also significantly higher and are based, at least in part, on productivity. By way of contrast, at many European universities and research institutes scientists are civil servants and receive the same (relatively low) pay regardless of performance. In France, for example, a professeur des universités with considerable seniority earns approximately $70,000.[19]

Relative salaries have an impact on who does science. The decline in the propensity (and for many years the number) of U.S. citizens to choose a career in science, particularly men, can be attributed in part to the low salaries scientists and engineers earn relative to the salaries in other occupations. Many of the best and brightest from Harvard routinely have gone to Wall Street. The $277,000 salary is not peanuts, neither is the $180,000 but these salaries come after years of training and hard work. Entry level jobs on Wall Street for freshly minted bachelor's degrees—especially before the crash—paid two-thirds of what the PhD at the top of his game was paid.[20] MBAs from a top program have the prospect of earning slightly more than three times the faculty salary—$559,802, to be precise—after they have been out 10 or more years and started their career in banking.[21]

Increased availability of fellowships for study, as well as an increase in the size of the fellowship, attracts more students into graduate programs. The widespread availability of research assistantships for study in the United States, and the possibility of working in the United States after completing graduate school, have proved to be powerful incentives in luring the foreign born to come to the United States to train.

Not all incentives are monetary. Non-monetary incentives are important to both faculty and institutions. Ask almost any scientist why they became a scientist, and the answer will almost invariably be an interest in solving puzzles. Most scientists derive considerable satisfaction from the "pleasure of finding things out." The enjoyment derived from puzzle solving is part of the reward of doing science. But scientists are also motivated to do science by an interest in recognition. Reputation matters in science. Reputation is built in science by being the first to communicate a finding, thereby establishing priority of discovery. A common way to measure the reputation of a scientist is to count the number of citations to an article or to the entire body of the scientist's work. The *h*-index, a citation-based method for measuring the impact of a scientist's work, has gained considerable use in recent years. Some scientists routinely include their *h*-index in their biographical sketches; others design webpages in which their *h*-index is prominently displayed on the screen.[22] Departments have been known to use the *h*-index to choose among job candidates when making hiring decisions.

The recognition that the scientific community bestows on priority has varied forms, depending on the importance the community attaches to the discovery. At the very top of the list is eponymy, the practice of attaching the scientist's name to the discovery. By way of example, the Richter scale is named for Charles Richter, who, along with Beno Gutenberg, devised the scale while working at Caltech in 1935.[23] The Hubble telescope is named for Edwin Hubble, the astronomer who discovered in 1929 that the universe is expanding. Other examples of eponymy include Haley's comet, the Salk vaccine, Planck's constant, and Hodgkin's disease.

Recognition also comes in the form of prizes. Among these, the Nobel is the best known. But hundreds of other prizes exist, and more are created every year. The Kavli Prize, for example, with its $1 million purse in each of three fields, was awarded for the first time in the fall of 2008 by the King of Norway.[24]

It is not only scientists and engineers who seek reputation. Universities strive to be highly rated, basing their position in the reputational hierarchy on metrics such as faculty research productivity (measured by citation counts or research dollars), number of Nobel laureates, or members of national academies. Their pursuit of status is undoubtedly one reason that, despite complaints that they routinely lose money on research grants, universities continue to urge (some would say pressure) faculty to bring in the grants.[25]

Knowledge as a Public Good

A reward structure that encourages scientists to share their discoveries in a timely manner is highly functional. The reason: knowledge has characteristics of what economists call a public good. It is nonexcludable and nonrivalrous. The classic example in economics of a public good is the lighthouse. It is nonexcludable: once built, anyone can use it. It is nonrivalrous: an additional user does not diminish the amount of light available for others. Parallels can be drawn with knowledge: once research findings are made public, it is difficult to exclude others from using the knowledge. And research findings are not depleted when shared.[26]

Economists have gone to considerable length to show that the market is not well suited for producing goods with such characteristics.[27] The incentives simply are not there. If one cannot limit access, it is difficult to make a profit. Public goods invite free ridership. Consumers can use the good without paying for it. Similar free ridership problems could exist for scientists. Unlike the wine maker, whose customers must pay if they wish to drink his wine, or the baseball team that can sell tickets to its games, the researcher has no way of excluding others from using his research if he makes it public through publication. He has no way of appropriating the monetary benefits. It is particularly difficult to appropriate the benefits of basic research, which at best is years away from contributing to products the market may or may not value. The lack of monetary incentives could lead to what economists refer to as "market failure," with society producing considerably less research than is socially desirable.

"Society, however, is more ingenious than the market."[28] The priority system has evolved in science to create a reward system that encourages the production and sharing of knowledge. The very act of staking a claim requires scientists to share their discoveries with others. By giving it away, scientists make the research findings their own. In the process, they also build their professional reputation, which indirectly leads to financial rewards in the form of higher salaries, consulting opportunities, and, in some instances, membership on scientific advisory committees of publicly traded firms.

This does not mean that scientists give everything away. One can have one's cake and eat it, too. Some research leads to patentable concepts; the findings of other research can be publicly shared while the techniques for doing the research remain somewhat clouded in mystery. Scientists also routinely fail to share materials with colleagues working in a similar area. Reputation is about being first: helping the competition could lead to second place.[29]

The Government's Incentive for Supporting Research

Priority may provide the incentive to do research, but it does not provide the wherewithal to do research. Thus, research, especially of a basic nature, has traditionally been supported by either the government or philanthropic institutions. The government's incentive for supporting scientific research rests partly on the argument that, due to market failure, private firms would not undertake a sufficient amount of research.[30] The public's incentive for supporting research also rests on the importance of research and development for specific outcomes deemed socially desirable and not directly provided by the market, such as better health and national defense. Life expectancy has increased by more than fourteen years since 1940 primarily because of advances in science, such as the development of antibiotics and effective treatments for cardiovascular disease.[31] The gains from increased longevity are substantial. Research suggests that citizens value the benefits associated with increased life expectancy to the tune of $3.2 trillion annually.[32]

Research plays an important role in national defense, as the Manhattan Project made abundantly clear. But there have been numerous other research breakthroughs, such as radar and the development of the electronic digital computer, that have contributed not only to national defense but also have had widespread commercial applications.[33]

Countries also support research because of a desire to win the "Scientific Olympics." Considerable bragging rights are involved in being the first to reach the moon or the first to create induced human pluripotent stem cells. Governments also support research because of humanity's quest for basic understanding. Numerous examples come to mind, but the spectacular images sent from the Hubble Space Telescope after it was repaired in the fall of 2009 are perhaps the best example in recent years. If and when the LHC succeeds in identifying the Higgs boson (what some physicists refer to as "God's particle"), science will have taken a considerable step forward toward knowing the origins of the universe.[34]

The case for public support of research is strengthened by the relationship between research and economic growth. The argument (which by now will sound familiar) goes something like this: economic growth is fueled by upstream research—research that is years away from leading to new products and processes. Moreover, basic research has the potential of having multiple uses, contributing to a large number of areas. Because of the multiuse nature of most basic research, as well as the long time lags between discovery and application, it unlikely that any one individual, company, or industry would support a sufficient amount of basic research to advance innovation at the desired pace. The economic incentives are not there. The findings would spill over, and others, including competitors, could use the knowledge at less than the original cost of producing it. Spillovers are great for growth, but they do not induce market-based institutions to conduct considerable amounts of upstream research. Hence, the government has a role in supporting research in the public sector.

Examples of how research in the public sector has contributed to new products and processes are plentiful. Global positioning devices, which have transformed the way we navigate, would not have been possible without the development of atomic clocks.[35] The idea of using atomic vibration to measure time was first suggested more than 130 years ago by Lord Kelvin in 1879; the practical method for doing so was developed in the 1930s by Isidor Rabi.[36] Hybrid corn, which did much to increase the food supply, was first produced by a faculty member at (what is now)

Michigan State University.[37] Lasers, which have had a profound impact on the fields of communication, entertainment, and surgery, as well as on defense, owe a substantial intellectual debt to the work of a graduate student at Columbia University in the 1950s.[38] Magnetic resonance imaging (MRI) technology, perhaps the most important advance in diagnostic techniques in over a century, had its origins in the work of Edward Purcell of Harvard and Felix Block of Stanford, who independently discovered nuclear magnetic resonance in 1946.[39] The two shared the Nobel Prize "for their development of new methods for nuclear magnetic precision measurements and discoveries in connection therewith" in 1952.[40] Modern high-capacity hard drives would not be possible were it not for the research of two European physicists, Albert Fert and Peter Gruenberg, who independently discovered giant magnetoresistance in the 1980s—the science behind the ability to store vast amounts of information in a small space. The two shared the Nobel Prize in physics in 2007. Nowhere is the contribution of public research more clear-cut than in the area of pharmaceuticals. Three quarters of the most important therapeutic drugs introduced between 1965 and 1992 had their origins in public sector research.[41]

And that is but prologue. Possibilities abound for new products and processes based on scientific research. If superconductors of sufficiently high temperature can be developed, the phenomenon of superconductivity could be harnessed to transmit electricity at no loss of efficiency.[42] (The current family of high-temperature superconductors operate in the range of 138 kelvin—far too cold to be used for the practical transmission of electricity; room temperature is at 300 kelvin.)[43] Wounds in fetal skin heal without a scar, suggesting that with sufficient research the underlying mechanism could be learned and a similar outcome could be accomplished after birth.[44] Gene therapy offers the possibility of restoring sight to those born with severe blindness.[45] The multi-billion-dollar investment in ITER is based on the hope that the fusion of hydrogen—the reaction that powers stars—inside the tokamak reactor can produce sufficient excess energy to be a viable source of energy.[46] Stem-cell research could lead to the ability to repair damaged organs. Advances in sensors, imaging tools, and the development of new software could create new ways to detect explosives.[47] Tiny transistors may be possible if researchers succeed in integrating carbon nanotubes into high-performance electronics.[48]

The relationship between research in the public sector and economic growth has been a rallying cry for resources for research in recent years. The 2007 report *Rising above the Gathering Storm,* issued in record time by the National Research Council, warned Americans that without substantial investments in research the nation would lag behind emerging economies. Science is the genie that will keep the country competitive, but the genie needs to be fed. University presidents routinely conjure up the economic contributions of universities in their quest for funds; local communities lobby for "research" universities in the belief that a research university will lead to economic growth.

The view that growth is built on public sector research is not incorrect. But it is too simplistic. Much of the research of universities and public research institutions cannot instantly be transformed into new products and processes. It can take time, as the examples of atomic clocks and hybrid corn clearly show. There are, of course, exceptions. The World Wide Web had a huge impact almost from its inception. The discovery of giant magnetoresistance transformed disk storage in a matter of years. There are also false hopes. Research that looks promising can fail to deliver on the predicted timeline. The discovery of the cystic fibrosis gene in 1989 brought the hope for gene-based treatments. To date, the "payoff remains just around the corner."[49]

It not only takes time; considerable investment and know-how are required to translate research into new products and processes. Industry, not academe, excels in doing this.[50] In singing the praises of academic research, one should not forget that innovations come from research and development—and development has long been the domain of industry.

Scientists and engineers working in the for-profit sector learn about research performed in the public sector by attending conferences and reading scholarly articles published by their university colleagues. They also engage in joint research with colleagues in academe. Relationships between universities and industry are fed by the constant supply of new talent that universities send to industry. In some fields, such as engineering and chemistry, universities place the majority of their newly trained PhDs in industry. University faculty also are hired as consultants to industry, and faculty receive about 6 percent of their research funds from industry.[51]

The flow of knowledge is not a one-way street from academe to industry. Faculty researchers with ties to firms report that their academic research problems frequently or predominately are developed out of consulting with industry.[52] Moreover, much of the technology that affects the rate of scientific advance in the public sector is developed in industry.

Economics and Science

Economics not only shapes science. Economics also provides a framework for studying science. One can draw on economic concepts in thinking about science and the research enterprise, such as that of the production function (which details the relationship between inputs and outputs) or the concept of public goods, as I have done above. One can also draw on the concept of economic efficiency, which asks whether it is possible to reallocate resources devoted to research in such a way as to get "more." It is not only a question of whether the amount invested is efficient; it is also a question of whether the allocation of resources among projects is efficient. The question also arises as to whether markets in science function efficiently. By way of example, are there special quirks in the PhD training model that lead to training more scientists than can effectively be employed in research? Is the market for scientific equipment so highly concentrated that sellers have extraordinary market power?

Economics also provides a tool bag that helps in analyzing the relationships between incentives and costs. It shares this tool bag with other fields. Certain concepts and approaches are especially key when studying science and scientists. Some are obvious, others less obvious. First, beware of attributing causality from correlation. Second, if at all possible, think of the counterfactual. Without a counterfactual, it is not possible to assess the impact of a policy on outcomes. The fact, for example, that the research that led to the MRI and the atomic clock originated in academic settings does not prove that the two would not have been invented elsewhere. Third, evidence from natural experiments is more convincing than most other kinds of evidence because natural experiments minimize effects caused by selectivity.[53] It is more powerful, for example, to see how patenting affects follow-on research if some exogenous event occurs that lifts restrictions resulting from a patent that has already been in place. Fourth, data that allow one to follow a panel of individuals over time have a distinct advantage over cross-sectional data—collected from individuals at a moment in time—in that such data allow one to control for what can be thought of as "fixed" effects—that is, individual characteristics that are unlikely to vary over time. The list could go on, but one gets the idea. The methodology underlying research findings provides some guidance concerning just how big the proverbial grain of salt should be.

The Focus of This Book

This book is primarily focused on the United States. This is the system that I know the most about, or to put it in economic terms, the area in which I have a comparative advantage. But the book is not exclusively about the United States. Comparisons are made with other countries, and alternative approaches for providing incentives as well as supporting scientific research are explored. Moreover, many of the underpinnings of science, such as the importance of priority and an interest in puzzle solving, transcend national borders. Science is also becoming increasingly international. A statistic frequently bandied about is that 50 percent of all the highly-cited PhD physicists in the world work in a different country than the one in which they were born.[54] Approximately 30 percent of papers published with one or more authors from a U.S. institution have as a minimum one international coauthor—more than double what it was 15 years ago.[55] Part of the increase reflects the fact that large-scale equipment is increasingly sponsored by a coalition of countries. Once again, money is a major factor. It is tricky in today's world for only one country to commit to a billion-dollar-plus piece of equipment that will provide insights for all. Part of the increase reflects the increased mobility of scientists and the widespread adoption of information technology that has dramatically changed the way in which scientists communicate with each other.

A fairly orthodox definition of science and engineering is employed in this book. To wit, the social sciences (including my discipline of economics) and psychology are not included in the analysis, despite the fact that the National Science Foundation includes these fields in its definition of science. This does not mean that the discussion is irrelevant to the social sciences. Many of the concepts developed here are relevant to the social sciences. By way of example, priority plays an important role in the social sciences as does the satisfaction derived from solving the puzzle. And research in the social sciences can require a substantial amount of resources, although usually not at the level required in science and engineering.

The book is particularly focused on research. Chapters 2 and 3 address the incentives for doing research, and Chapters 4 and 5 address how research is produced. Chapter 6 addresses how research is funded. In some of this discussion, the distinction is made between basic research and applied research. As used in this book, basic research refers to research directed at furthering fundamental understanding; applied research is directed at solving practical problems. Increasingly, and particularly in certain fields, such as the biomedical sciences, the distinction is somewhat moot. Researchers can have the dual goal of advancing fundamental understanding as well as solving practical problems. Donald Stokes referred to research directed at these dual goals as falling into Pasteur's Quadrant—in honor of Louis Pasteur and his research on bacteriology, which helped the wine and beer industry solve the problem of spoilage.[56] It also led to a fundamental understanding of the role that bacteria play in disease and provided a strong impetus for the investment in public water and sewer systems in the late nineteenth century—an investment that did more than anything else in human history to increase life expectancy.

The Plan of the Book

The book begins with a discussion of the intrinsic rewards of doing science. The enjoyment derived from puzzle solving, for example, is part of the reward of doing science. But scientists also strive for recognition. They are engaged in an enterprise that rewards the first to communicate a finding, thereby establishing their priority of discovery. The functionality of the priority system is also explored in Chapter 2, both in terms of the incentive to create and share new knowledge and in terms of the way priority solves what economists think of as the monitoring problem.

Science is often described as a winner-take-all contest, meaning that there are no rewards for being second or third. This is an extreme view. A more appropriate metaphor is to see science as following a tournament arrangement, much like those in tennis and golf. But science does share some characteristics of a winner-take-all contest—especially when it comes to inequality. Productivity in science is highly skewed: approximately 6 percent of scientists and engineers write 50 percent of all published articles. Chapter 2 examines the metrics for measuring research productivity as well as the highly unequal distribution of scientific output.

The financial rewards that accompany science include salary, royalties, and consulting fees as well as the considerable returns that a small number of scientists make from starting a company. These are examined in Chapter 3. Included in the analysis is a discussion of the degree to which faculty salary varies across individuals, depending upon rank, type of institution (public versus private), and field. The chapter also examines the degree to which salaries for researchers vary across countries and the implications this has for mobility of researchers.

Chapters 4 and 5 examine how research is produced. The focus of Chapter 4 is the people doing science and what they bring to the research enterprise. Chapter 5's focus is on equipment, materials, and space for research. The chapters examine not only the similarities in the way science is produced across disciplines but also the fact that no one model of production fits all fields of science and engineering. For example, the fields of mathematics, chemistry, biology, high energy physics, engineering, and oceanography all share certain common characteristics in terms of production. All require time and cognitive inputs. But in other dimensions there is considerable variability. A case in point is the way in which research is organized. Mathematicians and theoretical physicists rarely work in labs and often work alone, whereas most chemists, life scientists, engineers, and many experimental physicists collaborate on research, often working in labs. The chapter also examines how, in certain fields, research is organized and defined by equipment, as in the case of astronomy and high energy experimental physics. In other fields, the equipment required to do research is often minimal, as is the case in certain areas of mathematics, chemistry, and fluid physics.

Research costs money. An off-the-shelf mouse costs between $17 and $60; a postdoc can cost $40,000—more, when fringe benefits are included; a sequencer can cost $470,000; and a telescope can have a price tag in excess of a billion dollars. Chapter 6 examines public and private sources for supporting research and the mechanisms, such as peer review, prizes, administrative allocations, and earmarks, used to distribute research funds. The chapter also explores the benefits and costs associated with different mechanisms. Peer review, for example, has a number of pluses. It provides freedom of intellectual inquiry and encourages scientists to remain productive throughout their careers. It also promotes quality and the sharing of information. But peer review has its downside. The large amount of time required to apply for and administer grants diverts scientists from spending time doing research. The peer-review system also discourages risk taking. Failure is not rewarded.

Factors that play a role in determining who becomes a scientist or engineer are explored in Chapter 7. It is not all for "the love of knowledge," as some would suggest. The amount and availability of fellowship money influences the number of individuals choosing careers in science and engineering; high salaries in other fields, such as law and business, can discourage individuals from choosing careers in science and engineering. Pyramid schemes are not limited to Wall Street or to salesmen—they exist in science, especially in the biomedical sciences, where faculty persist in recruiting graduate students and postdocs to work in their labs despite strong evidence that a sufficient number of research jobs for those in training do not exist.

The foreign born play a substantial role in science and engineering today in almost every Western country. They are the focus of Chapter 8. Given the particularly large role that the foreign born play in the United States—where 44 percent of all PhDs in science and engineering are awarded to temporary residents, almost 60 percent of postdocs are temporary residents, and 35 percent of faculty were born outside the United States—the chapter primarily examines the foreign born in the United States. Once again, we see the important role that economics plays in determining who comes to study and who chooses to stay. We also see evidence that increased numbers of foreign born depress salaries, especially salaries of postdocs, and thereby may discourage U.S. citizens from choosing careers in science and engineering.

Chapter 9 explores further the relationship between science and economic growth introduced earlier. It also explores ways in which scientific knowledge diffuses between the public sector and the private sector.

Economics is not only about incentives and costs. It is also about the allocation of resources across competing wants and needs—or to use the jargon of the profession—economics is also about whether resources are allocated efficiently. The final chapter discusses issues of efficiency, and, where the evidence is sufficiently convincing, possible actions that could make the public research system—particularly in the United States—more effective. Where evidence is insufficient, I, in the tradition of other researchers, encourage further research.

CHAPTER TWO

Puzzles and Priority

ASK ALMOST any scientist what led him or her to become a scientist and the answer will be an interest in solving puzzles. The interest in puzzles persists throughout their career. It is not only the "hook" that attracts people to science, but it is also a key intrinsic reward for doing science. "The prize," to quote the Nobel-Prize winning physicist Richard Feynman, "is the pleasure of finding the thing out, the kick in the discovery."[1]

Scientists are not only motivated to do science by an interest in solving puzzles; they also are motivated by the recognition awarded to being first to communicate a discovery. The distinction between puzzles and recognition is that the satisfaction derived from puzzle solving occurs while doing the research; recognition comes from being the first to solve a particular puzzle and to communicate the findings to colleagues.

The rewards to a career in science also include money. Denials to the contrary, scientists take some interest in financial rewards. Although they do not choose careers in science with an eye to maximizing their income, they are not immune to the allure of monetary rewards. Such rewards come in a variety of forms, such as higher salaries, supplements associated with an endowed chair, royalties from patents, stock in start-up companies, and bonuses for receiving a grant. It is not just that money provides for greater material well-being; money is also a symbol of status.

This and the next chapter focus on the rewards to doing science. The discussion begins with the importance of puzzles and recognition. It continues in Chapter 3 looking at the role that money plays in science—not as a means to solve puzzles or to earn reputation (I do that in Chapter 6), but as an end in itself, a component of the extrinsic rewards that individuals receive from doing science.

Puzzles

The philosopher of science Thomas Kuhn describes normal science as a puzzle-solving activity. According to Kuhn, a primary motivation for engaging in normal science is an interest in solving the puzzle. Even though the outcome can be anticipated, the fascination with research is that "the way to achieve that outcome remains very much in doubt. Bringing a normal research problem to a conclusion is achieving the anticipated in a new way, and it requires the solution of all sorts of complex instrumental, conceptual, and mathematical puzzles. The man who succeeds proves himself to be an expert puzzle-solver, and the challenge of the puzzle is an important part of what usually drives him on."[2]

Warren Hagstrom, an early sociologist of science, picked up on the puzzle theme, noting that "research is in many ways a kind of game, a puzzle-solving operation in which the solution of the puzzle is its own reward."[3] The philosopher of science David Hull describes scientists as innately curious and suggests that science is "play behavior carried to adulthood."[4] He goes on to say, "The wow-feeling of discovery, whether it turns out to be veridical or not, is exhilarating. Like orgasm, it is something anyone who has experienced it wants to experience again—as often as possible."[5] The Nobel laureate Joshua Lederberg concurs with Hull, but sees the puzzle as too tepid an analogy: "But puzzle just doesn't capture the orgastic element of real discovery. As they say, if you haven't experienced it you can't convey it in words."[6]

The molecular biologist (and 1993 Nobel laureate) Richard J. Roberts recounts how it was his interest in puzzle solving that led him to a career in science. While Roberts was in elementary school, his headmaster encouraged his interest in math and provided him with problems and puzzles to solve. This led Roberts to want to be a detective, where "they paid you to solve puzzles." His ambition quickly changed when he received the present of a chemistry set and learned that science was full of puzzle-solving opportunities.[7] Jack Kilby, one of the inventors of the integrated circuit, is said to have fallen in love with the creative process of discovery. "I discovered the pure joy of inventing."[8] "The joy of discovery" is biochemist Steve Mc-Knight's answer to "why we choose to be scientists."[9]

Puzzle solving not only provides satisfaction. Puzzles are addictive. To quote Richard Feynman again, "Once I get on a puzzle, I can't get off."[10]

The satisfaction derived from puzzle solving is a first cousin to the "aha" moment associated with discovery that some scientists describe.[11] The biophysicist Don Ingber recounts such a moment when, as an undergraduate at Yale, he saw students walking around campus "holding sculptures that were made out of cardboard that looked like jewels," but also "looked very much like viruses to me in my textbooks."[12] The association led Ingber to enroll in a class where "tensegrity" was demonstrated—the word used to describe how the sculptor Kenneth Snelson used taut wires and stiff poles to make strong yet flexible monuments. In an interview, Ingber recounts how this experience changed the course of his professional life. The time was the late 1970s and researchers had just begun to publish papers describing how cells are held up by an internal scaffolding. Upon seeing the demonstration of tensegrity, Ingber reports, "I immediately thought: 'Oh, so cells must be tensegrity structures.'"[13]

Evidence concerning the importance of puzzles is more than anecdotal. Data collected by the National Science Foundation in the Survey of Doctorate Recipients (SDR) provide empirical support for the importance of the puzzle both as a motivating force and as a reward for doing research. When scientists were asked to score the importance of a number of job factors, they consistently gave the highest scores to intellectual challenge and independence. Not only do they see challenge as a key motivation for doing science, they also see the intellectual challenge as a reward. In the same survey, scientists working in academe reported that, among five job attributes, they were most satisfied with the intellectual challenge they received from their job as well as their ability to be independent on the job.[14]

Recognition

Many of life's tastes are acquired. Science is no exception. The 18-year-old physics major may have given little thought to the importance and kudos attached to publishing an article in *Science* or *Physical Review Letters*. But she quickly learns to value such a feat by seeing the importance others attach to the recognition that accompanies it and the way such recognition can be leveraged into resources for research. In this respect, scientists are no different from other human beings. "The pursuit of reputation in the eyes of others," according to philosopher and psychologist Rom Harré, "is the overriding preoccupation of human life."[15] "Give me enough ribbon," Napoleon reportedly said, "and I can conquer the world."[16] It is the form of recognition, not the interest in recognition, that varies from field to field.

Recognition is key in science, not only as an end in itself but also as a means for acquiring the resources to continue to engage in puzzle-solving activity. Here the focus is on recognition as an end in itself. Chapter 6 examines the importance that reputation plays in acquiring resources.

Reputation is built in science by being the first to communicate a finding—by establishing what the sociologist of science Robert Merton refers to as the priority of discovery. Merton further argues that the interest in priority and the intellectual property rights awarded to the scientist who is first are not a new phenomenon but have been an overriding characteristic of science for at least three hundred years.[17] Newton took extreme measures to establish that he, not Leibniz, was the inventor of the calculus.[18] Darwin was only convinced to publish *On the Origin of Species* when he realized that Wallace had reached similar conclusions and would be awarded priority of the discovery if he, Darwin, did not publish first. The importance of being first even made it into the vernacular in the 1950s Tom Lehrer song concerning a Russian mathematician—inspired by the nineteenth-century mathematician Nikolai Ivanovich Lobachevsky:

> And then I write
> By morning, night,
> And afternoon,
> And pretty soon
> My name in Dnepropetrovsk is cursed,
> When he finds out I publish first![19]

The interest in priority—and the knowledge within the scientific community that certain research questions are of particular importance—can lead to discoveries being made multiple times—as in the case of the calculus and natural selection, as already noted. In a speech delivered at the conference commemorating the 400th anniversary of the birth of Francis Bacon, Merton detailed the prevalence of what he called "multiples" in scientific discovery, giving, by way of example, twenty lists of multiples, compiled independently by various authors between 1828 and 1922. Moreover, Merton was quick to point out that the absence of a multiple does not mean that a multiple was not in the making at the time the discovery was made public. This is a classic case of censored data, where scooped scientists abandon their research after someone else is awarded the priority.[20]

Despite the censoring problem, examples of multiples abound. Hyperbolic geometry is a case in point, where the multiple involved is Lehrer's own Nikolai Ivanovich Lobachevsky (1830) and János Bolyai (1832). RSA, an algorithm for a public-key cryptosystem and the algorithm of choice for encrypting Internet credit-card transactions, was published in 1977 by Ron Rivest, Adi Shamir, and Leonard Adleman (hence the name RSA).[21] But Clifford Cocks, a mathematician working for the British intelligence agency GCHQ, described an equivalent methodology in a 1973 document that, due to its top-secret classification, was not revealed until 1997. Nanotubes provide another example: in 1993, Donald S. Bethune and his group at IBM and Sumio Iijima and his group at NEC independently discovered single-wall carbon nanotubes and methods to produce them using transition-metal catalysts.

Transgenic mice provide yet another classic example of a multiple: in the early 1980s, five independent teams published articles regarding the development of transgenic mice. In a remarkably short interval of time, the five teams described how the injection of foreign DNA (a so-called transgene) into mouse eggs, which were then transplanted into female mice, led to the incorporation of the genes into the offspring, creating a "transgenic" mouse.[22]

A necessary condition for establishing priority of discovery is to report one's research findings to the scientific community, usually through publication in a journal.[23] Indeed, the only way in which a discovery in science can be attributed to the scientist—and hence become the property of the scientist—is by publicly making the findings available. Later in this chapter we will return to properties of a reward system that is based on the premise of "making it yours by giving it away."

Fast turnaround can be important in establishing priority and building reputation. It is not unknown for scientists to write and submit an article the same day. Neither is it unknown to negotiate with the editor of a prestigious journal the timing of a publication or the addition of a "note added" so that work completed between the time of submission and publication can be reported, thus making the claim to priority all that more convincing.[24] *Science,* a leading if not the leading multidisciplinary journal in science, has the explicit policy of asking referees to return their reviews within seven days of receipt of the manuscript. Online publication has gained in popularity in recent years precisely because of the speed with which articles can be published. *Applied Physics Express (APEX)* promises, for example, rapid publication, with the online version appearing in the "record-shortest 15 days after submission."[25] The IEEE Engineering in Medicine and Biology Society recently announced *T-BME Letters,* promising two months from submission to publication.

The importance that scientists attach to establishing priority can be inferred by a variety of social conventions and practices in science. It is not unknown for scientists to argue about the order in which they appear on a program. Two issues are at stake: not wanting to be scooped and the prestige associated with being listed first. Scientists worry about the consequences of sharing data. The 2003 Nobel Laureate for Chemistry, Peter Agre, reports that he "lay awake at night worrying that my openness would cause us to be scooped."[26] Others take extreme measures to keep competitors at bay. Scientists have been known, for example, to collect class notes from students in an effort to stave off the competition or, in the case of mathematicians, to leave out a key point of a proof. In the two papers Paul Chu and Maw-Kuen Wu submitted to *Physical Review Letters,* describing their discovery of superconductivity above 77 Kelvin, the symbol Yb (ytterbium) was substituted for Y (yttrium). Chu claimed this was a "typographical error." Others claimed it was a deliberate effort on Chu's part to throw off the competition. Chu corrected the proofs in the final days that corrections could be made to the manuscript.[27]

Conflicts regarding the selection of Nobel Prize recipients provide another indication of the importance attached to priority and reputation. In 2003, the inventor Raymond Damadian, who was excluded from the list of winners for the invention of the MRI, took out full-page ads in the *Wall Street Journal* and the *New York Times* (with the banner "The Shameful Wrong That Must Be Righted") to protest his exclusion from the winners' circle. Money could not have been the issue—the ads cost far more than his share of the prize would have amounted to. The issue was reputation.[28] In 2008, considerable concern was expressed when Robert Gallo was excluded from the list of winners for identifying the HIV virus. No one contested that Francoise Barré-Sinoussi and Luc Montagnier were prizeworthy, but surprise was expressed that the third name on the prize was the German virologist Harald Zur Hausen rather than Robert Gallo. Gallo's public disappointment over being excluded was restrained.[29] But such was not the case with Jean-Claude Chermann who, rather than accepting the invitation of his former French colleagues to accompany them to Stockholm, invited journalists to lunch in order to explain why he should have shared the prize.[30]

Researchers can also manipulate where they stand in a hierarchy of prestige. By way of example, the Social Science Research Network (SSRN) website routinely generates a list of the top 10 downloaded papers by field. A recent study shows that individuals game the system, downloading their own papers when they are "close" to being in the top 10 or in danger of losing their top 10 status.[31] Whether this practice occurs in the natural sciences and engineering as well has not, to my knowledge, been studied.

Scientists can overstate the role they play in a discovery with an eye to augmenting their reputation. By way of example, a prominent engineer who had hosted a visiting scholar added his own name to an article that reported research done by the visiting scholar and a student in the engineer's lab when he realized the importance of the work and the attention that the research would garner. He subsequently gave interviews to the press that mentioned the visitor only in passing. Such honorary authorship is not uncommon but difficult to verify.[32]

Not all discoveries are equal. A common way to measure the importance of a scientist's contribution is to count the number of citations to an article or the number of citations to the entire body of work of an investigator. This used to be a laborious process, but changes in technology, as well as the incentives to create new products such as Google Scholar and SCOPUS, have meant that researchers, and those who evaluate them, can quickly (and sometimes erroneously) count citations to their work and thus judge where they stand relative to their peers.[33]

The growing obsession with measures and rankings has led to the creation of a variety of bibliometric indices and products. For example, Thomson Reuters, the company behind the large bibliometric database "Thomson Reuters Web of Knowledge" (formerly known as ISI Web of Science) markets a product that ranks scientists within a field in terms of citations. Scientists, their departments, or any other party that wants to know can use the Web of Knowledge to create a "Citation Report" for an individual or a group of individuals.

In 2005, Jorge Hirsch, a physicist at the University of California–San Diego, proposed the h-index to measure the productivity and impact of a scientist's research. The index became an instant success. Now, with only the click of a mouse, scientists can get one number that (supposedly) summarizes their productivity and the impact that their work has had. To be more precise, the h-index depends upon the number of papers published and the number of citations each paper has received. When papers are arrayed from the most highly cited to the least highly cited, the h-index measures the number of papers that have h or more citations. Thus, for example, if a scientist has authored 50 papers and 25 of them have 25 or more citations, she has an h-index of 25. A scientist who has published 35 papers, 30 of which have 30 or more citations has an h-index of 30.[34] In his original article, Hirsch suggested that an h value of about 10 to 12 might warrant tenure for a physicist; a value of 18, promotion to professor.[35] Despite its numerous limitations—the measure is sensitive to career stage, heavily discounts "blockbuster" articles, and can only increase with experience—the *h*-index enjoys considerable popularity. It is not unusual for scientists to list their *h*-index in their biography or on their webpage.[36]

Forms of Recognition

The recognition that the scientific community bestows on priority has varied forms, depending on the importance the scientific community attaches to the discovery. Heading the list is eponymy, the practice of attaching the name of the scientist to the discovery. The hunt for the Higgs particle, for example, is much in the news these days with the completion of the Large Hadron Collider (LHC) at CERN and the associated four detectors. The particle is named for the Scottish physicist Peter Higgs, who was the first to predict its existence (in 1964) as part of a theory that explains why fundamental particles have mass.[37] Many other examples of eponymy exist: Haley's comet, Planck's constant, Hodgkin's disease, the Kelvin scale, the Copernican system, Boyle's law, the RSA algorithm, to name but a few.[38]

Recognition also comes in the form of prizes—sometimes for a particular discovery, in other instances in recognition of a scientist's life work.[39] Among prizes, the Nobel is the best known, carrying the most prestige and a large—although not the largest—purse of approximately $1.3 million. But hundreds of other prizes exist, a handful of which have purses of $500,000 or more, such as the Lemelson-MIT Prize with an award of $500,000, the Crafoord Prize ($500,000), the Albany Medical Center Prize ($500,000), the Shaw Prize ($1 million), the Spinoza Prize (1.5 million euros), the Kyoto Prize ($460,000), and the Louis-Jeantet Prize (700,000 CHF), to name but a sampling. In some instances, the money that accompanies the prize is to support the winner's lab; in most instances, the award is given directly to the recipient.[40] How they choose to spend it is often a point of interest.

The number of prizes has grown in recent years. Zuckerman estimates that approximately 3,000 prizes in the sciences were available in North America alone in the early 1990s, five times the number awarded twenty years earlier (a rate of growth that outpaced growth in the number of scientists by a factor of two).[41] Although no systematic study of scientific prizes has been conducted since, anecdotal evidence suggests that the number continues to grow. *Science* regularly features recent recipients of prizes, many of which are awarded by companies and newly established foundations, and often have purses in excess of $250,000. Several very large prizes have been established recently. These include the Peter Gruber Genetics Prize, first awarded in 2000, with a value of $250,000; the Abel Prize in Mathematics, created in 2002 by the Norwegian government, with a monetary award of approximately $920,000; the Shaw prizes, referred to as the Asian Nobels, with a $1 million purse for each awardee, first awarded in 2004; the Kavli Foundation Award with a purse of $1 million, which was started in 2008; Joel Greenblatt and Robert Goldstein's Gotham Prize with a $1 million purse, first awarded in 2008; and the Frontiers of Knowledge Award, bestowed for the first time in 2009, with a monetary value of approximately $530,000 for each of eight prizes.

Not all prizes are large, and the size of the purse does not necessarily reflect the prestige associated with the prize. The Fields Medal, the closest equivalent to the Nobel Prize in mathematics, awarded only every four years, carries the nominal purse of around $15,000.[42] The Lasker Prizes in Basic Medical Research and Clinical Medical Research have a $50,000 monetary award, but are highly prestigious, having been awarded to seventy-five individuals who subsequently have gone on to win the Nobel Prize in physiology or medicine. Some prizes, especially targeted to young investigators, are in the $20,000 to $25,000 range. There is, for example, a Lemelson-MIT Student Prize with a monetary value of $30,000. In some instances, such as the Eppendorf and Science Prize for Neurobiology, the award involves not only a monetary reward ($25,000) but also the publication of the winner's article in *Science*.

Prizes are a two-way street. They bestow honor (and money) on the recipient; in return, the awarding group receives prestige through association with the distinguished recipients. It is not an accident that the Gairdner Foundation points out that 70 of its 288 awardees have gone on to win the Nobel Prize, that the Passano Foundation has a link on its webpage showing Passano scientists who have also won the Nobel Prize. The Lasker Prizes have a similar glow from association with the Nobel Prize.

Nor is it a surprise that in recent years many companies have created prizes. By way of example, Johnson & Johnson established the Dr. Paul Janssen Award for Biomedical Research in 2005 with a purse of $100,000; General Electric partnered with *Science* to create the Prize for Young Life Scientists in 1995 ($25,000); General Motors established the General Motors Cancer Research Prize ($250,000); and AstraZeneca created the Excellence in Chemistry award. The L'Oréal Foundation, whose parent company, L'Oréal, manufactures cosmetics for women, teamed up with UNESCO to make five awards "For Women in Science" annually.[43]

Other forms of recognition exist. Many countries, for example, have societies to which the luminaries are elected: the National Academies of Science, Engineering, and Medicine in the United States,[44] the Royal Society in England, the Académie des Sciences in France, and the Japan Academy. Membership in such societies is highly valued, and the invitation to join is rarely declined. Thus, eyebrows were raised in 2008 when Nancy Jenkins turned down the invitation to join the National Academy of Sciences.[45]

The Functional Nature of the Priority-Based Reward System

As noted in Chapter 1, scientific research has properties of what economists call a public good. Once it is made public, others cannot easily be excluded from its use.[46] It is also nonrivalrous in the sense that knowledge is not diminished with use and thus the cost of another user approaches zero. The market has special problems producing goods with such characteristics. Nonexcludability provides incentives for individuals to free ride, limiting the benefits the producer will receive if the good is provided and thus discouraging production. The fact that the cost of another user approaches zero means that an efficient price is zero. Clearly, the market cannot provide incentives at such a price to produce the good.

From an economist's point of view, an exceedingly appealing attribute of a reward system that is priority based is that it offers non-market-based incentives for the production of the public good "knowledge." Scientists are motivated to do research by a desire to establish priority of discovery.[47] But the only way that this can be done—that a scientist can establish ownership of an idea—is by giving the idea away. Thus, priority is another form of property rights, just as a patent is a form of property rights or a lease is a form of property rights. The interest in priority motivates scientists to produce and share knowledge in a timely fashion.

Merton deserves the priority for making the connection, doing so in the inaugural lecture of the George Sarton Leerstoel at the University of Ghent, October 28, 1986, which was published two years later in *Isis*. He describes the public nature of science, writing that "a fund of knowledge is not diminished through exceedingly intensive use by members of the scientific collectivity—indeed, it is presumably augmented . . ."[48] Merton not only recognized the public nature of science but went on to argue that the reward structure of priority in science functions to make the public good private: "I propose the seeming paradox that in science, private property is established by having its substance freely given to others who might want to make use of it." He continues, "Only when scientists have published their work and made it generally accessible, preferably in the public print of articles, monographs, and books that enter the archives, does it become legitimately established as more or less securely theirs."[49]

There are other socially desirable attributes of a reward system that is priority based. One relates to the monitoring of scientific effort. Economists have long been concerned about efficient ways to compensate individuals in jobs where monitoring is difficult. Science is a classic case: "Since effort cannot in general be monitored, reward cannot be based upon it. So a scientist is rewarded not for effort, but for achievement."[50] Priority also means that shirking is rarely an issue in science. The knowledge that multiple discoveries are commonplace makes scientists exert considerable effort.

The priority-based reward system also provides scientists the reassurance that they have the capacity for original thought and encourages scientists to acknowledge the roots of their own ideas, thereby reinforcing the social process.[51] Reputation also serves as a signal of "trustworthiness" to scientists wishing to use the findings of another in their own research without incurring the cost of reproducing and checking the results.

Priority also discourages plagiarism and fraud and helps to build consensus in science because the establishment of priority requires the sharing of information and evaluation by one's peers.[52] This is furthered by the small-world nature of scientific networks. The high degree of clustering characteristic of small worlds fosters monitoring, while the low degree of separation (estimated to be between five and seven) promotes the diffusion of scientific findings.[53]

Notwithstanding the public airing of scientific knowledge, fraud and misconduct do occur in science.[54] In recent years, there have been several high-profile cases involving misconduct and fraud. In the mid-2000s, Woo Suk Hwang, who claimed to have created human embryonic stem cells by cloning, was found to have fabricated data.[55] In 2010, Elizabeth Goodwin, an associate professor of genetics and medical genetics at the University of Wisconsin–Madison, was found to have falsified and fabricated data in grant applications.[56] The same year, Marc Hauser, a primate researcher at Harvard University, was found "solely responsible, after a thorough investigation by a faculty member investigating committee, for eight instances of scientific misconduct under FAS standards."[57] Later the same year, Mount Sinai School of Medicine fired two postdoctoral fellows working in the lab of Savio Woo for research misconduct. The university cleared Woo of any wrongdoing; four papers were retracted.[58] Earlier in the decade, many of the findings of Jan Hendrik Schön, a physicist working at

Bell Labs, regarding organic transistors were found to be fabricated. A number of papers (eight in the journal *Science,* seven in *Nature,* and six in *Physical Review*), were retracted. At the time, Schön was averaging one research paper every eight days.[59]

Economics provides some insight regarding who engages in fraud and the type of fraud most likely to be caught. Models predict, for example, that fraud is more likely to be caught in the case of radical research, such as that put forward by Woo Suk Hwang and Jan Hendrik Schön, but that fraud is more common in incremental research.[60] Economics, however, can only go so far. One is still left with the question of why a high-profile researcher would engage in fraudulent research of a radical nature that has a high probability of being scrutinized. One suspects that such behavior is irrational at its core and perpetrated by researchers who either seek to gratify their ego by making unsubstantiated claims or by researchers who are sufficiently irrational to drastically underestimate the probability of detection.[61]

One should not overstate the propensity of scientists to give all their discoveries freely away: scientists can have their cake and eat it, too—selectively publishing research findings while monopolizing other elements with the hope of realizing future returns. The legal scholar Rebecca Eisenberg argues that such behavior is more common among academic scientists than one might initially think because they can publish results and at the same time keep certain aspects of their research private by withholding data, failing to make strains available upon request, or restricting the exchange of research animals such as mice.[62] If such were the case in 1987, when Eisenberg made the argument, it would appear to be even more the case today, as academic scientists increasingly engage in patenting (see Chapter 3), which can restrict others from use of their research.[63]

A case in point is the "mouse that roared"—the OncoMouse—a transgenic mouse that carried specific cancer-promoting genes and opened up new areas for cancer research. The mouse, engineered by the Harvard scientist Philip Leder, was patented by the university in 1988 and then licensed exclusively to DuPont. DuPont took an aggressive stance regarding its patent rights, initiating "reach through" rights; this meant that DuPont owned a percentage share in any sales or proceeds from a product or process developed using the mouse, even if the mouse were not incorporated into the end product.[64] The research community was outraged; under National Institutes of Health (NIH) auspices in 1999, a memorandum of understanding (MOU) was signed allowing nonprofit researchers access to the OncoMouse, the only requirements being a material transfer agreement and a license.[65]

A clever piece of detective work on the part of Fiona Murray and her colleagues suggests that DuPont's practices had a chilling effect on related research. The study involves looking at what happened to citations to mouse articles before and after the NIH-issued MOU. Their research suggests that loosening property rights increases related research. They found that citations to OncoMouse research papers increased by 21 percent after the MOU was issued.[66] This finding is consistent with an earlier finding of Murray and her coauthor Scott Stern that knowledge that is embodied in both papers and patents—what are called patent-paper pairs—is cited less frequently once the patent has been issued.[67]

Litigation of patents is costly, which means that patent rights are not always enforceable and scientists are known to work around patents. But access to the research materials of others, such as cell lines, reagents, and antigens, depends upon the direct cooperation of one's colleagues—and here there is evidence that scientists have their cake and eat it too with some regularity. A survey of bioscientists regarding their experiences related to the sharing of materials finds that access is largely unaffected by patents. But access to the research materials of others is restricted: 19 percent of the material requests made by the sample were denied. Competition among researchers played a major role in refusal, as did the cost of providing the material. Whether the material in question was a drug or whether the potential supplier had a history of commercial activity were also relevant factors in refusal.[68]

The ability to keep certain findings and material for oneself (and one's students) is facilitated by the fact that publication is not synonymous with replicability. It is also facilitated by the fact that certain kinds of knowledge, especially knowledge that relates to techniques, can only be transferred at considerable cost. This is partly due to the fact that their tacit nature makes it difficult, if not impossible, to communicate in a written form. This "sticky" nature of tacit knowledge means that face-to-face contact is required for transmission.[69] It is one reason, as we will see in Chapter 9, why innovations are clustered in certain geographic areas, such as Silicon Valley. The tacit nature of knowledge also makes the location of where a scientist trains important: one cannot simply learn new techniques through reading published (codified) knowledge or by attendance at conferences. One must have hands-on experience to learn how to implement new techniques and use new instruments. Location is important.

Transgenic mice are a case in point. It is said that one needed "magic hands" to create such mice. Leder's lab at Harvard had not pioneered transgenic methods and had no such set of hands, but got them when Timothy Stewart (who had been a member of one of the five successful teams in early transgenic mouse developments) came to do a postdoc in Leder's lab.[70] With Stewart's expertise, Leder's group created a viable mouse that carried a *myc* oncogene and therefore had a predisposition for cancer. It is no surprise that during this time the director of one lab with transgenic expertise experienced "an uptick in applications from people wanting to do postdocs and learn the methods so they could take them elsewhere and gain fame and fortune."[71] Yes indeed—fame and fortune play a role.

The Nature of Scientific Contests

Science is sometimes described as a winner-take-all contest, meaning that there are no rewards for being second or third. This is an extreme view of the nature of scientific contests. Even those who describe scientific contests in such a way note that it is a somewhat inaccurate description, given that replication and verification have social value and are common in science. It is also inaccurate to the extent that it suggests that only a handful of contests exist. Yes, some contests are seen as world class, such as identification of the Higgs particle or the development of high temperature super-conductors. But many other contests have multiple parts, and the number of such contests may be increasing. By way of example, for many years it was thought that there would be "one" cure for cancer, but it is now realized that cancer takes multiple forms and that multiple approaches are needed to provide a cure. There won't be one winner—there will be many.

A more realistic metaphor is to see science as following a tournament arrangement, much like those in tennis and golf, where the losers get some rewards as well. This keeps individuals in the game, raises their skills, and enhances their chance of winning a future tournament. A similar type of competition exists in science. Dr. X is passed over for the Lasker Prize, but her work is sufficiently distinguished that she is invited to give named lectures, consistently receives support for her research, and is awarded an honorary degree from her alma mater. Dr. Y's lab is not the first to make a discovery, but Y's lab develops an instrument that contributes to break-throughs made by others, and he is credited with contributing to these discoveries.

Once one thinks of science in tournament terms, numerous analogies come to mind. First, there are classes of tournaments—or, more generally, tournaments are divided into leagues. Not every golfer plays in the Profes-sional Golf Association (PGA); some play in regional tournaments, others in more local tournaments. Or, to use a baseball analogy, not everyone plays in the major leagues.[72] Some researchers have the skill and good fortune to compete at the top, training and working at top research universities. Others, with perhaps less skill and good fortune, play in regional tournaments. They attend lower-ranked graduate programs, become postdoctoral fellows in less prestigious labs, and end up working in less prestigious universities. Occasionally, they are called to the major leagues. There is some mobility in science, but it is not that common. Sometimes those in the minor leagues make a discovery that is declared a home run by their peers. An interesting topic for future research is the career consequences enjoyed by regional players who achieve national and international attention.

Second, there are niches of tournaments such as tournaments for individuals younger than 35, or tournaments for individuals working in a special area. The NIH study sections are but one example of such "niche" tournaments.[73] Third, and related, funding for science does not follow a winner-take-all model but rather a tournament model with multiple winners. Panels at the National Science Foundation (NSF) make multiple awards to multiple principal investigators, even though the panel may view one proposal to be by far the best. In a similar manner, NIH study sections recommend multiple R01 awards (the bread-and-butter research grant in the biomedical sciences) to support research—not just one.[74]

The tournament nature of the reward system in science amplifies small differences in the underlying distribution of talent into much larger differences in recognition and economic rewards. Of the many who receive degrees in science and engineering, some win in the minors, some in the majors. The accoutrements of success involve independence in research, tenure, a named chair, reputation, awards. But some lose in the sense that they cannot find a position that enables them to play in *any* tournament. They drop out of science, following careers (or noncareers) in other areas, or work in the lab of a senior scientist who receives most of the glory and financial remuneration.[75]

Just such a person received considerable attention at the time the 2008 Nobel Prize was awarded in chemistry (to Osamu Shimomura, Martin Chalfie, and Roger Tsien) for the "discovery and development of the green fluorescent protein, GFP."[76] GFP, used as a tagging tool in research, allows scientists to watch processes involved in cancer, neural development, and more.

As is often the case, a fourth person, Douglas Prasher, was involved in the discovery. But in this case, the fourth person had left science and was driving a courtesy shuttle at the time the prize was awarded. He took the $8.50-an-hour job after a year of unemployment following the loss of a research position on a NASA-funded life science project. But it was Prasher who had cloned GFP, when hardly anyone understood its potential, and who had given it to Chalfie and Tsien when he realized that he might be leaving the field of bioluminescence. Tsien attributes Prasher as having played "a very important role." And, as Chalfie stated in numerous media reports, "They could've given the prize to Douglas and the other two and left me out." They did not, of course. Prasher became a face (the poster child?) for the inefficiencies that scientific tournaments can produce.

Inequality

A defining characteristic of contests that have winner-take-all characteristics—such as those that exist in science—is extreme inequality in the allocation of rewards. Science, also, has extreme inequality with regard to scientific productivity and the awarding of priority. One measure of this is the highly skewed nature of publications, first observed by Alfred Lotka after analyzing the publications of chemists listed in *Chemical Abstracts* for 1907–1916 and the contribution of physicists compiled by Felix Auerbach in 1910.[77] The distribution that Lotka found showed that approximately 6 percent of publishing scientists produce half of all papers. Lotka's "law" has since been found to fit data from several different disciplines and varying periods of time.[78]

Inequality in scientific productivity could be explained by differences among scientists in their ability and motivation to do creative research (to have the "right stuff"). But scientific productivity is not only characterized by extreme inequality at a point in time; it is also characterized by increasing inequality over the careers of a cohort of scientists, suggesting a casual process of state dependence, whereby current productivity—as measured by publications—relates to past success.[79]

There are several reasons that current productivity could be state dependent. First, the amount of recognition a scientist gets for a piece of work may be dependent upon the scientist's prestige. Merton christened this explanation the Matthew Effect, defining it as "the accruing of greater increments of recognition for particular scientific contributions to scientists of considerable repute and the withholding of such recognition from scientists who have not yet made their mark."[80]

One basis for the Matthew Effect, and the reason that Merton gave, is the vast volume of scientific material published each year, which encourages scientists to screen reading material on the basis of the author's reputation. Others argue that processes of cumulative advantage lead present productivity to be correlated with past success. Scientists who have enjoyed success, for example, may acquire a taste for more success and consequently work harder. Successful scientists may also find it easier to leverage past success into research funding.[81] A funding system such as NIH's that awards grants, at least in part, on past success clearly contributes to cumulative advantage (see Chapter 6). Moreover, scientists with a strong track record may find it easier to get their work accepted in top journals than do scientists without such a record.

Research productivity also relates to the current work environment (see Chapters 4 and 7). Facilities and equipment make a difference, as does the presence of research-active colleagues. Thus another reason for productivity to become increasingly unequal over the career of a cohort is that highly successful scientists are more likely to be recruited by strong departments and thus work in environments that promote productivity.

It is virtually impossible to determine what portion of success comes from having the right stuff and what portion can be attributed to state dependence. That's because it is impossible to randomly assign to people of comparable ability and motivation different packages of success. Even if one could, virtually no one has the budget or fortitude to observe how their careers would play out over thirty or forty years. But one could get some sense of the importance that past success plays by conducting an experiment in which identical research proposals which vary only in terms of the strength of the applicants' vitae are scored by experimental subjects. If proposals from those with stronger publication records consistently are rated higher, there is at least some evidence that success is in part state dependent.

Short of this we are left to sort things out by empirically analyzing career histories of scientists. One such approach controls for the right stuff by examining what happens to the careers of scientists who change institutions. If the productivity of movers is not correlated with the status of their new department, there is support for the right stuff. If the productivity of movers is correlated with the status of the department, factors other than the right stuff matter. At least one study which takes this approach finds productivity to be correlated with department prestige.[82] Another earlier study gives credence to state dependence processes without totally discrediting the right stuff.[83] Anecdotal evidence concerning unequal access in science also suggests that state dependence plays a role. For example, a physicist who has held academic positions at several institutions of different quality once wrote to me, saying, "I can tell you that there is a world of difference between writing a letter on Harvard stationary and writing on ____ stationary. In the former case, the door is opened immediately and you get a hearing. In the latter case you have to knock the door down."[84]

In the end, it is likely not a case of either or. Rather, it is highly probable that some sort of feedback mechanism is at work whereby able and motivated scientists leverage their initial success to greater success over their careers.[85] Such processes are characteristic of winner-take-all contests: "In all their manifestations, winner-take-all effects translate small differences in the underlying distribution of human capital into much larger differences in the distribution of economic reward."[86]

Policy Issues

The growth of prizes raises a number of interesting questions which, to the best of my knowledge, have yet to be investigated and are relevant for science policy. For example, what is the incentive nature of prizes: to what extent does the introduction of a new prize encourage individuals to work in a specific area? Second, is it more efficient to establish a prize that recognizes a particular piece of work or to award prizes toward the end of the career for a body of work? Third, does the introduction of yet another prize diminish the value of previously existing prizes? Fourth, are there too many prizes? Or, stated differently, does one more prize in an already prize-intensive field contribute in any way to research productivity—or does it merely bestow prestige, both to the recipient and to the foundation that awards the prize. If the latter is the case—and one suspects it may well be—surely more effective ways can be found to use the funds which meet the goal of conveying prestige while at the same time providing incentives for growing the stock of knowledge.

Another policy issue relates to the role state dependency plays in explaining productivity. To the extent that past success determines current success, scientists who are unlucky early in their career can be doomed throughout their career. By way of example, scientists who go on the job market in difficult economic times—such as those that the crisis of 2008 created—may find themselves working in environments that are not conducive to productivity. Lack of early success can severely hamper their future opportunities—even if and when the economy picks up. This suggests that funding agencies may want to have special grant programs geared particularly to individuals whose careers have been put on hold by such events. More generally, funding agencies may wish to pay more attention to the proposal, and less to the research record and preliminary data, especially for individuals who had the bad fortune to come of age at the wrong time, economically speaking.

Conclusion

Scientists are motivated to do science by an interest in puzzles and by the recognition awarded success—the ribbon. But it is not all about puzzles and ribbon; gold is also involved. Chapter 3 discusses the various types of financial rewards received by scientists working in the public sector.

CHAPTER THREE

Money

PUZZLE SOLVING and the recognition awarded to priority are not the only rewards to doing science. Money is also a reward, and scientists are, indeed, interested in money. They want, to quote Stephen Jay Gould, "status, wealth and power, like everyone else."[1] An eminent Harvard scientist said it well when asked by newly appointed Dean Henry Rosovsky the source of scientific inspiration. The reply (which "came without the slightest hesitation") was "money and flattery."[2]

What is remarkable about the two quotes is that they are now more than twenty-five years old and came during a time when opportunities for university scientists and engineers to augment their salaries were more limited than they are today. If money played a role in the 1980s, it plays a greater role today, since the opportunities for scientists and engineers working in academe to gain income and wealth from patenting and starting new companies have grown. Virtually none of Gould's colleagues or Rosovosky's scientists had earned millions at the time these statements were made. Today—although it is still rare—there are numerous examples of scientists and engineers working in academe who are, if not multimillionaires, very comfortably off.

The focus of this chapter is money as a reward to doing science. We look first at academic salaries, examining differences that exist between salaries for full professors and assistant professors, between those at top-ranked research institutions and baccalaureate institutions, between public and private institutions, and among fields. The chapter also examines the relationship between productivity (as measured by publications and citations) and salary and the ways by which academic scientists augment their income, focusing especially on the activities of patenting, starting companies, and consulting.

Before beginning, it is important to point out that money plays two other critical roles in science. First, and as will be developed in Chapter 7, money influences career choices. Salaries in science relative to salaries in other fields influence the number of individuals choosing to do advanced work in science and engineering (S&E), as does the amount of money available for graduate support. Second, research is expensive. Start-up packages are just that. Funds soon run out, and thereafter university-based researchers are expected to raise money to fund their research. This means that university scientists almost constantly think about money. I discuss the cost of research in more detail in Chapters 4 and 5 when I focus on the production of scientific research and again in Chapter 6 when I discuss paying for science.

Academic Salaries

Faculty pay varies considerably, depending upon academic rank, type of institution (public versus private, research intensive versus teaching intensive), and field. Full professors generally earn more than associate professors, and associate professors more than assistant professors. Faculty at Harvard earn more than faculty at the University of Michigan; faculty at the University of Michigan earn more than faculty at Central Michigan. Faculty in physics earn more than faculty in English but less than their colleagues in computer science.[3]

Pay also varies by characteristics associated with the individual, such as the number and quality of publications, the number of times the individual has moved, and gender. Some of these variables are highly correlated, making it difficult to distinguish causality. For example, highly productive faculty are more likely to be promoted and more likely to work at top-rated departments. Women, who often—especially in the past—face more family constraints than men, may be less mobile than men and thus have fewer job offers.[4]

The 2009–2010 American Association of University Professors (AAUP) salary survey provides some context for these generalizations. The salary (at the 60th percentile) for full professors at doctoral institutions was $120,867; that of associate professors was $84,931; and that of assistants was $72,672. Those who worked at master's institutions received considerably less: full professors earned $90,691, associates earned $71,326, and assistants earned $59,974. Those at baccalaureate institutions earned still less.[5] Private doctoral institutions paid 31.0 percent more than did publicly controlled doctoral institutions, a gap that has grown over time.[6] In the 2009–2010 academic year, only one public institution (UCLA) was among the top twenty research universities in terms of salaries paid to full professors—and it held the 20th position, $43,000 (or 25 percent) below top-paying Harvard.[7] Women full professors earned 91.8 percent of what men earned at doctorate-granting institutions, women associate professors earned 92.7 percent, and women assistant professors earned 91.9 percent. The gap, which is field dependent, has been narrowing over time.[8]

Variation by Field and by Rank

The AAUP data, while informative, are not available by field. But field matters. Faculty in law and finance, for example, generally earn much higher salaries than those in the humanities, science, or engineering. Within S&E there is also a definite hierarchy. Some sense of these differences is seen by examining data from the Annual Faculty Salary Survey by Discipline, commonly referred to (for the institution that collects the data) as the Oklahoma State University (OSU) survey.[9] The survey's intent is to collect information for institutions that are members of the National Association of State Universities and Land Grant Colleges, many of which are the "flagship" public doctorate-granting institution in the state. Thus, by design, almost all private institutions are excluded. This means that average salaries for research institutions are understated in the data, given that the privates, especially research-intensive private institutions, often pay higher salaries than the publics.

Table 3.1 reports summary data from the 2008–2009 OSU study for full-time employees. Means, as well as the highest salary reported by specific academic rank, are given by broad S&E discipline for the 117 institutions participating in the study. For purposes of comparison, average salaries for all disciplines excluding medicine as well as salaries in the two high-paying fields of law and finance are also reported. We see that computer scientists fare best among those in S&E, but engineering—especially at the rank of full professor—is not far behind. The biological and biomedical sciences pay almost the same as the physical sciences. The gap between salaries for these scientists and their higher paid colleagues in engineering and computer science is particularly noticeable at the lower ranks. Faculty in math and statistics receive the lowest salaries. These differences reflect market conditions. With the exception of the years immediately following the information

Table 3.1. Mean and high academic salaries in dollars, selected disciplines by rank, 2008, public research universities

	New assistant	Assistant	Associate	Full	Ratio full/ assistant
Computer and information sciences					
Mean	84,788	87,298	100,232	132,828	1.52
High	125,715	125,715	192,974	300,999	2.39
Biological and biomedical					
Mean	64,470	65,865	79,159	116,416	1.77
High	106,053	199,309	183,048	422,460	2.12
Engineering					
Mean	77,945	79,987	92,853	129,633	1.62
High	112,000	172,000	177,251	317,555	1.85
Mathematics and statistics					
Mean	61,979	65,684	76,654	110,889	1.69
High	86,000	103,000	131,950	328,200	3.18
Physical sciences					
Mean	64,670	67,161	78,728	116,557	1.74
High	99,000	99,000	140,000	382,945	3.87
Law					
Mean	90,892	97,714	113,380	164,070	1.70
High	130,000	190,000	175,000	318,600	1.68
Finance					
Mean	140,507	139,111	136,016	167,269	1.20
High	190,000	195,700	242,111	423,866	2.17
All disciplines except medical					
Mean	67,105	68,472	79,845	115,895	1.69
High	190,000	200,000	242,111	423,866	2.12

Source: 2008–2009 Faculty Salary Survey, Oklahoma State University.
Note: High salary: the highest salary reported for any full-time individual in a defined group.

technology bubble, academic institutions have had to compete with industry for engineers and computer scientists. But in the biomedical and physical sciences, demand from industry has been weak relative to supply.

Faculty in S&E generally earn about the same or more than the average faculty member does at these institutions—with the exception of faculty in math and statistics. But the S&E faculty are not the highest paid. Even those in computer science earn substantially less than those in law and finance.

The highest reported salary paid in S&E is in the biological and biomedical sciences: $422,460. This reflects the contributions that highly productive biomedical researchers make to the university—both in terms of external funding and, in some instances, royalties from licensing patents. There is also a considerable spread between the salary of top earners and that of average faculty, especially at the rank of full professors, where the spread ranges from 2.5 to 3.6 depending on field. The spread is characteristic of the tournament nature of science discussed in Chapter 2. Star scientists may not earn the megabucks that sports stars do, but they earn considerably more than their peers of equal rank and five to six times as much as rookies.

Comparable data are not collected from private institutions. However, the Survey of Doctorate Recipients (SDR) administered by the National Science Foundation (NSF) collects salary data from faculty who work at either private or public institutions. These data, reported for respondents working at doctorate-granting institutions in 2006 (the latest date for which data are available in 2010), are given in Table 3.2. The data are differentiated by those working at public institutions versus those working at private institutions. Confidentiality rules preclude reporting the "high" salary; instead, the salary received by the 90th percentile is reported.

The patterns are fairly similar to those seen in the OSU institutional data. Mathematicians receive the lowest salaries; engineering and computer scientists do relatively well. However, in the SDR data, for those working at private as well as at public institutions (at the rank of full professor), mean salaries are highest on average in the biological sciences. This likely reflects the fact that the salaries reported in Table 3.1 are for nine to ten months; those in Table 3.2 have been adjusted by the NSF to include summer pay, which adds a considerable amount to salaries in fields such as biology, where a large number of faculty receive summer support from research grants. The table also shows the salary gap that exists between public and private institutions—although it should be noted that the gap is not present in computer and information sciences, where the publics outpay (at least in terms of means) at every rank.

In many occupations, there is a large gap between what novices earn relative to what those who are well established earn. In the practice of law, for example, the differential can be of a magnitude of more than five. In medicine, similar gaps exist. Academe is somewhat different. The flat shape of the earnings profile is frequently noted, although over time the profile has become a bit steeper. To be more specific, full professors earned about 1.61 more than assistant professors in the physical sciences in 1974–1975;

Table 3.2. Mean and 90th percentile academic salaries, selected disciplines by rank, 2006, public and private PhD-granting institutions

	Public			Private		
	Assistant	Associate	Full	Assistant	Associate	Full
Biological sciences						
Mean	76,200	83,800	128,500	88,200	108,800	157,800
90th percentile	105,000	115,000	200,000	140,000	132,000	277,700
Computer and information sciences						
Mean	81,100	92,200	112,800	80,900	91,900	82,400
90th percentile	94,000	120,000	146,000	110,000	108,600	150,000
Engineering						
Mean	77,100	87,900	122,500	84,000	94,300	121,400
90th percentile	93,100	98,000	170,000	121,000	120,000	172,000
Math						
Mean	70,600	68,000	107,100	70,800	60,600	115,880
90th percentile	100,000	94,800	150,000	87,000	80,000	180,000
Physical sciences						
Mean	68,700	77,700	112,700	73,400	81,300	133,300
90th percentile	80,000	100,000	175,000	100,000	115,000	185,000

Source: 2006 Survey of Doctorate Recipients, National Science Foundation (2011b). The use of NSF data does not imply NSF endorsement of the research methods or conclusions contained in this book.

in 2008–2009, the ratio had grown to 1.74. In the life sciences, the ratio was 1.45 in the earlier period; it had increased to 1.76 by 2008–2009. These are significant increases, especially in the life sciences, and they undoubtedly reflect the effort of universities to recruit (or keep) highly productive faculty who bring in large external grants. This effort was particularly intense during the time that the NIH budget doubled.[10]

We see from both Table 3.1 and Table 3.2 that the gap between full and assistant professors is less noticeable in fields where newly minted PhDs have strong nonacademic options. In such markets, universities must ante up more competitive offers if they are to attract junior faculty.[11] Thus, for the public institutions reported in Table 3.1, the ratio in computer science is 1.52; in engineering it is 1.62, but in the biological and physical sciences it is over 1.7.

The gap between full and assistant professors is generally larger at highly prestigious research-intensive institutions than at less prestigious institutions.[12] This is not only because top institutions recruit and keep exceedingly productive senior faculty and thus pay high salaries to senior faculty;

it is also because prestigious institutions may not need to pay as much at the junior ranks, given the skills and status young faculty can acquire from working with illustrious colleagues.[13]

Inequality of Faculty Salaries

There has been considerable growth in income inequality in the United States over the past thirty to forty years. Academe, too, has experienced an increase in inequality, even among faculty working at doctorate-granting institutions. This can readily be seen from Table 3.3, which shows Gini coefficients by discipline and rank for the period 1975 to 2006 for faculty working at doctorate-granting institutions. (A Gini coefficient of 0 means that everyone receives the same salary; a coefficient of 1 means that all but one individual earn zero.)[14] With but few exceptions, in all fields and at all ranks, the Gini coefficient has more than doubled over the 33-year period. By way of comparison, over approximately the same time period, the Gini coefficient for full-time male earners in the United States grew by 35

Table 3.3. Inequality of salaries of faculty working at doctorate-granting institutions, 1973–2006, selected fields: Gini coefficient

	1973	1985	1995	2006
Engineering				
Assistant	0.072	0.079	0.106	0.164
Associate	0.064	0.082	0.118	0.152
Full	0.091	0.110	0.159	0.220
Math and computer science				
Assistant	0.071	0.115	0.119	0.164
Associate	0.079	0.095	0.143	0.184
Full	0.102	0.113	0.157	0.193
Physical sciences				
Assistant	0.070	0.099	0.132	0.142
Associate	0.091	0.104	0.141	0.146
Full	0.121	0.127	0.167	0.225
Life sciences				
Assistant	0.091	0.098	0.190	0.228
Associate	0.088	0.115	0.168	0.223
Full	0.120	0.128	0.206	0.250

Source: Survey of Doctorate Recipients, National Science Foundation (2011b). The use of NSF data does not imply NSF endorsement of the research methods or conclusions contained in this book.

percent, going from 0.314 to 0.424.[15] Salaries in academe may be more equally distributed than in the larger society, but income inequality has been growing at a much greater rate in academe.

The Relationship of Salary to Productivity

The relatively flat shape of the earnings profile arguably relates to monitoring problems discussed in Chapter 2 and the need to compensate scientists for the risky nature of pursuing research that may not be successful. To continue the tournament analogy of Chapter 2, not everyone who plays wins, and not everyone advances to the next tournament. One can thus think of compensation in science as being composed of two parts: one portion is paid regardless of the individual's success in tournaments; the other is priority-based and reflects the value of the scientist's contribution to science.

This clearly oversimplifies the compensation structure, but there is evidence—though most of it is extremely dated—that counts of publications and citations play a significant role in determining academic salary, directly as well as indirectly. One study found the salary of mathematicians employed at Berkeley between 1965 and 1977 (I said this work is dated!) to be positively related to career publications.[16] Another, based on data spanning the 1970s, found that an additional publication increased the salary for physicists, biochemists, and physiologists by about 0.30 percent.[17] There have been few studies of the relationship between publishing and salary in the ensuing years—perhaps because data are so difficult to assemble or because of ennui with the subject.[18] One exception is a study that uses data collected in the 1999 National Study of Postsecondary Faculty. The study finds, controlling for a large number of factors including region and the research intensity of the institution, that an additional publication increased salary by 0.24 percent—remarkably close to the earlier estimate of 0.30 percent. Although this is not a great deal at the margin, it suggests that a highly productive faculty member with fifty articles would earn about 10 percent more than a colleague with ten articles.[19]

There are other indications that salary bears a strong relationship to the publication record of scientists. For example, publications—and a funding record—play a key role in promotion and tenure decisions at research universities, and, as is clearly seen in Tables 3.1 and 3.2, salary bears a strong relationship to academic rank. Teaching and service matter in promotion, but it is the publication record that plays the key role. Institutions routinely seek letters from external reviewers, who are asked to comment on the contribution the individual has made to the field and to rank the individual on where he or she stands in the field.[20] A recent letter soliciting the opinion of an external reviewer for a tenure case at a top-ranked institution, for example, stated that "in making your evaluation, it would be helpful to us for you to rank Professor X within her peer group both with respect to the sub-field as well as the broader subject area or discipline, as the case may be."

Productivity also plays a key role in determining whether a scientist receives external funding for research. Grant proposals routinely require that the applicant submit a biographical sketch containing publication information. The NSF requires that the publication information be limited to ten articles and or books: the five most relevant to the proposed research and five "other significant publications." "Selected peer-reviewed publications" are a key component of the four-page biographical sketch that must accompany National Institutes of Health (NIH) applications, and they must be numbered and listed in chronological order. (The NIH used to not limit the number of publications that could be listed. As of January 25, 2010, "NIH encourages applicants to limit the number to 15.")[21] Reviewers and panelists routinely comment on the researcher's track record. In the case of the NIH, when faculty apply for a continuation of an R01 grant—the most common form of NIH research support—it is routine for reviewers to examine the quantity and quality of articles published during the previous funding period. Comments such as "excellent productivity of PI," "outstanding record," "very productive researcher," "published x number of papers in funding period" are common.

Funding levels, in turn, affect salary. For medical schools, the relationship can be direct, even for tenured faculty: no grant (from which to charge off salary), no pay (or reduced pay). To be more specific, in 35 percent of medical schools, tenure is accompanied by no financial guarantee for basic science faculty. In 52 percent, tenure is accompanied by a specific financial guarantee, but only in 13 percent is this guarantee for total institutional salary.[22] Thus, increasing amounts of risk are being shifted from universities to faculty, at least in medical schools. Some medical schools also have begun to adopt the practice of awarding bonuses to faculty who have received external funding. In 2004, for example, 59 percent of basic science faculty at medical schools were eligible for bonus pay; 20 percent reported having received bonus pay.[23]

There is also the nontrivial issue of summer pay. Most academic scientists in the United States are hired on nine-to ten-month contracts. It is the grant that pays for their summer, not the institution. Grants are crucial—not only to support one's research, but also to support oneself.

In Europe, salaries of university faculty have been less clearly linked to productivity. In countries such as Belgium, France, and Italy, university faculty are considered civil servants, and salaries are determined at the national level; the local university has virtually no say in negotiating or determining pay, and there is very little mobility between universities. Hence, it is only through promotion—which is based in part on productivity—that individuals are able to leverage publications into higher salaries. Salaries for a specific rank are determined nationally.[24]

This is not the case everywhere, however. In Spain, a special agency (Agencia Nacional de Evaluación de la Calidad y Acreditación, or AN-ECA) was recently set up to evaluate Spanish university faculty for positions leading to tenure, based on their track record of publications; a review process has been in place for more than eighteen years that evaluates tenured individuals for a "sexenio," which is accompanied by a 3 percent raise.[25] Universities in the United Kingdom have developed considerable autonomy when it comes to setting salaries. The Research Assessment Exercise, which allocates resources to university departments and places considerable weight on publications, led to "just-in-time" hiring as universities attempted to build up their fire power in advance of the evaluation exercise.[26] Between 2002 and 2006, the number of faculty earning more than £100,000 in the United Kingdom grew by 169 percent.[27]

In a handful of countries, national policies have been implemented that award cash bonuses to individuals who publish in top international journals. The Chinese Academy of Sciences adopted such a policy in 2001. Rewards vary by institute, but they represent a large amount of cash compared with the standard salary of researchers. Bonuses are particularly high for publications in journals such as *Science* and *Nature* and, depending upon the institute, can be as high as 50 percent of salary. The Korean government inaugurated a similar policy in 2006 whereby 3 million won (U.S. $3,000) or approximately 5 percent of salary is paid to the first and corresponding authors on papers in key journals such as *Science, Nature,* and *Cell.* When bonuses awarded by the university are included, the value can easily exceed 20 percent. In 2008, Turkey introduced a national agency that collects publication data and for each article pays a cash bonus equivalent to approximately 7.5 percent of the average faculty annual salary.[28]

Royalties from Licensing/Patenting

Isolated instances of faculty patenting in the United States go back more than 100 years. In 1907, for example, Frederic Cottrel of the University of

California–Berkeley received the first of six patents for the electrostatic precipator, a device for removing fumes from smoke stacks.[29] Sixteen years later, in 1923, Harry Steenbock and James Cockwell of the University of Wisconsin discovered that exposure to ultraviolet light increased the vitamin D concentration in food; they applied for a patent. In 1935, Robert R. Williams and Robert E. Waterman developed a process for the synthesis of vitamin B_1 in Williams's lab at the University of California–Berkeley and received a patent for the process in 1935. In 1956, Donald F. Jones and Paul C. Mangelsdorf received a patent for what is known as the Jones-Mangelsdorf hybrid seed corn. Mangelsdorf, who was on the faculty of Harvard University, subsequently used his share of royalties to found the Mangelsdorf Chair of Economic Botany at Harvard.[30]

Thus, there is nothing new about faculty patenting. What is new is the rate at which faculty are patenting, the amount of revenues universities and faculty receive from patents, and the direct involvement of universities in managing patents. Cottrel's patents are a case in point. They were managed by the Research Corporation, a corporation set up specifically to manage Cottrel's patents—and to provide a seemly distance between the university and the commercial activity of licensing the patents. Royalties from licensing patents were paid to the corporation and distributed to support university research. (Later, the Research Corporation managed income from patents and licensing for a number of universities.) What is of particular interest is that Cottrel himself chose to receive no royalties from the inventions. The case of the vitamins is similar. Steenbock assigned his patent to the newly created Wisconsin Alumni Research Foundation (WARF), which then licensed the technology to Quaker Oats for breakfast cereals. Steenbock also chose to receive no royalties. WARF subsequently used the proceeds to support research at the University of Wisconsin, where one of its projects was the creation of a library named in honor of Steenbock. The inventors of the vitamin B_1 process followed Cottrel's lead and assigned their patent to the Research Corporation, as did Jones and Mangelsdorf.

The university patent landscape has changed significantly in the ensuing years. In terms of mere volume, between 1969 and 1995 the number of patents issued to universities grew by a factor of 10, going from slightly less than 200 per year to slightly more than 2000. The university share of all patents issued by the U.S. Patent and Trademark Office (USPTO) went from 0.3 percent to approximately 2.0 percent. In the next thirteen years, the number of university patents grew by an additional 50 percent, and by 2008 slightly more than 3,000 patents were issued to universities. The university share of U.S. patents remained at about 2.0 percent.[31]

It is not only that the same faculty are patenting more, but more faculty are patenting. In 1995, only 9.6 percent of faculty reported having been named an inventor on a patent application in the past five years. In 2001, the figure was 11.7 percent; by 2003 (the latest year for which we have reliable data), it was 13.7 percent.[32]

It is common to attribute the dramatic increase in university patenting and licensing to the passage in 1980 of the Bayh-Dole Act, which gave U.S. universities intellectual-property control over inventions resulting from research funded by the federal government, and which universities extended to intellectual property developed from other sources of support.[33] But attributing the increase exclusively to Bayh-Dole is far too simplistic. It ignores dramatic changes that occurred in molecular biology during these years, which opened up opportunities for scientists to conduct research that has the possibility not only of advancing basic understanding but also of being "use" oriented—that is, for scientists to work in what is commonly referred to as Pasteur's Quadrant.[34] It also ignores important court decisions which played a key role in the 1980s in increasing the range of what could be patented and, consequently, the number of patents.[35]

Universities did not stand by passively during the debate that accompanied Bayh-Dole. Rather, a number of universities, including Harvard, Stanford, the University of California, and MIT, actively lobbied for passage of the act.[36] From the national perspective, Bayh-Dole was seen as a way of fostering U.S. competitiveness by clarifying intellectual-property rights arising from federally sponsored research. From a university perspective, it was a matter of economics: licenses could provide needed revenue. It might be incremental, but "it could make a substantial difference" in light of the plateau in federal funding for research that universities experienced in the 1970s. (See the discussion in Chapter 6.)[37]

By the early 1960s, the notion of keeping a discrete distance between the university and commercial operations was in decline. Universities had begun to develop their own offices for technology transfer. One of the most successful was that at Stanford, developed by Neils Reimers in 1968. "I looked up the income we had from Research Corporation from '54 to '67," Reimers said, "and it was something like $4,500. I thought Stanford could do a lot better licensing directly, so I proposed a technology licensing program."[38] Other universities followed suit. By the mid-1990s, almost all research universities had an office of technology transfer.

The Research Corporation seldom had annual gross income of more than $9 million during the years that it was the primary representative of U.S. universities in the patenting and licensing arena.[39] By 1989–1990, U.S. universities reported licensing revenue of $82 million; by 2007, the sum had increased to $1,880 million (excluding the extraordinarily large payment received by New York University in 2007—see the discussion below).[40]

Long gone is the faculty practice of declining a share of the royalties, and the growth in royalties means that there is more to share. Faculty now routinely receive a portion of the net royalty income, although the "sharing" formula varies across universities. In slightly more than 60 percent of all universities, faculty get the same percentage of royalties regardless of whether the sum is $5,000 or $50 million. The average for such arrangements is 42 percent, but there is some variation. For about a third of the universities that pay out a fixed percentage, the rate is at or below 33 percent. But four out of ten universities share fifty-fifty, and a handful share more than 50 percent with faculty. Northwestern University has one of the least favorable sharing rates, paying only 25 percent of the licensing income to faculty inventors; the University of Akron has the most generous rate, paying the inventors 65 percent.[41]

The other 40 percent of universities have chosen to structure the rate regressively, paying out a smaller percentage the larger the amount of royalties received from the patent.[42] For these universities, the rate paid on the first $50,000 is about 49 percent.[43] Because approximately 96 percent of university patents result in royalty payments of less than $50,000, the 49 percent is a close approximation of the average rate that faculty who patent can expect to get in these universities; for faculty working in universities with a fixed share, the average percentage, as previously noted, is 42 percent. These are not, however, the average rates paid on all royalty income because the royalty distribution is heavily skewed.[44] By far the largest percentage of the royalty income comes from the small number of patents paying over $50,000, some of which are of the blockbuster variety, bringing in more than $100 million to the university in royalties. Faculty in universities with fixed sharing formulas receive on average 42 percent on such blockbuster patents as well as on "semi-blockbusters." But for faculty at institutions with regressive formulas, the average rate for the distribution of royalty shares over $1 million (which in most instances is the last rate on the schedule) is 32 percent.

Examples of Blockbuster Patents

The Cohen-Boyer patent for the technique of recombinant DNA (gene splicing) was the first major blockbuster patent to come out of university research in recent years. The patent takes its name from its coinventors,

Stanley Cohen and Herbert Boyer. The pair met at a conference in Hawaii in 1972 and became interested in each other's work. Four months later, they successfully cloned predetermined patterns of DNA.[45] The patent, which was applied for in 1974, was issued in December 1980 after the Supreme Court ruling in June of that year that made the patenting of life forms possible.[46] Two other patents followed, reflecting the fact that during the process the initial patent application was split into three applications. The three related patents were assigned to Stanford where Cohen was an associate professor of medicine at the time of the discovery. Royalties were shared with the University of California–San Francisco, where Boyer was a biochemist and genetic engineer. The first patent expired in 1997, the second in 2001, and the third in 2005. By 2001, the patents had generated $255 million in licensing royalties. The two inventors' share was in the neighborhood of $85 million.[47]

The Cohen-Boyer patent may have been the first blockbuster, but it by no means has generated the largest royalties for universities and faculty. Much larger sums lay down the road. In 2005, Atlantans woke up to find that three Emory faculty had just divided more than $200 million, the result of the sale by Emory of its royalty interest in emtricitabine (Emtriva), used in the treatment of human immunodeficiency virus, to Gilead Sciences and Royalty Pharma. To be more precise, Emory received $525 million in cash. The share of the three inventors, Dennis C. Liotta, Raymond Schinazi, and Woo-Baeg Choi, amounted to 40 percent. These were not the only payments that the university or the professors had received. Emory had been receiving royalty income since licensing the drug in 1996.

Similar deals followed in 2007, first for New York University (NYU) and then for Northwestern University. In the former case, the university sold an undisclosed portion of its worldwide royalty interest in the anti-inflammatory drug infliximab (Remicade) to Royalty Pharma for $650 million in cash. Under the terms of the agreement, NYU retained the portion of the royalty interest payable to the two NYU faculty inventors Jan T. Vilcek and Junming Le, who in collaboration with the company Centacor (which was started by Vilcek), had developed the drug as a treatment for rheumatoid arthritis, Crohn disease, ankylosing spondylitis, psoriatic arthritis, and other inflammatory diseases.[48] Vilcek's share made him sufficiently rich to enable him to announce a gift of $105 million to NYU in 2005. Five years earlier, he and his wife had set up the Vilcek Foundation, with the goal of honoring the contributions of immigrants to science and the arts.[49]

The case of Northwestern University was similar. In late 2007, Royalty Pharma paid Northwestern $700 million for an undisclosed share of Northwestern's royalties from the drug pregabalin (Lyrica). The drug was originally developed to treat diabetes and later epilepsy. Its fortunes rose when, in June 2007, it won U.S. Food and Drug Administration (FDA) approval to treat the common chronic condition of fibromyalgia. The drug was developed by a chemistry professor at Northwestern, Richard B. Silverman, and a postdoctoral fellow at the time, Ryszard Andruszkiewicz. Northwestern's technology transfer policy calls for sharing 25 percent of royalty payments with inventors. Silverman recently made an undisclosed gift to Northwestern to help fund a new research center, which will bear his name.

Paclitaxel (Taxol) is another example of a drug that has generated millions for the university—in this case, Florida State—and for the inventor, Robert Holton, who succeeded in synthesizing it. To be more precise, before Bristol-Myers found another (and cheaper) method for making the drug, which treats certain kinds of breast cancer and ovarian cancer, the university took in more than $350 million in royalty income. Holton's share is reportedly 40 percent of this, or $140 million.[50]

One should not conclude that faculty gold comes exclusively from medically related patents and licenses. Less than a third of all patents issued to universities in the last twenty years have been in the technology areas of pharmaceuticals and biotechnology. Other areas with strong university patent activity are chemicals (19 percent), semiconductors and electronics (6 percent), computers and peripherals (5 percent), and measurement and control equipment (5 percent).[51] But the lion's share of revenue comes from medically related patents. The last year university licensing revenues were reported by field was in 1996, and in that year, among U.S. institutions reporting revenue by field, 76.7 percent of the royalty income came from patents in the life sciences.[52]

Universities also benefit from intellectual property that is not patented. The University of Florida has made millions off trademarked Gatorade. The University of Chicago receives over $4.5 million annually in royalties from the Everyday Mathematics curriculum developed by faculty at the university. Stanford University benefited handsomely from its arrangement with Google, which was started by Larry Page and Sergey Brin while they were PhD students at Stanford in the 1990s.[53]

The Financial Fruits of Inventive Activity

The above discussion makes clear that faculty—albeit a limited number—are enjoying the financial fruits of inventive activity. Excluding the $650 million that NYU received for the previously mentioned sale, net royalties to universities in 2007 equaled $1,880 million.[54] Given that 91 percent of university licensing revenue comes from universities with licenses earning more than 1 million a year in revenue[55] and that on average faculty receive 38 percent of the royalties from licensing mega-agreements, we may conclude that faculty received about $650 million in royalties from megalicenses in 2007.[56]

The number of faculty with such earnings is limited; only fifty-three universities, university systems (in the case of the University of California and the SUNY system), and medical schools reported licenses generating in excess of $1 million in 2004; the average number of licenses earning more than a million was 2.5 for those reporting one or more licenses earning a million or more. If one makes the further assumption that there is a one-to-one correspondence between licensing and patenting (admittedly a bit of a stretch),[57] that the average number of faculty inventors on a patent is three,[58] and that few faculty hold more than one blockbuster patent, one concludes that the $650 million is being shared by approximately 400 faculty. Although this is but a miniscule number, it is sufficiently large—and the amount shared sufficiently impressive—to make other faculty aware that the possibility for receiving "big bucks" from inventive activity clearly exists. Indeed, on more than half of the research-intensive campuses in the United States, there are a handful of faculty who earn more than their salary each year from royalties. For every one of these there are at least thirty times as many faculty who have applied for a patent in the past five years.[59]

Incentives for Patenting

Do faculty patent for the money? Research by Lack and Schankerman finds a positive and significant relationship between the royalty share going to faculty and the revenue a university receives from licensing. The relationship is stronger for private universities than for publics.[60] But when one examines data at the individual level rather than at the university level, the evidence is not as strong. There is no evidence, for example, that faculty are more likely to patent on campuses that provide a more generous share of net royalties with faculty.[61] Moreover, when the number of patents a faculty member has applied for is related to a set of monetary and non-monetary motives, financial motives only prove statistically significant in explaining patenting activity for those in the physical sciences. In engineering, the motives of intellectual challenge and advancement are related to patenting. For those in the biomedical sciences, the motive of having an impact on society trumps all others in the regression analysis, consistent with Raymond Schinazi's view: "Saving lives is what motivates us. Some people can make a beautiful painting, and I can make a beautiful drug. That's enough for me."[62] That is easy for him to say, perhaps, after earning over $70 million in royalty payments from Emory—but others, who have earned considerably less, appear to share his view as well.

Technology transfer offices (TTOs) on university campuses, however, see things differently.[63] When TTO offices were surveyed regarding the perceived importance to faculty of five outcomes (license revenue, license agreements executed, inventions commercialized, sponsored research, and patents), they listed license revenue as the second most important outcome, taking second place to sponsored research. Funds for research are sacred, but royalties are not to be taken lightly.[64] And virtually no faculty turn down the royalty income they are awarded.

The TTOs may be right; the data just are not up to teasing out the relationship between patenting and financial incentives of faculty. One reason is that patenting is a noisy measure of faculty inventive activity. University policy requires faculty to disclose to the TTO if they have a discovery, but it is the TTO that decides if and when to apply for a patent. A large number of disclosures are never patented. Moreover, in a number of instances, the invention is licensed but never patented. It is also important to remember that the financial rewards from inventive activity are highly skewed, and even if realized, occur ten to twenty years down the road. Almost twenty years elapsed between the time that the Emory faculty disclosed to the TTO and the point when they realized their share of the $520 million. The present expected value of a highly unlikely large sum twenty years down the road may provide little incentive compared with other, more immediate rewards.[65]

Faculty Patenting in Other Countries

Patenting is not the exclusive domain of U.S. faculty. European faculty were patenting earlier than U.S. faculty. Lord Kelvin, for example, filed numerous patents in the nineteenth century, and patent royalties contributed to the considerable wealth that he accumulated.[66] Despite their early start, it is far harder to track the patent activity of European faculty because, until recently, "professor privilege"—the assignment of the patent to the faculty inventor and not to the university—has been common. In practice, this means that faculty in many European countries assign the patent to the firm that sponsored their research or for whom they consult. For example, 60 percent of patent applications having a faculty inventor in France are owned by a firm. The comparable figure for Italy is 72 percent and for Sweden 81 percent.[67] In Germany, 79 percent of the patents identified as having an inventor with the title "Prof. Dr." were assigned to industry.[68] In the United States, faculty also patent outside the university, but the vast majority of their patents go through the university. One study, for example, estimates that 67 percent of U.S. faculty patents are assigned to the university; another finds that 74 percent are.[69]

Start-Ups

Robert Tjian, the president of the Howard Hughes Medical Institute, did it when he was on the faculty of the University of California–Berkeley in 1991.[70] Susan Lindquist, a highly productive researcher at MIT and the former director of the Whitehead Institute, who studies protein folding, did it in 2003. Leonard Adleman, Ronald Rivest, and Adi Shamir, the inventors of the encryption algorithm RSA, did it in 1982. Dean Pomerleau, a roboticist at Carnegie Mellon University, did it in 1995. John Kelsoe, a psychiatric geneticist at the University of California–San Diego whose research focuses on looking for the gene behind bipolar disorder, did it in 2007.[71] Elizabeth Blackburn, who shared the Nobel Prize in physiology or medicine in 2009 did it in 2011. James Thomson, the University of Wisconsin professor who developed the first human embryonic stem cells and went on to lead the team that showed that human somatic cells could be reprogrammed to pluripotent stem cells in 2008, has done it twice. Robert Langer, the director of the MIT Technology Lab, has done it thirteen times.[72] Leroy Hood, the 2003 winner of the MIT-Lemelson Prize and a member of all three national academies, has done it more than fourteen times—first while a professor at the California Institute of Technology and then at the University of Washington and the Institute of Systems Biology.[73] Stephen D. H. Hsu, a professor of physics at the University of Oregon, has done it twice.[74] In the late 1990s, over one-third of the forty-five professors in the Stanford University Department of Computer Science were thought to have done it at least once.[75]

"It" refers to starting a company while on the faculty or on leave from a faculty position—another way faculty earn income and wealth. The most profitable scenario for the faculty member generally arises when the company they founded makes an initial public offering (IPO) and a market develops for their equity shares. Sometimes the rewards are of staggering proportions, at least on paper. Eric Brewer, a computer scientist at the University of California–Berkeley, landed on *Fortune* magazine's list of the forty richest Americans under age 40 in October 2000 when the company he founded, Inktomi Corporation, went public and his net worth was reported to be $800 million.[76] (The company subsequently made it onto the Nasdaq 100 before it was bought by Yahoo in 2003.)[77] Leroy Hood has received substantial, although undisclosed, amounts when some of the companies that he helped found—such as Amgen and Applied Biosystems—went public. So, too, has the Harvard professor George Whitesides, who helped start Genzyme Corporation. The amounts involved are not miniscule: it is estimated that academic founders of biotechnology firms that made an IPO during the period 1997 to 2004 held equities with a median value of $3.4 million to $8.7 million (depending on the date of the offering) based on the closing price of the stock the day the IPO was issued.[78]

The rewards can be significant when realized. A study of fifty-two IPOs in biotechnology in the early 1990s, for example, followed forty faculty who had sufficient options or stock to require disclosure at the time of the IPO until early in January 1994. Fourteen of the faculty exercised options and then made a sale that realized a profit. The minimum was $34,285, the maximum was $11,760,000, the median was approximately $250,000, and the mean was $1,237,598.[79]

Faculty also realize substantial gains when the company they founded is sold. David Sinclair, a Harvard professor and founder of Sirtris Pharmaceuticals, held shares in Sirtris worth more than $3.4 million when Glaxo acquired the company in 2008.[80] Robert Tjian received millions in 2004 when Tularik, the company he cofounded when he was a faculty member at University of California–Berkeley, was sold to Amgen for $1.3 billion.[81] Robert Langer received a considerable amount of stock when Advanced Inhalation Research was acquired in 1999 by Alkermes for 3.68 million shares of stock in the company. Stephen Hsu received a substantial amount when Symantec bought one of the two software companies that he had founded for $26 million in cash in 2003—a fact that Hsu lists on his curriculum vitae.[82] John Hennesy, a computer scientist and the tenth president of Stanford University, realized considerable gains when MIPS Technology, a company he cofounded while on sabbatical from Stanford during the academic term 1984–1985, was acquired by Silicon Graphics for $333 million in 1992.[83] He was not alone. A third of the Stanford computer science department were millionaires in 2000, although it is unclear how much of their wealth persisted after the dot-com bubble.[84]

It is also not uncommon for start-up companies to license intellectual property belonging to the university, often based on the invention of the founder. Thus, a number of the scientists involved in start-up companies share in the licensing revenue the university receives and realize additional payments when the firm begins to sell a product and the university receives royalties for the license.

One need not be a founder to enjoy the benefits. There is also a role in start-up companies for faculty colleagues who serve on scientific advisory boards (SABs) of the start-up companies. In biotechnology, the number of faculty involved is substantial; one study, for example, identified 785 unique academic members of SABs for companies that made an initial public offering between 1972 and 2002.[85] The pay is not that high―$500 to $2,500 per meeting―but it is steady, and the majority of SABs offer stock options to members.[86] Faculty who serve as directors―but not as members of the SAB―also frequently receive stock options. It is also not uncommon for faculty members of SABs as well as faculty directors to serve as consultants to the new companies.[87]

Just how common is it for a faculty member to be involved in a start-up? And what percentage of the start-ups survive long enough to yield substantial rewards? That is considerably more difficult to determine. The Association of University Technology Managers (AUTM) estimates that 3,376 academic start-ups were created in the United States from 1980 to 2000. Not all of these were the doing of faculty―some were the brainchildren of students. Google is by far the best known of these, having been started by Sergey Brin and Larry Page in 1995 when they were graduate students at Stanford (and disclosed to Stanford in 1996), but other examples clearly exist. Some university start-ups do not survive long enough to yield substantial rewards; 68 percent of the 3,376 start-ups remained operational in 2001.[88] A considerably smaller number go public: one study puts the lower bound at 8 percent.[89] Although this is a healthy percentage (perhaps 114 times the "going public rate" for U.S. companies generally), it is small and suggests that only a handful of university scientists hit it big through starting companies. Nevertheless, the number is sufficiently large that at many research-intensive universities―and not just the Harvards or Stanfords of the world―at least one or two faculty members have made millions through holding equity in a company that goes public or through a buyout prior to an IPO, and others have benefited, albeit in a more limited way, as directors and members of the SABs. The amounts may pale compared with those received by investment bankers and hedge fund executives in the early 2000s, but by academic standards they represent a fortune.[90]

University-based scientists, of course, have other than financial incentives for becoming involved with start-up firms. In some instances they write joint-authored articles with scientists employed by the firm.[91] They also place graduate students in start-ups. There is also the motive of wanting to contribute to society. John Criscione, a bioengineer at Texas A&M, is a case in point.[92] Criscione, who founded CorInnova in 2004, said, "My goal has always been to provide these technologies to patients that need it, so at that point this became the only route to take—there really wasn't an alternative."[93] But financial rewards clearly provide an incentive for involvement, and, as the previous examples show, the amounts some faculty receive through start-up activity can be considerable.

Consulting

Faculty also augment their income through consulting, a practice that has a long tradition in academe and grew out of the commitment of universities—in many instances since the earliest days of their founding—to provide useful knowledge to the local and regional economy. Although this was often done through the establishment of research and extension programs or by creating courses designed to meet the needs of local industry (such as the University of Akron's research in the processing of rubber, which developed the university's expertise in polymer chemistry),[94] the faculty also consulted with industry. It was, for example, somewhat common for engineering faculty at MIT to serve as consultants to such companies as Standard Oil of New Jersey.[95] The shaping of closer ties between the university and industry is considered to be one of the great legacies Frederick Terman ("the father of Silicon Valley") left to Stanford University. Although these ties took a variety of forms, consulting was one of the activities that Terman actively encouraged while dean of the School of Engineering and later as provost of the university.[96]

Considerable anecdotal evidence exists concerning consulting activity, but there has been little systematic study of how pervasive consulting is among faculty. Indeed, much of what is known comes from surveying the firms, not from surveying faculty. By way of example, a survey of U.S. research and development (R&D) managers found that approximately a third listed "consulting" to be moderately or very important to industrial R&D.[97] An earlier survey asked firms to identify five academic researchers whose work had contributed the most to the development of new products or processes by the firm. A follow-up survey of the faculty firms identified found that 90 percent of the researchers had been consultants to industry; the median amount of time they spent consulting annually was 30 days.[98]

The interaction between firms and faculty is reciprocal: relationships with firms enhance not only income but also the productivity of faculty. Academic researchers with ties to firms report that their academic research problems frequently or predominately are developed out of their industrial consulting, and that this consulting also influences the nature of the work they propose for government-funded research.[99] In the words of an MIT engineer, "it is useful to talk to industry people with real problems because they often reveal interesting research questions."[100]

Additional insight regarding the prevalence of consulting among faculty can be gathered by studying patents that have a university faculty member as an inventor but are assigned to a firm rather than to the university. The practice is somewhat common: one study (as noted earlier) estimates that 33 percent of faculty patents are assigned to industry; another estimates the number to be slightly lower, at 26 percent.[101]

One might initially think that such activity represents nefarious behavior on the part of faculty—as universities almost universally have the policy that inventions belong to the university—but interviews with faculty, technology transfer personnel, and firm R&D managers strongly suggest that the majority of such patents evolve through the consulting activity of faculty.[102] Additional confirmation that patents assigned to firms arise from consulting activity comes from examining the characteristics of the patents, which have been found to be considerably more "incremental" in nature than are patents assigned to universities. This is consistent with studies that find that faculty consulting projects are generally more incremental than projects originating in university labs, which are of a more basic nature.[103]

Some consulting arrangements are extensions of start-up activity. As noted earlier, it is not uncommon for faculty founders, members of SABs, and directors of start-up companies to have a consulting arrangement with the start-up firm. Sometimes, and as part of this arrangement, new patents are filed. While the first patent—often the founding piece of intellectual property—belongs to the university and is licensed to the firm, subsequent inventions made at the start-up belong to the firm.

Some of the consulting activity comes with the active encouragement of the TTO. For example, the university may pass on a faculty disclosure—choosing not to file for a patent—and leave it up to the faculty member to seek a patent. Or the university may patent an invention and, if it decides not to license it, turn the invention over to the faculty member. Or firms may request faculty involvement at the time of licensing because the intellectual property that the firm is licensing is so undeveloped that it is but a "proof of concept" and requires considerable faculty involvement to successfully develop.[104]

Consulting is not the only formal tie that faculty have with industry. Other mechanisms exist. The most common is the practice of industry support for faculty research—what is called sponsored research—which constituted 5.8 percent of all R&D funding at universities in 2009.[105] I will examine this practice in Chapter 6 when I focus on funding for science; for now, suffice it to say that the amount of sponsored research grew dramatically in the 1980s and 1990s, although it tapered off in the early part of the next decade.

Policy Issues

Do increased opportunities for faculty to earn money, be it through patenting, starting a company, or consulting, impede science? Does increased patenting activity, for example, affect the character and quantity of knowledge available in the public domain? Do patents limit academic scientists' access to materials and instruments?

These, and other related questions, belong to the wider debate regarding what is happening to the scientific commons.[106] Some argue, for example, that the financial rewards associated with inventive activity encourage faculty to substitute applied research for basic research.[107] Others argue that patenting diverts faculty from doing research that is published and hence made publicly available.

The evidence suggests otherwise. Research shows that patenting and publishing go hand in hand: the number of patents a faculty member has relates to the number of articles the faculty member has published, and the number of articles published relates to the number of patents.[108] This could, of course, result from unobserved characteristics among researchers, but the research is relatively robust to controlling for such effects. One reason for the high correlation is that patents are often a by-product of a line of research that is published. The large number of patent-paper pairs that have been documented is consistent with this.

The complementarity between patents and publications arises in part because scientists increasingly work in Pasteur's Quadrant, generating both fundamental insights and solutions to problems.[109] The dual nature of research also helps explain why there is little evidence to suggest that the incentives associated with inventive activity have diverted faculty from doing basic research.[110] One can do fundamental research that provides answers to specific questions and has commercial value.

The research and the entrepreneurial activity of Susan Lindquist concerning protein folding provide an excellent example of the dual nature of research occasioned by what Lindquist calls the "blooming of knowledge" in her field.[111] Since her first patent application in 1994, Lindquist has been listed as an inventor on twenty-one other U.S. utility patent applications, and she cofounded a company in 2003. She sees these activities as necessary for "her life's work to make a difference." Along the way, there has been no apparent decline in her production of published research nor in the scientific significance of that work. Since the first patent application, she has authored 143 papers, which have received over 10,622 citations in journals tracked by Thomson Reuters Web of Knowledge. All but one are in journals that bibliometricians classify as "basic."[112]

This does not mean that patenting by universities does not impede research. If managed poorly, patents on materials and instruments can cast a chill on the future research of others, as the work of Murray and her colleagues regarding mice (discussed in Chapter 2) so aptly demonstrates.

It also does not mean that universities are as effective as they could be in the transfer of knowledge to industry. Indeed, some would argue that universities, in an effort to raise revenues, have become overly aggressive in negotiations with industry, thus discouraging the diffusion of knowledge.[113]

The question also arises as to whether the close connection between industry and academe slows the production of public knowledge by discouraging or delaying publication, as well as by discouraging the practice of the open discussion of research within the university community. Numerous studies have looked at the issue—particularly in the biomedical sciences, where the practice of forging close ties between industry and academe became more common in the 1990s. Most find that industry sponsorship comes with the price of delayed publication. I return to this in Chapter 6.

A serious problem for the scientific commons is that some researchers do not make their close and lucrative involvement with industry known, as is generally required by universities and funding agencies. One study found that fully one-third of all articles published in fourteen leading biology and medical journals in 1992 had at least one lead author with a financial interest in a company related to the published research, but virtually none of the authors disclosed the relationship.[114]

The amount of money can be considerable. In 2008, Charles Nemeroff—an Emory psychiatrist—failed to report at least $1.2 million in outside income that he received from drug companies—often for speeches he had made regarding the efficacy of the drugs that he was studying. The NIH responded by initially transferring the $9.3 million study comparing depression treatments to another faculty member. A month later, the NIH halted funding of the grant.[115] The NIH subsequently investigated twenty other faculty members for taking income from drug companies without reporting it.[116]

There is also the concern that faculty put their name on articles that have been "ghosted" for them by industry. In the case of the drug rofecoxib (Vioxx), Merck employees prepared manuscripts and subsequently recruited academics to serve as coauthors. Although 92 percent (22 of 24) of the clinical trial articles included a disclosure of Merck's financial support, only 50 percent (36 of 72) of the review articles contained either a disclosure of sponsorship or a disclosure indicating that the author had received financial compensation from Merck.[117] Such unethical practices diminish the credibility of science and lower public trust in research.

Conclusion

No one would become a scientist solely for the money. There are too many other, more lucrative careers that require fewer years of training and fewer hours of work and pay higher salaries. Nonetheless, success in science is accompanied by monetary rewards, and scientists are not immune to their allure. Just as the prizes attached to tournaments are larger the more skill the tournament requires, the rewards in science depend in part on the level of competition. For example, scientists employed at top research universities earn considerably more than those employed at master's-level institutions. Tournaments also exist within departments: those who are professors almost always earn more than those who are assistant professors, regardless of the institution.[118] But the salaries of scientists and engineers are not entirely linked to research performance. They also depend on contributions to teaching and service within the university.

Scientists and engineers can augment their salary by consulting, an activity that has a long tradition in academe. Moreover, consulting is not the exclusive domain of highly productive scientists. Proximity matters. Research shows that many firms seek out local scientists and engineers as consultants, especially when working on applied problems. The "big guns" are only brought in when the problem is of a more basic nature.[119] Scientists can also augment their income by serving as an expert witness.

Many of these rewards are within the grasp of most scientists and engineers, be they journeymen or stars. Most can hope to accumulate a sufficient record of research to be promoted to full professor. Many will seek out—or be sought by—industry and will earn additional income through consulting.

For a few, the rewards are significantly greater. Some will receive prizes, which, in addition to the honor they bestow, are accompanied by a substantial amount of cash. Possibilities for great wealth also arise through patenting and starting a company. I have taken care to demonstrate that, although the rewards to such activities can be extremely large, few scientists and engineers participate at the megalevel. On the other hand, the rewards associated with patenting and starting up companies are not the exclusive domain of those who strike it big. A significant portion of faculty are associated with a patent application, and a significant number serve on advisory boards and as directors of their colleagues' companies. Although only a small percentage of the inventors will strike it rich, more can expect to earn $10,000 or more a year in royalties. Those on boards often hold equity in their colleagues' companies and also receive compensation for serving as consultants.

One final note: Wealthy a handful may be but there is little evidence that wealthy scientists slow down. Robert Tjian, who earned millions when the company he cofounded was bought by Amgen, has a reputation for the long hours he works at the Howard Hughes Medical Institute. John Hennesy became president of Stanford after the company he founded went public and was eventually acquired by another. LeRoy Hood has continued to work and be productive into his 70s, twenty-five-plus years after founding the first of many companies.

The Production of Research:
People and Patterns of Collaboration

THE LAB OF KATHY GIACOMINI, professor and co-chair of the Department of Bioengineering and Therapeutic Sciences at the University of California–San Francisco (USCF), studies how genes affect the response to medication. The particular focus of the group is how genetic variation in transporter genes across ethnically diverse groups is associated with variation in therapeutic and adverse drug response. The lab also studies novel anticancer platinum agents. In addition to herself, the Giacomini group includes a medical doctor (who directs the clinical studies), a laboratory manager, four postdoctoral fellows (postdocs), five graduate students, and a visiting scientist from Japan.[1] The majority of the funding for the Giacomini lab comes from the National Institutes of Health (NIH). The lab occupies approximately 2,500 square feet at the Mission Bay Campus of UCSF. It uses a variety of equipment and materials in its research, including genetically modified mouse models, cofocal microscopy, and Applied Biosystems (ABI) equipment for sequencing and genotyping. The microscopy and sequencing equipment is "core," and is housed outside the Giacomini lab and used by others. The equipment for genotyping is housed in the lab, but it is also used by other researchers in the building who help pay for the service contract.

The IceCube Nutrino Observatory sits underneath the South Pole. The telescope is the brainchild of Francis Halzen of the University of Wisconsin–Madison, and involves sixty-seven faculty, sixty-two PhD research scientists and postdocs, and ninety-five students, drawn from thirty-three institutions, approximately half of which are located outside the United States. The project was conceived more than twenty years ago; the actual construction of the IceCube Observatory began in 2005. The observatory is designed to detect high-energy neutrinos by capturing the charged particles they create when they interact with nuclei in the ice. The goal is to solve the puzzle of the origin of cosmic rays. The array is a cubic kilometer in size and is composed of eighty-six holes in the ice, varying in depth from 1,450 to 2,450 meters, into which specially designed photomultipliers have been placed to detect neutrino activity. Each hole takes approximately two days to complete. The project deployed the last string of photomultiplers in late December 2010. During the construction period, 170 people worked on the project, although less than 40 could be on the ice at any one time, causing serious scheduling challenges. IceCube also employs a number of technicians and administrators off site. Approximately 85 percent of the $280 million project has been paid for by the National Science Foundation (NSF); other agencies and countries have contributed the other 15 percent.[2]

The fluid physicist David Quéré has two labs, one at the École Superieure de Physique et Chimie Industrielles of France (ESCPI) and the other at the École Polytechnique.[3] Quéré, who is a professor at the École Polytechnique, also teaches at ESPCI and is a research director at the French National Center for Scientific Research (CNRS). The research interests of the group Quéré helps lead cover "systems with liquids in which interfaces play a predominant role." The group calls itself Interfaces & Co.[4] The group is composed of Quéré, another CNRS research director, nine graduate students, three postdocs, and a visitor being hosted from the Tokyo Institute of Technology. Quéré received considerable attention in September 2010 for a paper he published with three members of the group, which used a tank of water, a slingshot, a high-speed camera, and a computer to examine the behavior of projectiles in fluids. The paper concluded by discussing how their research could explain what has become known as "the impossible goal," scored by Brazilian soccer player Roberto Carlos on June 3, 1997, against the French team.[5]

Zhong Lin (ZL) Wang's Nano Research Group in the College of Engineering at the Georgia Institute of Technology works in a wide variety of areas, including the development of nanogenerators for converting mechanical energy into electricity. The group occupies 7,500 square feet of space in the Institute of Paper Science and Technology building at Georgia Tech. Including Wang, the group's size, which is constantly changing, stood at thirty-three in the spring of 2011: seven postdocs, one visiting student from China, eleven graduate students, four research scientists (one of whom is the coordinator of electron microscopy), two research technicians and seven visiting scientists. Funding for Wang's group comes from a number of sources, including the NSF, NIH, Department of Defense (DOE), National Aeronautics and Space Administration (NASA), Defense Advanced Research Project Agency (DARPA), and industry. The group uses a variety of specialized equipment, including a transmission electron microscope, an atomic force microscope, and a field emission gun scanning electron microscope (FEG-SEM), all of which can be seen by clicking on the laboratory tour link on the group's website.[6]

All of the above groups combine inputs, such as effort, knowledge, equipment, materials, and space, to produce research.[7] They do not, however, use the inputs in the same proportion. The importance of equipment, for example, and the way the research is structured, varies considerably. More generally, one model of the production of scientific research does not fit all fields of science and engineering (S&E). Mathematicians, chemists, biologists, high energy physicists, engineers, and oceanographers share certain similarities in terms of the production of scientific research. All, for example, require effort and cognitive inputs. In other dimensions, however, there is considerable variability across fields in the way research is produced.

The way research is organized is a case in point. Mathematicians and theoretical physicists rarely work in labs (although they may identify with a group and work with coauthors), but most chemists, life scientists, engineers, and many experimental physicists do. The role of equipment provides another dimension. In some fields, the equipment required to do research is fairly minimal, as in the case of certain areas of math, chemistry, and fluid physics. In others, research is almost entirely organized and defined by equipment, as in the case of astronomy and high-energy experimental physics. Materials also play a role. In vivo experiments require access to living organisms. For many biomedical researchers this means having—and taking care of—large numbers of mice, and, in more recent years, zebrafish.

Thinking of research as a production process raises several questions. Is there, for example, any evidence of diminishing returns? Are certain inputs complements while others are substitutes for each other? Does a change in the cost of one input, such as the cost of employing a graduate research assistant, lead principal investigators (PIs) to hire more postdocs and cut the number of doctoral students they support? Does an increase in the technological prowess of an instrument, such as that used in sequencing genes, lead to a substitution of equipment for people?

This chapter examines how research is produced, focusing on the people doing science, attributes they possess and patterns of collaboration. The discussion begins by looking at the contributions that scientists make to the process of discovery, in terms of time and cognitive inputs. It continues by examining the important role that labs play in many areas of science. It concludes by examining the substantial and increasing role that collaboration is playing in science. Chapter 5 continues the discussion of production, focusing on the inputs of equipment, materials, and research space.

Time and Cognitive Inputs

Although it is popular to characterize scientists as having instant insight—eureka moments—science takes time and persistence. Productive scientists—and eminent scientists especially—are described as highly motivated, with "stamina" or the capacity to work hard and persist in the pursuit of long-range goals.

Persistence

Persistence is especially important. Slightly over half of the physicists questioned in a study of what it takes to succeed in their field chose persistence from the list of twenty-five adjectives. No other quality came close.[8] The persistence of the cancer researcher Judah Folkman was legendary. It took years before the scientific community accepted his idea that tumors can be choked by blocking blood-vessel growth.[9] Edward Norton Lorenz, the father of chaos theory—sometimes referred to as the third scientific revolution of the twentieth century—is described as being persistent.[10] The inventor Zalman Shapiro, who in June 2009 at age 89 received his fifteenth patent, attributes his success to that quality as well: "Persistence is absolutely essential. You have to be persistent, otherwise you can't come up with anything . . ."[11]

Persistence is closely related to practice, as in "practice makes perfect." Recent work suggests that it is practice—more than talent—that leads to success in fields as diverse as writing, tennis, and music.[12] Persistence also relates to creativity. If creativity occurs, as some would argue, through the chance combining or recombining of two or more ideas, then the more one works, the more likely is one to achieve a creative outcome.[13]

Persistence translates into long hours of work. According to a NSF survey, scientists and engineers in academe for whom research is either the most important or second most important work activity spend 52.6 hours in a typical week working on their main job.[14] Many scientists work even longer hours; the standard deviation was 9.1, and the maximum number of weekly hours reported was 96.[15] One reason for the long hours is that research is not just work—satisfaction is derived from doing research. But the long hours also reflect the need to continue being productive in order to remain competitive and the tournament nature of research, where "the slightest edge can make the difference between success and failure."[16] The amount of time spent on administrative details also contributes to the long hours. A 2006 survey of U.S. scientists found that scientists spend 42 percent of their research time filling out forms and in meetings, tasks split almost evenly between pre-grant (22 percent) and post-grant work (20 percent). The tasks cited as the most burdensome were filling out grant progress reports, hiring personnel, and managing laboratory finances.[17]

Knowledge and Ability

Several dimensions of cognitive resources are associated with discovery. One aspect is ability. Although persistence may trump talent, ability matters. Lorenz not only was persistent, he possessed "plain old intelligence."[18] It is generally believed that a high level of intelligence is required to do science, and several studies have documented that, as a group, scientists have above average IQs.[19] There is also a general consensus that certain people are particularly good at doing science and that a handful are superb.

In recent years, and particularly after Lawrence Summers's presentation at a 2005 National Bureau of Economic Research conference, considerable attention has focused on the relationship between mathematical aptitude and success in science—especially the relationship between success and being in the extreme right-tail of the math distribution.[20] Two questions arise. First, to what extent is there is a relationship? And second, how much does mathematical ability vary by gender—especially at the right-tail of the distribution? Summers (and his critics) focused on the latter, assuming the former to be affirmative—even though the verdict on that is not yet in. Even psychologists Stephen Ceci and Wendy Williams, who have studied the subject extensively, acknowledge that one need not be in the top 1.0 percent or top 0.1 percent of math performance to be successful in science, engineering, or math.[21]

Another dimension of cognitive inputs is the knowledge that a scientist possesses, knowledge that is used not only to solve problems but also to select problems and the sequence in which the problem is addressed.

The importance that knowledge plays in discovery leads to several observations. First, it intensifies races, because the public nature of knowledge means that multiple investigators working in the same field have access to the same underlying knowledge. Work in the area of high-temperature superconductors and induced pluripotent cells are but two cases in point.[22]

Second, knowledge can either be embodied in the scientist(s) working on the research or disembodied but available in the literature (or from discussions with others). Different types of research rely more heavily on one than the other. The nuclear physicist Leo Szilard, who left physics to work in biology, once told the biologist Sydney Brenner that he could never have a comfortable bath after he left physics. "When he was a physicist he could lie in the bath and think for hours, but in biology he was always having to get up to look up another fact."[23]

Third, certain forms of knowledge are tacit, meaning that they cannot readily be written down and codified. The only way to acquire such knowledge is by working directly with individuals knowledgeable in the area. For example, creating transgenic mice, as we have seen in Chapter 2, was not something that one could pick up by reading an article—one needed to train in the lab of someone who had the expertise. Likewise, the new technology of microfluidics requires hands-on training. The importance of tacit knowledge is one reason why scientists and engineers visit other labs—or send their students to visit them. A biomedical researcher reported that a postdoc from Japan expressed no need to find a job after she had completed her training because a job was waiting for her in her mentor's lab in Japan. Her sole purpose in coming (and the reason she had been sent) was to learn specific techniques in which the lab excelled. The honey bees—as graduate students who enhance a lab's productivity are sometimes called— often do so by describing how a problem was approached in a previous lab in which they worked. Although not all of this is tacit, a component is.

Fourth, the knowledge base of a scientist can become obsolete if the scientist fails to keep up with changes occurring in the discipline. Certain fields move so rapidly that an absence of two or three months from the field can prove disastrous. Work with induced pluripotent cells is a case in point; organic synthesis is not. The need to stave off obsolescence is undoubtedly one reason why liberal arts colleges—as well as master's institutions—do not discourage research on the part of their faculty.[24] On the other hand, the presence of fads in science (which are somewhat common in theoretical particle physics) means that the latest educated are not always the best educated.[25] Vintage matters in science, but the latest knowledge is not always the "best" knowledge.

Fifth, there is anecdotal evidence that "too much" knowledge can occasionally be a bad thing in discovery in the sense that it encumbers the researcher. There is the suggestion, for example, that exceptional research may at times be done by the young because the young "know" less than their elders and hence are less encumbered in their choice of problems and in the way they approach a question. This is one of several reasons that exceptional contributions are often more likely to be made by younger persons.[26]

Sixth, and perhaps most important, "many problems in science require an array of cognitive resources that no single scientist is liable to possess."[27] Scientists can augment the knowledge available for addressing a problem by drawing on the cognitive resources of others—by becoming part of a team. Research is rarely done in isolation.

Labs

Collaboration in science often occurs in a lab. The lab environment not only facilitates the exchange of ideas. It also encourages specialization, with individuals working on specific projects or with specific pieces of equipment, materials, or animals. By way of example, there are researchers who are electron microscopists and researchers who are electrophysiologists and use micromanipulators to measure single ion channel activity.

How labs are staffed varies across countries. In Europe, research labs are often staffed by scientists holding permanent positions, although increasingly these positions are held by temporary employees.[28] In the United States, although positions such as staff scientists and research associates exist, the majority of scientists working in the lab are doctoral students and postdocs, as the examples in the introduction to this chapter suggest.

Labs at U.S. universities "belong" to the faculty PI, if not in fact, at least in name, as is readily seen by the common practice of naming the lab for the faculty member. A mere click of the mouse, for example, reveals that all of the twenty-six faculty at MIT in biochemistry and biophysics use their name in referring to their lab.[29] Sometimes, as in the case of the Nobel laureate Philip Sharp, lab members and former members are referred to using a play on the PI's name—in this case "Sharpies."[30] In a similar vein, graduate students and postdocs working in Alexander Pines's lab at Berkeley are known as "pinenuts," and alumni are referred to as "old pinenuts."[31]

It is common practice for labs to maintain webpages, with links to research focus, publications, funding, the PI's curriculum vitae, and members of the research group. Most pages provide pictures of people who work in the lab, sometimes in group shots, other times as individual pictures. Most pictures are of a traditional nature, but it is not uncommon for photos to be on the humorous side. Susan Lindquist's lab at the Whitehead Institute, for example, features a poodle on its webpage. Sometimes the photos are more daring. The webpage for chemist Christine White's lab depicts White seated on a stone throne, engulfed in flames and surrounded by graduate students, one of whom sports horns. Two celebrities have been added to the picture.[32]

Staffing Labs

The mix of personnel, as well as the number of personnel, in U.S. university labs varies by field. The biomedical sciences rely on a considerable number of postdocs. Twenty of the thirty-nine scientists working in Lindquist's lab, for example, are postdocs.[33] But in other labs and other fields, graduate students can outnumber postdocs. A study of 415 labs affiliated with a nanotechnology center, and drawn from departments of chemistry, engineering, and physics, found, for example, the average lab to have twelve scientists, excluding the principal investigator (PI). Fifty percent were graduate students, 16 percent were postdocs, and 8 percent were undergrads.[34]

Populating labs with graduate students and postdocs has been embraced in the United States for a variety of reasons. Pedagogically, it is an efficient training model. It is also an inexpensive way to staff laboratories. The average postdoc earns half to two-thirds of what a staff scientist—the closest substitute to a postdoc in the lab—earns.[35] Moreover, as faculty are not abashed to note, it provides a source of "new" ideas, especially given the relatively young age of doctoral students and postdocs. Trevor Penning, while serving as the Associate Dean for Postdoctoral Research Training at the University of Pennsylvania School of Medicine, was quoted as saying, "A faculty member is only as good as his or her best postdoc."[36]

In addition, funding is often readily available for predoctoral and postdoctoral students. The typical NIH grant, for example, supports both graduate research assistantship and postdoc positions, as do many other forms of grants. The NSF has had the explicit policy of supporting students for many years. According to Rita Colwell, the Director of the NSF from 1998 to 2004, "In the 1980s, NSF asked investigators to put graduate students on their research budgets, saying it preferred to fund graduate students rather than technicians."[37] There is also the added advantage that postdocs and graduate students, with their short tenure, provide for more flexibility in the staffing of laboratories than do permanent technicians.

The mix between postdocs and graduate students depends in part on cost. At first blush, graduate students, who can receive as much as $28,000 a year in certain fields but as little as $16,000 in others, may seem like a bargain compared with a postdoc, who can cost $38,000 or more plus fringe benefits.[38] But the cost advantage can quickly vanish—especially at private universities—once tuition (which can exceed $30,000 and is paid for in part from the PI's grant) is added into the equation.[39]

The cost advantage also depends on the number of hours worked. The average postdoc in 2006 reported working approximately 2,650 hours a year in the life and physical sciences. Postdocs worked about 100 hours less in engineering and about 150 hours less in math and computer science. Contrast this with first- and second-year graduate research assistants, who, while taking classes, often work around thirty or so hours a week in the laboratory. One quickly concludes that before fringe benefits the hourly rate for a postdoc is about half the rate for a graduate student at a private institution in a relatively high-paying field such as the life sciences. And this says nothing of the skill and knowledge advantage the postdoc brings to the lab nor that postdocs can work independently while graduate students, especially in the first years of their program, require supervision.[40] The cost advantage, however, declines as graduate students become more advanced and begin to log in the same, if not more, hours a week in the lab as the postdoc.

Some postdocs are supported on fellowships rather than on the faculty member's grants, providing another cost advantage to populating labs with postdocs. In some labs this is the norm, not the exception. For example, Lindquist's lab page explicitly states that "postdoctoral fellows in the laboratory generally secure independent funding through grants and fellowships."[41] This is not to say that the faculty member plays no role in helping the trainee get the funding. Postdocs can come without a fellowship in hand but with a project in mind, and the PI will help the aspiring candidate write the proposal for the fellowship. It is not all altruism on the part of the PI.[42] The resulting publications come out of the PI's lab (with the PI's name as a coauthor).

Fellowships also play a role in graduate education, as we will see in Chapter 7. However, it is rare for a fellowship to pay for more than three years of study, and it is common for students on a fellowship to work in a lab. Some graduate students in the biomedical sciences are supported on NIH training grants for the first one or two years of study before becoming a graduate research assistant. Rotation through a number of labs is a requirement of the training grant. The bottom line: regardless of the source of support, most graduate students in the United States in experimental fields and in engineering work in labs.

The Number of Graduate Students and Postdocs

The number of graduate students and postdocs involved in university research is considerable. For example, approximately 36,500 postdoctoral scientists and engineers were working in academe in graduate departments in the United States in 2008—more than twice as many as in 1985.[43] Almost 60 percent of the 36,500 postdocs were in the life sciences; the next most likely field for postdocs to be working in was the physical sciences.[44] There is reason to believe that the 36,500 is an undercount of postdocs in academe. Identifying exactly who holds a postdoc position is challenging, given the creative titles that are often bestowed on individuals who are technically postdocs.

Considerably more graduate students than postdocs work with faculty on research. In 2008, for example, approximately 95,000 graduate students worked as research assistants in S&E departments in the United States.[45] An additional 22,500 graduate students in S&E were supported on a fellowship, which often involves work of a research nature; another 7,615 were supported on a traineeship grant, which generally requires work in a lab.

Authorship patterns in the journal *Science* provide one way of examining the role that graduate students and postdocs play in research at U.S. universities. Applying such a lens to articles with strong ties to a U.S. university, one finds that 26 percent of the articles had a graduate student as the first author, and 36 percent had a postdoc as the first author. If one looks at all authors rather than just the first author (on articles having ten or fewer authors), one finds that 22 percent of the authors are postdocs and 20 percent are graduate students.[46]

The United State's reliance on staffing labs with postdocs and graduate students has contributed to its eminence as a training center for foreign-born students. It provides not only a hands-on learning experience but also financial support for graduate study and postdoctoral work, something that many other countries cannot provide. In 2008, almost 60 percent of postdocs in the United States were temporary residents. Forty-four percent of all PhD recipients in S&E were temporary residents.[47] The heavy reliance on foreign talent to staff labs is a topic that we will return to in Chapter 8.

The Pyramid Structure of U.S. Labs

Organizationally, labs in the United States are structured as pyramids. At the pinnacle is the faculty PI—"God in his realm," as one researcher put it.[48] Below the PI are the postdocs, below the postdocs are graduate students, and below them are the lowly undergraduates. Some labs, as already noted, also have scientists who have completed postdoctoral training in the lab or in another lab, and who have been hired in non-tenure-track positions as staff scientists or research scientists.

The pyramid analogy does not stop there, however—in certain ways, the research enterprise itself at U.S. universities resembles a pyramid scheme. In order to staff their labs, faculty recruit PhD students into their graduate programs with funding and the implicit assurance of interesting research careers.[49] They look especially for students who have academic aspirations because such aspirations make them especially good worker bees in the PI's lab. Upon receiving their degree, it is mandatory in most fields for students who aspire to a faculty position to first take an appointment as a postdoc. Postdocs then seek to move on to tenure-track positions in academe. The Sigma Xi study of postdocs, for example, found that 72.7 percent of postdocs looking for a job were "very interested" in a job at a research university and 23.0 percent were "somewhat interested."[50] Such a system of staffing labs with temporary workers—who aspire to the same types of jobs—only works as long as the number of jobs grows quickly enough to absorb the newly trained. In recent years, however, the transition from postdoc to tenure track has proved difficult in many fields because, not surprisingly, the number of tenure-track positions has failed to keep pace with the large number of newly minted PhDs.

It is not uncommon for recent graduates to feel that the system has not delivered what it promised. The inherent problems of a system that relies on young temporary workers to staff labs—and continues to recruit students despite the difficulties recent graduates experience in finding research jobs—is a topic that we return to in Chapters 7 and 10.

Collaboration and Coauthors

A number of factors promote collaboration in science. One, as we have already noted, is the advantage that arises from sharing knowledge with others. Data and material sharing also promote collaboration. A recent paper in *Nature Genetics* concerning how "protein trafficking" contributes to the development of Alzheimer's disease provides a good example. Forty-one researchers working at fourteen different institutions looked at the association between Alzheimer's disease and gene variations in people of varying ethnic backgrounds.[51] Collaboration is also facilitated when scientists conduct research requiring large equipment, such as a telescope or a collider, or, in the case of oceanography and certain areas of geology and marine biology, a vessel.

Coauthorship patterns provide a way of studying the important role that collaboration plays in discovery. They also show the substantial growth in collaboration that has been occurring over time. Papers written by teams increasingly outnumber those written by solo authors. An analysis of approximately 13 million published papers in S&E over the 45-year period 1955 to 2000 found that team size had increased in virtually every one of the 172 subfields studied and that, on average, team size had nearly doubled, going from 1.9 to 3.5 authors per paper. Team size even increased in mathematics, generally seen as the domain of individuals and the field least dependent on capital equipment: during the same period, the fraction of articles written in mathematics with more than one author went from 19 percent to 57 percent, with the mean team size rising from 1.22 to 1.84.[52]

Collaboration patterns are even more striking when one focuses on papers with one or more authors from a research-intensive U.S. university—a group of institutions for which exceptionally good data exist for the period 1981 to 1999.[53] During this nineteen-year period, the average number of coauthors of articles rose from 2.77 to 4.24. Teams were largest in physics (7.26) and smallest in mathematics (1.91). The large number of coauthors associated with physics papers reflects the pattern in high-energy physics of granting authorship to all individuals participating in an experiment. There are reports of physics papers that are shorter than the author list! A recent article on the emission of high-energy gamma rays had more than 250 coauthors, affiliated with sixty-five institutions.[54] Patterns by field are given in Table 4.1.

Growth in the number of authors per paper is due both to a rise in collaboration within a university—and an increase in lab size—and to an increase in the number of labs and institutions collaborating on a research project. A study of publications from 662 U.S. institutions that received NSF funding found that collaboration across institutions, which was rare in 1975, had grown every year; by 2005, the last year of the study, one out of three articles involved scientists or engineers coming from different institutions.[55] During the same time period, the incidence of solo authors declined, as did the incidence of writing exclusively with colleagues at one's own institution.

Scientists and engineers also increasingly collaborate with colleagues in other countries. The foreign share of addresses on papers with one or more authors from a top research university in 1981 (measured by the ratio of foreign affiliations to all affiliations) in the United States was 0.036. By 1999, it was 0.111 (see Table 4.1). The field with the largest share of foreign addresses on papers is astronomy (one in four), followed by physics (one in five). The field with the smallest share is medicine, where only

Table 4.1. Coauthorship patterns at U.S. research institutions by field, 1981 and 1999

	Team size		Ratio of foreign affiliations to all affiliations	
	1981	1999	1981	1999
Agriculture	2.41	3.31	0.028	0.104
Astronomy	2.65	4.95	0.086	0.245
Biology	2.81	4.27	0.034	0.110
Chemistry	2.82	3.60	0.046	0.108
Computer science	1.86	2.64	0.043	0.113
Earth sciences	2.29	3.62	0.052	0.161
Economics	1.57	1.94	0.041	0.094
Engineering	2.29	2.98	0.040	0.105
Mathematics	1.53	1.91	0.071	0.168
Medicine	3.26	4.58	0.021	0.077
Physics	3.09	7.26	0.070	0.196
Psychology	2.21	3.14	0.016	0.059
All fields	2.77	4.24	0.036	0.111

Source: Adams et al. (2005).

0.077 of the addresses are foreign. The considerably higher incidence of international collaboration in physics and astronomy reflects the fact that some major instruments are located outside the United States. For example, the La Silla Paranal Observatory in Chile has eleven instruments and plays a key role in observing the southern skies. The largest particle physics laboratory in the world is located at CERN, in Switzerland. Its newest collider, the Large Hadron Collider (LHC), came online (for a second time) during the fall of 2009.

A recent survey of scientists and engineers working at U.S. academic institutions found that slightly more than a quarter—26.8 percent, to be precise—were collaborating with someone outside the United States on research. The percentage was highest in the physical sciences and computer and information sciences (almost 30 percent) and lowest in math and statistics (23.7 percent).[56] Almost all reported using the telephone or e-mail to collaborate (98 percent); about half of those who collaborated internationally also traveled to do so. A slightly higher percentage of U.S. collaborators worked with someone who traveled to the United States to work on the project. About 40 percent of those who collaborated with someone outside the United States did so by "web-based or virtual" technology.

Authorship, of course, does not necessarily correlate with contribution. Individuals who make a contribution may be excluded (for example, ghost authors) and those who did not may be included in the list of authors. The latter are sometimes referred to as gift, guest, or honorary authors. We have already noted instances of ghost authors in Chapter 3, where scientists in industry write the articles and then recruit faculty to be the named author, with the goal of giving credibility to the work. But ghost authorship can also occur when individuals who work on a project (such as graduate students or junior faculty) are intentionally excluded from the list of coauthors.

It is difficult to know just how common these practices are. A survey of six peer-reviewed medical journals found that 26 percent of review articles contained evidence of honorary authorship and 10 percent contained evidence of ghost authorship.[57] A more recent survey found that 39 percent of Cochrane reviews showed evidence of honorary authors, and 9 percent showed evidence of ghost authors.[58]

Sufficient concern existed in the biomedical community regarding the attribution of authorship to warrant the crafting of criteria for authorship; some journals now require coauthors to list their specific contributions.[59] Most journals in the field have adopted the criteria. The criteria, however, are sufficiently ambiguous to allow considerable variation in what constitutes authorship. In the United States, for example, it is common practice for the PI to be the last author on articles coming out of the PI's lab, regardless of the level of contribution: My lab, my article.

In some fields, everyone who is involved in the larger project is listed as a coauthor, regardless of whether they contributed to the specific piece of research. Articles coming out of the IceCube project, for example, list all project members—256, most recently—as authors in alphabetical order.[60] In other fields, such as in the biomedical sciences, authorship order generally relates, at least to some extent, to the level of contribution. The first author did the heavy lifting; the last author contributed the lab, assembled the team, and set the research agenda. It is less obvious how authorship order is established in between.

Inventorship is more closely guarded, not only because the criteria for inventorship is defined by law, but also because more is at stake.[61] In the case of authorship, it is reputation; in the case of inventorship, it is reputation and money. A study of 680 patent-paper pairs for a sample of Italian academic scientists found the number of coauthors to be higher than the number of coinventors on the patent. First and last authors of articles were less likely to be excluded from patents; the probability of exclusion also decreased with seniority. Although the authorship order finding is congruent with contribution, especially with regard to the first-author finding, the seniority finding suggests that status may affect the outcome.[62]

Factors Contributing to Increased Collaboration

Several factors contribute to the increased role that collaboration plays in research. First, the importance of interdisciplinary research and the major breakthroughs that often occur in emerging disciplines encourage collaboration. Systems biology, which involves the intersection of biology, engineering, and physical sciences, is a case in point.[63] By definition, no one has all the requisite skills required to work in the area; researchers must rely on working with others.

The importance of collaborating with someone with a different skill set is described eloquently by Rita Levi-Montalcini, who found her lack of training in biomedical techniques to be an impediment in trying to identify the "nerve-growth-promoting agent." Then she met Stan Cohen, a biochemist, and "the complementarity of our competences gave us good reason to rejoice instead of causing us inferiority complexes." She recalls Cohen as saying, "Rita, you and I are good, but together we are wonderful."[64] For their collaborative work, the two won the Nobel Prize in Physiology or Medicine in 1986.

Second, researchers arguably are acquiring narrower expertise over time. To some extent, this is a necessary adaptation to the increased educational demands associated with the growth of knowledge over time.[65] But it also reflects the benefits accruing to the group when members specialize.

The evidence supports the gains arising from collaboration: teams produce better science. Team-authored articles receive more citations than sole-authored articles in virtually all fields of S&E, and a team-authored paper is 6.3 times more likely than a solo-authored paper to receive 1,000 or more citations.[66] Articles coauthored with a scientist at another institution (in the United States) are more highly cited—especially if the scientists come from different elite institutions. For example, authors working together at Harvard tend to produce lower impact papers than do authors working together from both Stanford and Harvard.[67]

Third, the rapid diffusion of connectivity has decreased the costs of collaboration across institutions. Twenty-five years ago, the only way to work with someone at another institution was to talk with them on the phone, visit them in person, fax them material, or communicate by mail. Phone calls and travel were expensive. The cheapest ticket to Europe cost approximately $1,800 in today's dollars. Mail required patience. The Internet as we know it today did not exist—nor did e-mail. Data arrived on tape; offsite equipment had to be visited to be operated. The information technology (IT) revolution has changed all of this, making it possible to communicate online, share databases online, and (as we will see in the next chapter) operate equipment online.

The IT revolution can be dated to the creation of ARPANET by the Department of Defense in 1969. Restricted access to ARPANET, however, meant that most researchers could not use it. This led to the development of other networks. Among these, BITNET emerged as the leader. Conceptualized by the Vice Chancellor of University Systems at the City University of New York (CUNY), BITNET was first adopted by CUNY and Yale in May 1981. At its peak in 1991–1992, BITNET connected about 1,400 organizations in forty-nine countries; almost 700 of these were academic institutions.

The speed with which BITNET was adopted by research universities and medical schools (tier 1) is seen in Figure 4.1. Master's institutions (tier 2) and liberal arts colleges (tier 3) were much slower to adopt the new technology. By 1992 (the last year that data on its use were collected), over 80 percent of all research institutions had adopted BITNET, approximately a quarter of master's institutions had adopted it, and slightly more than 10 percent of liberal arts colleges had access to the technology.[68]

By the mid-1990s, BITNET had been replaced by the Internet. A key requirement for efficient communication on the Internet was the development of the domain name system (DNS)—such as harvard.edu. Figure 4.2 uses data regarding the adoption of domain names to plot the speed with which use of the Internet diffused among U.S. institutions of higher education. Particularly noteworthy is the rapidity with which the system diffused and the fact that, although research institutions adopted more quickly, by 2001 almost all institutions that granted a BA degree in the United States had access to the Internet.

When the productivity of biomedical scientists is related to the availability of IT, some support is provided for the idea that the productivity of individuals who worked at institutions that had access to IT, especially early on, increased. The data also support the hypothesis that IT enhances collaboration. There is also evidence that connectivity has differential effects on productivity, depending on a scientist's individual characteristics and position in academe. Specifically, women scientists benefit more than their male colleagues in terms of overall output and an increase in new coauthors. This is consistent with the idea that IT is especially beneficial to individuals who face greater mobility constraints.

There is also evidence that the tier of the research organization matters.[69] The availability of IT has a greater effect on the productivity of scientists at nonelite institutions than it does for scientists at elite institutions.

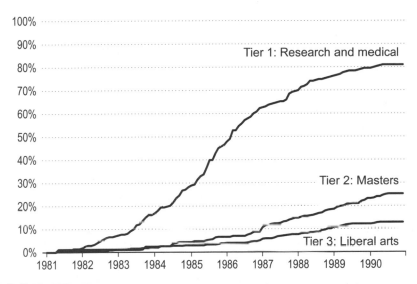

Figure 4.1. Cumulative percentage of institutions adopting BITNET, by tier.
Source: Winkler, Levin, and Stephan (2010).

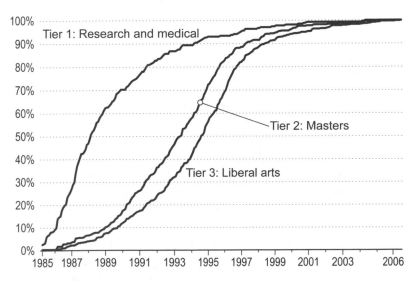

Figure 4.2. Cumulative percentage of institutions adopting a domain name,
by tier. *Source:* Winkler, Levin, and Stephan (2010).

The finding is consistent with the idea that faculty at nonelite institutions, with fewer in-house colleagues and resources, have relatively more to gain from the availability of IT.

The gender and research-tier results suggest that IT has been an equalizing force, at least in terms of the number of publications and the gains in coauthorship, enabling scientists outside the inner circle to participate more fully. A study of engineers found somewhat similar results: those who worked at medium-ranked research universities benefitted the most—in terms of increased publishing—by the adoption of BITNET.[70]

The increasing complexity of equipment also fosters collaboration. At the very extreme are the teams assembled to work at colliders. The LHC's four detectors have a combined team size of just under 6,000: 2,520 for the Compact Muon Detector (CMS), 1,800 for the Atlas, 1,000 for ALICE, and 663 for LHCb.[71] The IceCube Project, with approximately 250 associated scientists, is small by comparison.

The vast amount of data that is becoming available also fosters collaboration by enhancing the proclivity of researchers to work together in solving "large" problems. The Alzheimer's research discussed earlier is but one example. Among other examples from recent years, probably the best known is the Human Genome Project and its associated GenBank database. Many other large databases have recently come online, such as PubChem, which as of April 2009 contained 48 million recorded substances,[72] and the Worldwide Protein Data Bank (wwPDB), a depository of information regarding protein structures. And this is the tip of the iceberg. It is estimated that if all the data produced by the LHC at CERN were burned onto disks, "the stack would rise at the rate of a mile a month."[73]

At least one other factor leads researchers to seek coauthors. That is the desire to minimize risk by diversifying one's research portfolio through collaboration, just as one can minimize financial risk by holding a diversified portfolio.

Some of the factors encouraging collaboration are new (such as greatly enhanced connectivity, the creation of large databases, and the increasing complexity of equipment), but growth in the number of authors on a paper is not new. As noted previously, team size has grown in all but one of the 171 S&E fields studied from 1955 to 2000.

Government Support for Collaborative Research

Governments have bought heavily into the importance of collaborative research. The rationale, although not always explicitly stated, is based on the idea that collaborative research produces better research and creates incentives for labs to share data and materials. Consequently, governments actively foster collaboration within institutions, across institutions, and, in the case of the European Union, across countries. For example, the National Institute of General Medical Sciences (NIGMS) at the NIH has encouraged collaboration within universities by creating initiatives to promote quantitative, interdisciplinary approaches to problems of biomedical significance. In practice, this has led to the funding of centers in systems biology. Another way that the NIH fosters interdisciplinary, collaborative work is by creating training grants in fields that span disciplines and that require departments at the same university to work together in the training of students.

In an effort to foster collaborative research across institutions, the NIH funds large-project grants, called P01s. The grant mechanism is "designed to support research in which the funding of several interdependent projects as a group offers significant scientific advantages over support of these same projects as individual regular research grants."[74] Budgets for P01s are often in the $6 million (direct cost) range.

On a larger scale, NIGMS supports "Glue Grants," with the purpose of making "resources available for currently funded scientists to form research teams to tackle complex problems that are of central importance to biomedical science and to the mission of NIGMS, but that are beyond the means of any one research group." The amount of resources involved can be quite large—on the order of $25 million in direct costs. The goal is to provide sufficient resources "to allow participating investigators to form a consortium to address the research problem in a comprehensive and highly integrated fashion."[75]

The NIH also supports large networked groups—such as the Pharmacogeneics Research Network, made up of groups from twelve different institutions. Each group has ten or more associated researchers; in many instances groups have more than twenty.[76] In the 2010 competition for funds, the UCSF team, led by Giacomini, received $11.9 million for research into the genetics behind membrane transporters. Giacomini will also oversee a $3.2 million NIH grant to continue and expand work with other countries concerning variation in drug responses.[77] All in all, the NIH is spending $161.3 million on the effort. That sum is dwarfed by the more than $700 million that the NIH has spent supporting groups of researchers studying protein structure under two consecutive protein structure initiatives.

On the other side of the Atlantic, the European Union bought heavily into the gains arising from collaboration as a rationale for supporting research at the European level. The great majority of funding under the various Framework Programmes requires research consortiums composed (in most cases) of at least three legal entities based in three different European Union member states.[78] Although such programs create incentives for individuals to work together, research has yet to show their effectiveness relative to other forms of funding. Clearly this is a topic that warrants further research.

It is not just governments that allocate resources with the goal of fostering collaboration. A primary motive behind Harvard's decision to create a new campus in Allston, Massachusetts, was to foster collaboration. The idea is that the new campus will help to connect basic research—which has been done primarily among faculty in Arts and Sciences on the Cambridge campus—with faculty doing applied research, located at the Medical School across the river in the Longwood area of Boston as well as at other hospitals. A major impetus for the creation of the new campus was the realization that Harvard lagged significantly behind such peer institutions as the Massachusetts Institute of Technology (MIT) and Stanford in bringing together faculty doing basic research with faculty doing applied research.[79] The 2008 financial crisis, however, caused Harvard to put plans for the Allston campus on hold (or, as the University said, to "pause in the construction") in December 2009.[80]

Policy Issues

Intellectual property rights, whether in the legal form of being awarded inventorship on a patent or in the symbolic form of being awarded priority of discovery, are still largely conceived of as rights of the individual. As such, they are functional, motivating scientists to do research and share their research with others. But intellectual property rights are more difficult to determine as the number of collaborators working on a problem grows.[81] This presents challenges for organizations. For example, as the number of coauthors grows, it becomes increasingly difficult to evaluate curriculum vitae at tenure and promotion time. It also has become increasingly difficult to maintain the tradition that penalized young scholars for publishing with their mentor subsequent to completing a postdoctoral appointment.[82]

Increased collaboration can also present challenges for individuals. When does one join a team? When does one become a team leader? When does one join a large, multi-institutional collaboration? The U.S. system has, in a sense, made some of these decisions for the individual. One's role on the team is assigned while in graduate school, first as a worker bee, then at dissertation time as a lead researcher on a project for which the student is often the first author, giving the last-author position to the PI. As a postdoc, the young scientist may be lucky enough to lead a small research project in a PI's laboratory. The hope is to move on and have a lab of one's own—to be the one who sets the research agenda and shares in the intellectual property rights of the research coming out of the lab. But as we will see in Chapter 7, a smaller and smaller percentage of scientists are able to make this transition. This means that, if they choose to continue doing research, scientists are likely to play supporting roles for life, and the intellectual property rights that proved so motivating may become farther from their reach.[83]

Growth in collaboration also challenges nonprofit organizations to rethink the awards they bestow in science. Prizes are not awarded to groups; they are handed out one by one (or at most three at a time). Typical are the Nobel, the Kyoto, and the Lemelson-MIT prizes. But pathbreaking work is being done by scientists working together. Choosing a winner among so many is difficult, and it also can be dysfunctional. There is clearly a need to rethink the way in which prizes are structured. Status, as the Nobel Peace Prize so aptly demonstrates, need not be conferred on one person at a time. It is time to think about creating prizes that can be shared by a team.[84]

Conclusion

This chapter has focused on the people doing science, the attributes they possess and their patterns of collaboration. The goal has been to convey several facets of the production process. First, research requires persistence and hard work. Brains help, but science is not all about brains. Second, collaboration plays an important and growing role in science. We see the pattern at the lab level—which in the United States has more or less a pyramid structure and is heavily reliant on the input of graduate students and postdocs. Or we can observe the pattern across labs and institutions, looking at both domestic and international collaborations. Third, fields differ in the production of research in a variety of dimensions, including collaborative patterns, the location in which research is conducted, and the importance of materials and equipment. We turn to a discussion of materials and equipment in Chapter 5.

CHAPTER FIVE

The Production of Research: Equipment and Materials

B IOPHYSICIST LILA GIERASCH was "wooed by an NMR machine" to the University of Texas Southwestern Medical Center after she repeatedly had difficulty obtaining funds to purchase a high-field nuclear magnetic resonance (NMR) machine in an environment where her lab would be the only major user.[1] It's no wonder: high-field NMRs are not cheap. Depending on strength, they currently run anywhere from $2 million to $16 million. The McLaughlin Research Institute in Great Falls, Montana, successfully recruited a researcher when they offered him a mouse package with a mouse per diem that was more than 50 percent less than what he had been paying.[2] Access to equipment and materials matter to researchers and greatly affect productivity. Scientists and engineers know this; so do deans. Start-up packages for faculty consistently contain funds for equipment and materials.

This chapter discusses the importance of equipment and materials in the production of research. It also focuses on the cost of equipment and how the development of new equipment can affect the pace of discovery. It closes with a discussion of the physical space used for academic research.

Several themes emerge from the discussion. First, a technological revolution is occurring in the speed (and associated unit cost) with which discoveries can be made. One consequence of this is that the amount of scientific data that are available is growing at an exceptionally fast pace. Another theme is that the new technologies have the potential to affect the ratio of equipment to people used in research (what economists call the capital/labor ratio). Another is that a considerable market exists for the new technologies that are emerging—and companies are adroit at marketing the new technologies. A recent ad for the Maxwell 16 System touts, "Releasing good research first often leads to a lot of better things—better results, better publications and a better chance for your next grant."[3] Still another theme is that access to equipment and materials can affect stratification in science, in terms of where research is performed. Here, not all forces work in the same direction. For example, increased specialization of equipment, and the associated increase in price, can further stratify the scientific research community in terms of where research is performed. But increased access to materials can have a democratizing effect. The latter effect is reminiscent of the finding, discussed in Chapter 4, that the diffusion of information technology boosted the publications of individuals working at lower tier institutions more than those at higher tier institutions.

Equipment

The important role that equipment plays in scientific research is reported again and again in accounts of scientific discovery. Galileo had his telescope. Boyle had an air pump. X-ray diffraction was key to uncovering the double helix. Einstein could not have redefined "simultaneity" in his 1905 paper on relativity without the technology that led to synchronized clocks. The human genome was successfully mapped because of the development of automated sequencers.[4] The goal to sequence a human genome for $10,000 or less can only be achieved with the development of next-generation sequencers. Perhaps nowhere is the role of equipment more obvious than in particle physics, where accelerators operating at higher and higher levels of energy are opening an inward world that scientists only dreamed of in the not so distant past. To quote Wolfgang Panofsky, the first director of SLAC (formerly known as the Stanford Linear Accelerator), "Physics is generally paced by technology and not by the physical laws. We always seem to ask more questions than we have tools to answer."[5]

The historian of science Derek de Solla Price writes, "If you did not know about the technological opportunities that created the new science, you would understandably think that it all happened by people putting on some sort of new thinking cap . . . The changes of paradigm that accompany great and revolutionary changes may sometimes be caused by inspired thought, but much more commonly they seem due to the application of technology to science."[6]

The key role that equipment plays is one reason to stress what is sometimes referred to as the nonlinear model: scientific research leads to advances in technology, but it is new technology that often brings about advances in science. Peter Galison's account of Einstein's pathbreaking work more than a century ago provides an excellent example, showing that "the new theoretical physics in any age is just as likely to be stimulated by the technologies of the moment as to be spun out platonically from the abstractions of the past."[7] Or consider astronomy, where new technologies allow astronomers to detect electromagnetic radiation of various wavelengths that come from stars and galaxies and facilitate precision studies of the microwaves lingering from the big bang.[8]

In some instances, the scientist is both the researcher and the inventor of the new technology. The biologist Leroy Hood, author of more than 500 papers and inventor of "four instruments that have unlocked much of the mystery of human biology, including the automated DNA sequencer," is an excellent example of such an academic researcher.[9] But numerous other examples exist, and it is common in scientific publications to report on the development of a new tool, such as fluorescent markers or time-lapse microscopes, that permit the detection or observation of things heretofore not observed. It is also common practice to identify the company that manufactured the equipment; this facilitates the reproduction of the research by others.

Costs of Equipment

Some of the equipment and materials used in science are cheap. Gregor Mendel used peas. T. H. Morgan used fruit flies. Alejandro Sánchez Alvarado uses planarian. Susan Lindquist uses yeast. Early researchers in the science of chaos used Apple computers. The lab of fluid physicist David Quéré measures with paper rulers that Ikea freely distributes to anyone who walks in the door. The physics lab of the late Bill Nelson of Georgia State University "scrounged for parts" to build the K-band EPR/ENDOR spectrometer that they used in their research.

But most equipment does not carry bargain-basement prices. Even Quéré's lab, with its reliance on readily available products such as shaving cream, slingshots and a toy gun, requires expensive cameras to capture the experiments. And the spectrometer that Nelson and his group built incorporated a magnet bought for about $125,000 in 1997.

In the United States, it is not uncommon for a scientist to have a lab with a quarter of a million dollars of equipment and materials. And this is toward the lower end; the equipment in a lab can easily exceed $1 million.

More expensive equipment—such as an NMR that costs millions of dollars, or equipment for sequencing—is often shared by scientists working in different labs at the same institution and housed in a core facility.

These expenditures add up. In 2008, U.S. universities spent nearly $1.9 billion on equipment out of current funds.[10] Of this, 41 percent was spent in the life sciences, 17 percent in the physical sciences, and 23 percent in engineering. Johns Hopkins University headed the list in terms of equipment expenditures ($69.8 million); other universities at or consistently near the top in recent years are the University of Wisconsin–Madison, the Massachusetts Institute of Technology (MIT), and the University of California–San Diego.[11]

Exceedingly expensive equipment is generally shared among members of a consortium. The Large Hadron Collider (LHC), which came on line at CERN in 2009 for the second time (and at half its maximum energy), cost $8 billion. The Gemini 8-Meter Telescopes Project (one for the southern skies and one for the northern skies) cost approximately $184 million and has an annual operating budget of $20 million.[12] Chikyu, the Japanese ocean-drilling vessel used in research, cost approximately $550 million.[13] Alvin, the U.S. Navy-owned deep-submergence vehicle, operated by Woods Hole Oceanographic Institution, recently was refitted at a cost of $40 million.

Some scientists and engineers do not require equipment for their research, but most do. Even theorists have become increasingly dependent on computers in modeling mathematical systems for which the required calculations are too complex to compute with paper and pencil.

Not all equipment is located in the lab of the scientist or near the scientist's university. Telescopes are a case in point. The telescope that the California Institute of Technology (Caltech) helps manage is not located in or near Pasadena, California. Instead, the telescope is located in Mauna Kea, Hawaii, where viewing conditions are optimal.[14] Nor, in the determination of protein structure, is the diffraction equipment generally located in the lab of the scientist doing the study. The crystals that B. C. Wang (and other scientists) analyze to determine protein structure are bombarded at Argonne National Laboratory, outside Chicago. And virtually no one has an accelerator in their backyard—especially since SLAC shut down the PEP-II in 2008.[15] Two nuclear physicists from Georgia State who work on the Relativistic Heavy Ion Collider (RHIC) project at Brookhaven National Laboratory in Upton, New York, are part of a 400-plus team of physicists. Some of their work is done at Georgia State, some on site. Many, many more physicists who do experimental particle research will depend on the LHC. Some will do their research at CERN, some as members of virtual communities, others as visiting scientists.

Access to Equipment

There are a variety of ways that scientists and engineers gain access to equipment, but it usually starts with a dean or a department chair who provides space and a start-up package at the time of hiring. Although the packages also include stipends for graduate research assistants and postdoctoral positions, a key component is funds for equipment.[16] In 2003, the average start-up package for an assistant professor in chemistry was $489,000; in biology, it was $403,071. These are not modest sums—they represent four to five times the starting salary that the institution paid a junior faculty member at the time.[17] At the high end, it was $580,000 in chemistry, and $437,000 in biology.[18] For senior faculty, start-up packages averaged $983,929 in chemistry (high end: $1,172,222) and $957,143 in biology (high end: $1,575,000). Start-up packages usually have a life of three years; thereafter, faculty are on their own in raising the funds for equipment (and the funds for other expenses related to running a lab, such as the stipends for postdocs and graduate students).

A major component of grant proposals is the request for funds to buy equipment for one's lab. More expensive equipment, such as an NMR or a magnetic resonance imaging (MRI) machine, is often shared across labs, and institutions commonly submit proposals to foundations to support the purchase of such equipment.[19] Supercomputers, which are typically one of a kind and can cost from $10 to $65 million, are acquired either though a national competition initiated by the National Science Foundation (NSF) or though local initiatives.[20] Access to extremely expensive equipment, such as a telescope, an accelerator, or an underwater vehicle, is often obtained by writing a proposal to a review panel. Time on an NSF-funded supercomputer is allocated in a similar manner.

Lack of access to equipment can affect one's research, as some young physicists learned all too well in 2009. They had planned to use data coming out of the LHC in 2008 and 2009 for their dissertations, but the "accident" that closed the LHC on September 19, 2008 put an end to those plans. The students were forced to lower their sights in terms of available data. They also lost precious time waiting for the LHC to come back online. Astronomers who fail to land jobs at institutions with ready access to a telescope have traditionally had more difficulty getting telescope time and producing research.

More generally, access to equipment is not evenly distributed across universities. There are universities that have funds for the purchase of equipment and those that do not. By way of example, the equipment expenditures of the top five universities (as noted previously) constitute almost

12 percent of the total spent by all U.S. universities on equipment. Some scientists attend graduate schools with state-of-the-art equipment. Some land jobs at institutions that provide strong start-up packages. Some have minimal trouble getting grants that provide funds for the purchase of equipment, but others do not. Where one works makes a difference in terms of career outcomes. We return to this in Chapter 7.

Access to equipment also plays a role in priority of discovery and the recognition that accompanies it, as discussed in Chapter 2. Once the equipment that is required to understand a phenomenon becomes readily available, others can make the discovery as well. It is no wonder that a recent advertisement for the Genome Sequencer FLX system showed a racing horse with the caption "More applications lead to more publications."[21]

The discussion that follows provides examples of equipment that plays a key role in certain fields. It starts with a discussion of sequencing, moves on to a discussion of the role that equipment plays in protein structure determination, and ends with a discussion of telescopes.

Sequencers

The Human Genome Project (HGP) was the first large-scale international project to demonstrate the important role that equipment could play in the biological sciences.[22] The challenge was to sequence the 3 billion base pairs of the human genome and to do so in fifteen years. The sequencing method used to elucidate the genome employs the chain-termination method or Sanger method developed by Frederick Sanger and colleagues at the University of Cambridge in the mid-1970s (another case of eponymy— see Chapter 2). For his seminal work, Sanger was awarded his second Nobel Prize in chemistry in 1980 (which he shared with Walter Gilbert and Paul Berg).[23]

The Sanger method uses dideoxynucleotide triphosphates (ddNTPs) as DNA-chain terminators. It relies on radioactivity to detect the sequence of the four nucleotides (ATGC) of the genetic code. Scaling up the procedure had limitations, both from a hazard point of view and from the fact that it was person intensive: "The whole procedure [was] manual by its very nature and worse, the interpretation of the data was subjective."[24] The sequencing process became safer when fluorescent dyes replaced radioactivity as the means of detection. The dyes produce a chromatogram in which each color represents a different letter in the DNA code.[25]

The procedure became less labor intensive with the invention in 1986 of the DNA sequencer by Leroy Hood and colleagues Michael Hunkapiller and Lloyd Smith. The machine "rapidly determines the order of the four letters across the 24 strings of DNA by labeling the letters with laser-activated fluorescent dyes in red, green, blue or orange."[26] The machine was sold by Applied Biosystems, one of the companies that Hood helped found (see the discussion in Chapter 3).

The machine is one of the four inventions for which Hood won the Lemelson-MIT Prize in 2003. In 2011, Hood was awarded the Fritz J. and Delores H. Russ Prize ($500,000) "for automating DNA sequencing, which has revolutionized biomedicine and forensic science."[27] His inventions (and his interest in invention) were not greatly appreciated by his department at Caltech; their attitude was one of the reasons, according to Hood, that he left Caltech for the University of Washington. The invention, which with incremental improvements made DNA sequencing 3,000 times faster, helped to usher in the genomics revolution, where speed and cost play a key role.

A simple chronology tells the story. When the HGP began in 1990, the best-equipped lab could sequence 1,000 base pairs a day. By January 2000, the twenty laboratories involved in mapping the human genome were collectively sequencing 1,000 base pairs a second, 24/7. The cost per finished base pair fell from $10.00 in 1990[28] to under $0.05 in 2003[29] and was roughly $.01 in 2007.[30] This is now ancient history. Measured in terms of base pairs sequenced per person per day, the productivity of a researcher operating multiple machines increased more than 20,000-fold from the early 1990s to 2007, doubling approximately every 12 months.[31] In terms of overall expenditures, including administrative costs, the HGP cost $3 billion. It is a commentary on the cost reduction resulting from the continued improvement of the equipment that the genome would have been sequenced at a cost of only $25 to $50 million had it been possible from the beginning of the project to use the equipment available in 2006.[32]

Machines were widely acknowledged as playing an important role in bringing the HGP to fruition at the time it was completed. For example, in an article which appeared in June 2000, soon after it had been announced that a working draft of the genome had been compiled, *The New York Times* discussed the key role that sequencing machines played, reporting that machines "reached their zenith in the latest generation of the machines known as capillary sequencers, like PE Biosytems' Prism 3700 and Amersham Pharmacia's excellent though less widely used Megabace." The *Times* went on to say, "If the human genome project were allowed a robotic hero, it would be the Prism 3700."[33] Francis Collins, Michael Morgan, and Aristides Patrinos, the major figures leading the HGP, in their 2003 article regarding lessons from large-scale biology, described the important role that equipment played in the HGP effort, heading the section

"Technology Matters." According to the three, "The advent of capillary sequencing machines from Amersham and Applied Biosystems provided a much-needed boost in efficiency, enhancing the gains already being made due to the use of better enzymes and dyes."[34] Sequencers were not the only technology that made the HGP a reality. Computers played a key role. Without advances in computer technology and software, it would never have been possible to evaluate the quality of the raw data and piece it together.[35]

A new generation of sequencing machines began entering the market in 2005, rendering the earlier machines increasingly obsolete. Rather than read a hundred or fewer different DNA base pairs at a time, these "next generation" machines read millions of sequences at once, although the "length" of the base that is read is substantially shorter. It is not just that the machines themselves are faster; new reagents for the machines and new software also make them faster.

The first of these next generation sequencers was invented by Jonathan Rothberg and marketed by the company he helped found, 454, now a subsidiary of Roche.[36] The initial sequencer they sold had a read length of 100 bases and could sequence 20 million bases in less than five hours. In 2010, the company had an instrument on the market with a read length of 400 to 500 bases and the ability to generate more than 1 million sequencing reads per ten-hour run; they were hyping that longer read lengths would be forthcoming in 2011. With a wink and a nod to readers, the company promoted the longer length in advertisements for the FLX system that proclaimed "length really matters."[37]

The company—and Rothberg in particular—was extremely creative in getting the word out about its FLX instruments. For example, in 2006 they approached James Watson (of double helix fame) regarding the possibility of mapping his genome; early in 2007, they made the announcement that they had mapped Watson's genome at a cost of $200,000, using their technology.[38] They succeeded in getting the 454 equipment installed at the Broad Institute, a leader in sequencing, and they successfully partnered with a researcher in Germany (Svante Paabo) to sequence the first million base pairs of the Neanderthal genome.[39] The research was reported in a cover article of *Nature* in 2006. The company was awarded *The Wall Street Journal* Gold Medal for Innovation in 2005 for their method for low-cost gene sequencing.[40]

Rothberg himself is an interesting example of the entrepreneurial scientist discussed in Chapter 3. He started his first company, CuraGen, while completing his PhD in biology at Yale University and since then has founded or cofounded three other science-based companies: 454,

RainDance, and Ion Torrent Systems. He attributes his motivation for inventing a faster sequencer (and eventually founding 454) to his son's visit to an emergency room. In 2002, he established the Rothberg Institute for Childhood Diseases, dedicated to finding a cure for children suffering from tuberous sclerosis complex, a genetic disorder that his oldest daughter has.[41]

At least three other next-generation machines rapidly entered the market, one from Helicos, one from Applied Biosystems, and one from Illumina. Helicos's cofounder Stephen Quake made headlines in 2009 when he announced in his *New York Times* blog that he had successfully mapped himself using the Helicos equipment. Four months later, Quake published an article in *Nature Biotechnology* that showed the amount of overlap between his genome and the genomes of Watson and Craig Venter (whose genome had been mapped in 2007). The publication was followed by an article in *The New York Times* that reported that the mapping had taken four weeks and a staff of three and had cost $50,000.[42] This is notable given that just two years before it had taken 454 something like two months to map Watson's genome, and this was for only three "passes"— nine passes were required to produce the final draft of the HGP.[43] Despite the Helicos hype, it was Illumina's machine that captured the second-generation market.

New-generation machines are changing the location of the work and the number of researchers who have access to sequencing technology. Just how it will sort out is still up in the air as new equipment and new business models come online. The next-generation sequencing equipment introduced in 2007 was not cheap. Illumina's Genome Analyzer System, for example, costs $470,000 (about $170,000 more than the cost of Applied Biosystems' model 3730 sequencer) and the Helicos Single Molecule Sequencer costs about $1 million "depending on how hard you bargain."[44] But the speed and associated lower unit cost mean that the equipment has the potential of being used in a large number of labs and hospitals to address a number of research and clinical questions. This is in contrast to first-generation equipment, which eventually was being run in a small number of highly specialized labs. Illumina uses access as a selling point, noting on its website that the Genome Analyzer System "enables even the smallest lab to have the sequencing capabilities of the largest genome centers."[45] Despite this, equipment and access remain highly concentrated. One estimate put half the world's 1,400 sequencing machines in just twenty academic and research settings in 2010.[46]

The business model for sequencing is also in flux. Just when it looked like the next-generation equipment might increase the number of locations doing sequencing, the consolidated model of sequencing got a big boost when Complete Genomics, of Mountain View, California, successfully sequenced material supplied by the Institute of Systems Biology in Seattle.[47] One of the coauthors of the resulting publication was no other than Leroy Hood, the father of the original sequencing machine, who serves on the scientific advisory board of Complete Genomics. The project: to decode the genomes of two children with rare genetic diseases and compare their genomes to those of their parents. The research was published via *Science-Xpress* in March 2010. Complete Genomics reports that they have "perfected a low-cost, high-quality sequencing method that will cut time and reduce the cost for researchers from as much as $250,000 to as little as $5,000."[48] It has the goal of opening ten sequencing centers around the world with the capacity of sequencing 1 million human genomes annually. If they have their way, sequencing will become a service industry, and researchers, regardless of location, will have access to the technology.

If Jonathan Rothberg has his way, sequencing technology is more likely to remain in house—and in more houses. In March 2010, he demonstrated a silicon-chip sequencer, manufactured by his latest company, Ion Torrent Systems, which directly translates chemical information into digital data. The sequencer became available at the bargain basement price of $50,000 in January 2011. Rothberg's goal is to open the sequencing field to hundreds of smaller research groups that currently lack access to sequencing technology at their research facilities. He also envisions putting the small machines (the size of a desktop printer) in doctors' offices. The name he chose for the machine, Personal Genome Machine (PGM), reflects this ambition. In its current form, however, the machine only sequences 10 million bases per run, making the cost per base pair extremely high and inappropriate for sequencing the entire genome.[49] Other competitors are actively pursuing third-generation alternatives. Pacific Biosciences introduced the first machine to scan a single DNA molecule in real time in 2010. The machine (known as the *RS*) was awarded the "top invention of 2010" by *The Scientist*.[50] One of the three judges was no other than Jonathan Rothberg!

One thing is for sure: the new sequencing technologies require fewer technicians. This became abundantly clear when the Venter Institute eliminated twenty-nine sequencing-center jobs in December 2008, announcing that the staff reduction "is a direct result of a technology shift and is not a reflection of the tough economic times that we are all facing in the United States today."[51] The Broad Institute followed about seven weeks later, firing twenty-four staff, saying once again that the layoffs were due to a shift in technology and were not related to the recession.[52] The layoffs come as no surprise to economists, whose models predict that a change in relative prices will lead to a substitution of the relatively cheaper input for the relatively more expensive input.

The decline in the cost has led to the goal of sequencing personal genomes for $1,000 or less. To incentivize the race, the Archon X Prize for Genomics was established in March 2007 with the goal of awarding $10 million to the first group that can "build a device and use it to sequence 100 human genomes within 10 days or less, with an accuracy of no more than one error in every 100,000 bases sequenced, with sequences accurately covering at least 98 percent of the genome, and at a recurring cost of no more than $10,000 per genome."[53]

Should the HGP and the sequencing technology that has evolved be viewed as a major step forward in addressing human disease? The answer depends upon whom one talks to, and their time horizon. For Eric Lander (the first author on the first published draft of the human genome and the head of the Broad Institute, a leader in genome medicine, in Cambridge, Massachusetts) the answer is "yes." According to Lander, speed (and associated lower costs) mean that sequencing "can be applied to about any problem." The new instruments offer, for example, a better understanding of diseases associated with problematic genes, as well as the prospect of personal genomics.

Francis Collins, the leader of the HGP, sees the glass as half-full. The HGP and sequencing technology is "helping to piece together many of [medicine's biggest] puzzles." But not at the rate Collins predicted in 2000. "The First Law of Technology," according to Collins, "says we invariably overestimate the short-term impact of a truly transformational discovery, while underestimating its longer-term effects."[54]

Others see it differently. Despite advances in new drugs for a few cancers, and genetic tests that can predict the efficacy of a handful of drugs or whether people with breast cancer need chemotherapy, the "original hope that close study of the genome would identify mutations or variants that cause diseases like cancer, Alzheimer's and heart ailments and generate treatments for them has given way to the realization that the causes of most diseases are enormously complex and not easily traced to a single mutation or two."[55] By way of example, a 2010 study led by Nina P. Paynter of Brigham and Women's Hospital in Boston found that 101 genetic variants that had been statistically linked to heart disease had no value in predicting who among 19,000 women had gotten heart disease. Family history, on the other hand, was a significant predictor.[56]

Protein Structure Determination

Proteins, which are present in all biological organisms, fold into spatial conformations in order to perform their biological function. Determination of the three-dimensional structure of a protein is important in under-standing protein function at a molecular level and is a major component of the field of structural biology.[57] Structural determination has generally been a difficult, time-intensive procedure. The protein must first be crystal-lized, then the crystal must be successfully mounted for an X-ray diffrac-tion study, and finally the resulting data must be analyzed to determine the structure. Crystals play such a key role in determining structure and are sufficiently difficult to grow that a common saying in the grants commu-nity used to be "no crystal, no grant."

In recent years, structural determination has been greatly expedited through the development of new technologies and software. Much of the funding for this has come from the National Institute of General Medical Sciences (NIGMS) at the NIH, which has funded a series of Protein Struc-ture Initiatives (PSI). Although in some ways the PSI project has been a disappointment, providing (to date) "structures that are by and large di-vorced from biological function," the technological progress that has evolved has been considered a major success. The same assessment report that spoke of concerns and disappointments regarding the initiative also concluded that "the PSI has been highly successful in establishing an auto-mated pipeline for protein production and structure determination."[58]

One important technological advance has been in the use of robotics to grow and screen crystals. For example, a robot can set up multiple crystal-lization experiments simultaneously and can automatically screen whether a crystal is being grown and the quality and size of a crystal if crystalliza-tion occurs. One such system is produced by Thermo Scientific. Such ro-botic systems, with accessories, cost on the magnitude of $57,000.[59]

Technological advances also play an important role in the actual diffrac-tion studies conducted at a synchrotron. A visit to the lab of Bi Cheng Wang (who prefers to be called B. C.) at the University of Georgia pro-vides a good example of their role and the evolution of the technology.

Wang was recruited to the University of Georgia in 1995 as a Georgia Research Alliance Eminent Scholar in an effort to build up the program in structural biology at the university. At approximately the same time, Ar-gonne National Laboratory in Illinois announced that it would be opening a new facility, called the Advanced Photon Source (APS). The national laboratory was looking for groups or consortia to build one or more of the thirty-six available sectors.

At the time that Argonne made the announcement, a number of research-ers in the southeast were using the facilities at the Brookhaven National Laboratory (New York), Lawrence Berkeley National Laboratory (Califor-nia), or the Stanford Synchrotron Radiation Lightsource at SLAC (Califor-nia), but there were no formal groups or consortia in the southeast. In June

1997, Wang called a meeting of regional researchers to see if they were interested in forming a consortium. Thirty people from a number of institutions attended, and the Southeast Regional Collaborative Access Team (SER-CAT) consortium was formed. Initially, a share in SER-CAT cost $250,000; several institutions bought more than one share, including Wang's group, which purchased four. Other universities in the state that joined were Emory University, Georgia State University, and the Georgia Institute of Technology. At the time I spoke with Wang in 2008, SER-CAT had sold fifty-four of the seventy available shares. (Membership is not limited to southeastern institutions—the University of Illinois at Chicago, for example, is a member.) In addition to the initial membership fee, an annual operational maintenance fee is also assessed, which in 2008 was approximately $38,000.

Each synchrotron beamline costs approximately $7 million to construct; each sector has two or more beamlines, with individual detectors, plus a possible small backup detector. The first SER-CAT beamline was finished in 2002. At that time, the standard procedure was for a researcher to go to Argonne to conduct the diffraction study.

As early as 1999, Wang and his group began looking into the idea of building a robot, with the goal of increasing efficiency at the SER-CAT sector. The SER-CAT Board, however, was not enthusiastic about the idea and preferred to focus on beamline development rather than robotics. By 2002, robotics were being used elsewhere in diffraction studies. Another group at the APS bought a Rigaku robot, and the University of California–Berkeley designed their own robot. At this point, the SER-CAT group realized that they, too, needed a robot. They used the Berkeley design (which was publicly available) as a template to build a modified robot at Argonne.

One of the ways in which SER-CAT increased efficiency was to reduce the amount of machine time allocated for a run. By 2002, other facilities were typically allocating two days for each user group to visit their facilities. SER-CAT was able to reduce this by one day yet enable users to collect their needed data. By the time I visited Wang in 2008, the goal was to whittle the run time down to six hours in the near future. They were also able to increase efficiency by creating software for high-throughput structure determination on site. In 2004, the group succeeded in determining five structures in twenty-four hours. In 2008, a researcher on the SER-CAT beamline got five structures in six hours. At the time of this writing, SER-CAT has not yet implemented the six-hour runs, but since the summer of 2009 they have been allocating twelve-hour-run shifts. To make this possible, they hired two additional staff members and extended their on-site user-support services from eight hours to sixteen hours a day.

A complementary innovation that further increased efficiency is that members no longer need go to Argonne to collect data. They can control the robot off site from a home or lab computer with software that takes a minute to mount, center, and start the data collection process. (This was cautiously termed by many synchrotron facilities "remote control data collection." It was later called "remote participation," and is now more commonly referred to as "remote access.") Wang initially became intrigued with the remote access idea while visiting a National Aeronautics and Space Administration (NASA) facility in 1999. If NASA could control equipment in outer space from a computer, why couldn't he (and others) mount a crystal and do diffraction studies remotely?

Argonne's rules require that 25 percent of the operating time be used by nonconsortium members, so non members can use the facility as well, including remote access, for data collection. The crystals are shipped to Argonne by mail. Long before remote access became available at SER-CAT, SER-CAT instituted the practice called "FedEx crystallography" or "mail-in crystallography," by which researchers mail their crystals to SER-CAT. As a special service to members who needed the data quickly or preferred not to travel, the staff would collect the data for them personally, a practice that continues at SER-CAT today but only as a special perk for its institutional members. The FedEx crystallography service SER-CAT pioneered has been adopted by others in the protein structure community.

The closest competing method for determining protein structure is NMR spectroscopy, which has produced slightly more than 7,800 structures.[60] Kurt Wüthrich, the first to have used the method for determining structure, shared the 2002 Nobel Prize in chemistry for this work.[61] The major advantage of NMR over X-ray crystallography is the ability to determine protein structures in solution under near physiological conditions, without the need to crystallize proteins into an ordered lattice. However, NMR is labor intensive and is largely limited to proteins of smaller size, disadvantages that are currently being overcome. The other emerging method for the characterization of proteins is mass spectrometry.

Protein structures, once determined, are deposited in the Protein Data Base (PDB), a repository for three-dimensional structural data of proteins and nucleic acids. In 1971, when it was created at the Brookhaven National Laboratory, it contained seven structures. By the summer of 2009, it contained over 59,000 structures. It is currently headquartered at Rutgers University.[62] Over 3,500 of the structures deposited by the summer of 2009 were identified by researchers supported through the Protein Structure Initiative of NIGMS at the NIH.

Telescopes

The telescope—which celebrated its 400th anniversary in 2008, is one of the oldest instruments used in the study of science. Without it, Galileo would not have observed the moons of Jupiter or refuted the established view that the universe revolves around the earth. Although Galileo's telescope was small and portable (and he fiercely guarded the knowledge of its workings), within a fairly short time telescopes became considerably larger. They also began to be supported by governments. By 1675, for example, England had established the Royal Observatory at Greenwich. A major rationale for royal support was that the telescope was thought to be key to solving the "longitude problem," crucial for a seafaring nation such as England that routinely lost ships because of the inability to determine longitude.[63]

Today, a variety of types of scopes are in use, including optical, radio, and space, as well as instruments for detecting cosmic neutrinos produced by violent astrophysical sources and for detecting high-energy gamma-rays. Historically, optical scopes in the United States have belonged to a university or a consortium of universities, dividing U.S. astronomers into "the haves and the have-nots." Considerable animosity exists between the two communities. As one have-not astronomer said, "They [the haves] don't give a flying fuck about the rest of us."[64] The instruments controlled by Caltech are a case in point. For forty-five years (1948–1993), the university operated the world's largest optical telescope (200 inch or 5.1 meter) at the Palomar Observatory in California. It was surpassed only when Caltech joined with the University of California to build a 10-meter telescope at Mauna Kai, funded by the W. M. Keck Foundation and named the W. M. Keck Center.[65]

Not all scopes "belong" to an institution (or consortium of institutions), and, as the cost of building and running telescopes has increased and demand for time on telescopes has grown, the trend has been toward building national and international telescopes, made up of consortia of universities and/or nations. Kitt Peak, built with NSF funds and operated by a consortium of U.S. universities, is one such example. But by the end of the 1970s, the demand for observing time at Kitt Peak's two largest scopes outnumbered the available nights by a factor of three, and pressure began to mount to build another optical telescope.[66] As a consequence, a portion of the U.S. astronomy community (especially the have-nots) began to explore building a new telescope with government support. Eventually this effort evolved into the Gemini 8-Meter Telescopes Project, and two scopes were built: one for the southern skies in Chile and one for the northern skies in Mauna Kea, Hawaii. Along the way, it became an international consortium, with partners drawn from a number of countries, including Brazil, Argentina, Chili, Australia, Canada, and the United Kingdom. The project initially cost $184 million, and currently costs $20 million annually to operate. New instruments have been added over time.

Twenty-five percent of the time on the Gemini is allocated to engineers working on the telescope and to the host country and local staff. The rest of the time is allocated to countries by a formula, depending on the amount of support (the United States gets approximately 35 percent of the time) and through peer review, with proposals submitted to the National Time Allocation Committees.

Some telescopes are dedicated to specific projects. A 2.5-meter telescope on Apache Point in New Mexico, for example, has been dedicated to surveying the skies since 2000. The project, known as the Sloan Digital Sky Survey (SDSS), is equipped with a 120-megapixel camera.[67] It has generated millions of images of galaxies that can be viewed by volunteers online in an outreach program called Galaxy Zoo. The $150 million effort is named for the Alfred P. Sloan Foundation, which has provided support for the project. Papers coming out of the project are a team effort. According to Michael Strauss of Princeton University, "People giggled when we put out papers with 100 authors. But we showed that many astronomers could get along without killing each other and [that] a large survey could be enormously scientifically productive."[68]

In recent years, competition in the optical community intensified considerably when Caltech and the University of California announced plans to build a 30-meter telescope (TMT). Much of the funding for the $77 million design-development phase was provided from the Gordon and Betty Moore Foundation (of Moore's Law fame); additional funds were provided by Canadian partners. Part of the funds for building the $1 billion telescope will come from the Moore Foundation, which made a $200 million gift, and from Caltech and the University of California, which have made a joint pledge of $100 million.[69] Canadian partners will provide the enclosure, the telescope structure, and the first light-adaptive optics.[70] But the cost is so substantial that the project could well stumble before its projected completion in 2018. The location for the telescope was announced in the summer of 2009. Once again Mauna Kai was selected.

Rather than rely on one giant mirror, the TMT uses technology developed by Jerry Nelson, an applied physicist at the Lawrence Berkeley National Laboratory, to join together 492 thin, hexagonal mirror segments to form a smooth parabolic surface.

The TMT is not the only large optical telescope on the drawing board in the United States. The Giant Magellan Telescope (GMT) is also in the design stage, nipping at the heels of the TMT project. The project, led by Carnegie Observatories and the University of Arizona, is based on using seven monolithic 8.4-meter mirrors, arranged like flower petals, to function as a mirror 24.5 meters in diameter.[71] The chosen location is Las Campanas in Chile. The estimated cost is $700 million.[72] Considerable rivalry exists between Jerry Nelson and Roger Angel, the designer of the monolithic mirrors for the GMT, as well as between the two projects.

Astronomy, as already noted, is highly competitive. Both the TMT and the GMT will be dwarfed if European astronomers have their way and succeed in building the European Extremely Large Telescope (E-ELT). The scope is planned to have a 42-meter segmented-mirror—almost half the length of a football field.[73] Europe had to "settle" for the 42-meter telescope after plans to build a 100-meter scope—known, appropriately, as the "Overwhelmingly Large Telescope" (OWL)—proved too expensive and overly complex. The 42-meter telescope in all likelihood will be located in either Chile or the Canary Islands. Planning is still in preliminary stages; the earliest that the $1.5 billion facility could open is 2016. The current plan is to build the primary mirror with hexagonal panels about the same size as those used in the TMT design.[74]

Optical telescopes can be configured for different purposes. Not all configurations carry huge price tags. One such example is the astronomical interferometer operated by the Center for High Angular Resolution Astronomy (CHARA) at Mount Wilson in California. The array was the brainchild of the astronomer Harold McAlister at Georgia State University and consists of six 1-meter telescopes. It was built with funds from the NSF, Georgia State University, the W. M. Keck Foundation, and the David and Lucile Packard Foundation. McAlister began the search for funding in the early 1980s and received initial seed money from the NSF in 1985. Ground was broken at Mount Wilson in 1996. The telescope became fully functional in 2004. Excluding Georgia State's contribution, the telescope cost slightly over $8 million to construct. The array can be operated remotely from Georgia State, 2,000 miles away.[75]

Radio astronomy also provides key insights into the universe.[76] The largest of the radio telescope facilities currently on the drawing board is the Square Kilometer Array (SKA), with a projected construction cost of $1.5 billion and an annual operating budget of $100 million. If and when SKA is built, it will dwarf the 305-meter-diameter Arecibo radio telescope, which was opened in 1963 in the Puerto Rican city it is named for and is funded by the NSF and managed by Cornell University.[77] Observations made by Joseph Taylor and Russell Husle at Arecibo provided the first proof that gravity waves, predicted by Einstein's General Theory of Relativity, actually exist. The two won the Nobel Prize in 1993.[78] A 2006 review instigated by the NSF recommended that the NSF stop funding Arecibo in 2011.[79]

Fifty-five institutes and nineteen countries are involved in the planning and funding of the SKA, which will have 3000 dish antennas as well as two other types of radio wave receptors.[80] The main aim of the SKA is to "search for faint radio signals from the most distant reaches of the universe, helping scientists examine clues to what existed before the first stars were born and to probe the nature of dark matter and dark energy."[81] But there are many obstacles to overcome before it can be completed. Selection of a location is one of these: unlike optical scopes, where the number of appropriate locations is limited by the clarity of night skies and the number of days in the year with clear nights, there are a number of places where the SKA could be constructed. And, just like the Olympics, there is considerable competition: China wanted it, as did Australia, South Africa, and Argentina. By the spring of 2011 selection had been narrowed to sites in either South Africa or Australia.[82]

The SKA provides an excellent example of the extremely long horizon required to create a new instrument. In this case, planning first started in the early 1990s; it is unlikely that the SKA will be finished before 2022. The instrument is clearly for the use of the next generation of radio astronomers. This generation's reward is to design and create it, much like planting an olive tree for one's child or grandchild.[83]

But it is not all about one's children or grandchildren. There are rewards along the way: many of the instruments are conceived by an individual or a group of individuals who gain status and a sense of accomplishment by watching "their" instrument be built. There are also papers that are generated along the way. Francis Halzen, the physicist "father" of the Ice-Cube project—the $280 million neutrino observatory built in the ice in Antarctica—became an expert on glaciers as the project developed and has coauthored papers in the area.

Telescopes are not restricted to the earth. The Hubble Space Telescope, launched by NASA in 1990, is the best example to date of a telescope that operates in outer space. It is also an "open-use" facility in the sense that anyone can apply for observing time without restriction to nationality or academic affiliation. Competition, however, is intense: only about one in six of the proposals for observation are selected. Furthermore, unlike earth scopes, Hubble's days are numbered; NASA expects that it will be out of commission by 2019 if not earlier.[84]

Hubble is controlled remotely; given its location, this is a necessity. But as telescopes get larger—and more expensive—it is likely that most optical scopes will be run remotely as well. "With thousands of astronomers clamoring for observation time, the scheduling of observations and steering of the telescope are likely to be fully automated to squeeze out every useful second."[85] Moreover, the competition for time is likely to force astronomers into larger and larger collaborations.

Living Organisms

Genetic model organisms such as budding yeast (*Saccharomyces cerevisiase*), fruit flies (*Drosophila melanogaster*), and round worms (*Caenorhabditis elegans*) have been used in biological research for over 150 years. They are ideal genetic models for a number of reasons, including their small size, rapid growth, and the ease with which their genome can be manipulated. They are also inexpensive. Examining spontaneously occurring or induced mutations of these organisms has facilitated the identification of a number of important proteins, including those required for cell growth and proliferation, protein synthesis and processing, and signal transduction.[86]

Other model organisms are also used in research. Planaria, for example, whose regenerative powers were first studied by scientists in the nineteenth century, have proved to be an excellent model for Alejandro Sánchez Alvarado's work examining the molecular components underlying regeneration.[87] Zebrafish, originally collected to populate aquariums, have become widely used for research as well; they are cheap and reproduce quickly; their eggs are easily studied and manipulated, being fertilized externally. They can also be genetically modified to "glow in the dark," allowing researchers to study development at its earliest stages.[88]

But mice are king. They have been used as a research tool at least since the days of Gregor Mendel, who preferred mice to peas and only switched after the Church forbade their use. The grounds, among other things: the study of mice involved copulation. (Mendel later gloated, "You see, the Bishop did not understand that plants also have sex.")[89] Fifty years later, the Harvard biologist Clarence Little read Mendel's recently rediscovered work, became interested in using mice for research, and began breeding mice at Harvard. The fact that mice can be inbred to remove genetic variation makes them especially desirable as a research model. In 1929, Little went on, with the help of several benefactors, including Edsel Ford, to found the Jackson Laboratory (commonly known as JAX).[90] It is now the largest nonprofit mouse facility in the world. In 2008, it supplied more than 2.5 million mice.

The use of mice as a research tool accelerated in the late 1980s as a result of dramatic breakthroughs in genetic engineering. No longer did one need to use "spontaneous mice" (naturally occurring sick animals with specific recognizable symptoms) for disease studies; it was now possible to engineer mice with specific diseases or susceptibility to specific diseases, using one of three new technologies. Knockout methods deleted specific genes in a mouse; transgenic methods inserted novel genes into a mouse; Cre-lox technology allowed the "conditional" deletion of gene regions at specific times or in specific tissues. Some transgenic (e.g., the OncoMouse) and Cre-lox mice were patented; the knockout mouse was not.[91] Three researchers who played a key role in creating knockout mice were awarded the Nobel Prize in 2007 in physiology or medicine.[92]

As a result of these technologies, mice models are now available for almost all common diseases. There are mice that develop Alzheimer's disease, mice with diabetes, obese mice, mice with heart disease, blind mice, deaf mice, and mice who show the symptoms of obsessive-compulsive disorder, schizophrenia, alcoholism, or drug addiction. And mice with all varieties of cancer. You name it, a mouse model is available. And if a mouse model does not exist, one can be ordered. Johns Hopkins University, for example, has a lab designed to do precisely this for Hopkins researchers.[93]

It is estimated that mice constitute 90 percent of all animal models used in labs today.[94] Just how many mice are in use is difficult to estimate. Some say as many as 80 million; others say between 20 and 30 million.[95] Regardless of the disparity, everyone agrees that there are "a lot." Hopkins alone had approximately 200,000 mice at ten facilities in 2008; ten years earlier Hopkins had but 42,000 mice.[96]

Several factors lead the mouse to be the preferred vertebrate research model.[97] Mice are "close" cousins; the mouse and human genomes have about 99 percent similarity; mice reproduce cheaply and quickly; and mice, unlike other animals, have very few human advocates. For a variety of reasons, they are not high on the list of animal rights advocates.[98]

One inbred off-the-shelf mouse costs between $17 and $60; mutant strains begin around $40 and can go to more than $500. The prices are for mice supplied from live-breeding colonies. But more than 67 percent of JAX's 4,000 strains are only available from cyropreserved material. Such mice cost considerably more: the cost to recover any strain from cryopreservation (either from cryopreserved sperm or embryos) is $1,900. For this, investigators receive at least two breeding pairs of animals in order to establish their own breeding colony.[99] Custom-made mice can cost considerably more. Hopkins, for example, estimates that it costs $3,500 to engineer a mouse to order.

With such a large number of mice in use, the cost of mouse upkeep becomes a significant factor in doing research. Johns Hopkins, for example, employs ninety people, including seven veterinarians, to care for their

200,000 mice. The university estimates that mice costs represent about 75 percent of its annual $10 million animal-care budget.[100] It is common for U.S. universities to charge principal investigators a mouse per diem. Boston University, for example, charged a cage per diem of $0.91 (a cage generally holds five mice) in 2009.[101] By comparison, the University of Iowa's $0.52 per diem is a real bargain.[102] Such charges can rapidly add up. Irving Weissman of Stanford University reports that before Stanford changed its cage rates he was paying between $800,000 and $1 million a year to keep the 2,000 to 3,000 cages he was using for research.[103] Costs for keeping immune deficient mice are far greater (on the order of $0.65 per day per mouse) because their susceptibility to disease generally requires that they be housed separately.

Male mice are more commonly studied than female mice. Indeed, only in reproductive studies is the ratio of female subjects to male subjects greater than one.[104] Costs are a factor: the four-day ovarian cycle of female mice means that researchers must monitor females daily in experiments where hormones may play a role. As many as four times the number of females to males may also be required if researchers wish to ensure that their subjects cycle in sync.[105] But female mice have at least one cost advantage over males: they are less aggressive, and thus more females can be kept in the same cage.[106]

The equipment for mouse care is big business; 30 million mice require at least 6 million cages. Moreover, specialized robotic equipment has been developed to move cages for cleaning and feeding. One also needs equipment to study mice, such as surgical instruments. Observational equipment is also important. The titanium dorsal skinfold chamber (designed to fit under the skin on a mouse's back) allows the researcher to "nondestructively record and visualize microvascular functions."[107] One of the most remarkable pieces of equipment to come on the market recently is designed to conduct mouse ultrasound studies. The high-frequency machines go for $150,000 to $400,000, depending on the system and the configuration of hardware and software options.[108] The market is reported to be brisk.

Access to Research Materials

Research materials such as cell lines, reagents, and antigens also play a major role in research. Some of these materials are purchased from labs, but many scientists gain access to materials through a process of exchange, which has a long tradition in science and plays a considerable role in fostering research and in creating incentives for scientists to behave in certain ways.[109] For example, scientists routinely share information and access to research materials and expertise in exchange for citations and coauthorship.[110]

John Walsh, Charlene Cho, and Wes Cohen examined the practice of sharing materials among academic biomedical researchers and found that 75 percent of the academic respondents in their sample made at least one request for material in a two-year period, with an average of seven requests for materials to other academics and two requests for materials from an industrial laboratory.[111] Scientists don't always get what they want: 19 percent of the material requests made by the sample were denied. At least 8 percent of respondents had to delay a project due to the inability to obtain access to research materials in a timely fashion. The likelihood of compliance depended on the costs and benefits. Competition among researchers (and hence the intensity of the race for discovery) played a major role in refusal, as did the cost of providing the material. Whether the material in question was a drug or whether the potential supplier had a history of commercial activity was also relevant in refusal, suggesting that the prospect of financial gain contributed to refusal.[112]

Access to materials has been fostered in recent years by the establishment of biological research centers (BRCs) whose stated purpose is to preserve, certify, and disseminate material deposited by researchers. These centers often receive their funding from government or nonprofit organizations. Sometimes the collections they receive had been languishing in a researcher's refrigerator and were transferred by the institution at the time the researcher moved, retired, or died; in other instances, the transfer is made because the institution can no longer afford to maintain the collection. Deposits can also be mandated by funding agencies.

Certification of noncontamination is not a trivial concern. Contaminated cell lines can lead researchers to draw faulty conclusions. A particularly famous case of contamination was documented by Walter Nelson-Rees and colleagues, who were able to show that an extraordinarily robust cell line known as the HeLA (named after the cervical cancer donor Henrietta Lacks) had contaminated dozens of cell lines widely used in the 1970s.[113] Their research called into question a considerable body of cancer research, including the work of Nobel laureates. More recently, three research groups found that their earlier findings that mesenchymal stem cells (MSCs) could become cancerlike were caused by contamination of the MSCs by tumor cells used for other studies.[114]

Recent work by Furman and Stern uses citation patterns to study the effect that deposit (and hence availability to others) of research materials at biological research centers has on research practices. The authors focus exclusively on material that was transferred by an exogenous event, such as the death of a researcher, in order to ensure that the sample material had not been deposited solely because of its research importance or the prominence of the researcher. The methodology involves matching citations to the root paper that originally described the material's characterization and application. The authors find that the exogenous deposit of materials has increased the breadth of the research community: postdeposit citations to root papers grow faster from authors at new institutions and new countries, measured by not having cited the root article in the previous periods. Citations also grow faster in journals that had not published work related to the material in the previous periods.[115]

The tremendous increase in patenting among academics (see Chapter 3) raises the logical question of the degree to which patents affect the sharing of material. Walsh, Cohen, and Cho also examine how patenting affects access to material and find that it is largely unaffected, primarily because of issues related to lack of enforceability.[116] Only 1 percent of academic researchers reported that they had delayed a project due to the patents of others; none reported abandoning a project. Moreover, only 5 percent reported that they regularly checked to see if their research could be affected by relevant patents, suggesting that infringement is of little concern. But not all institutions wink and look the other way when infringement occurs. Several cases of strong patent enforcement have been widely documented that have affected research. A recent example concerns human embryonic stem cells. The University of Wisconsin, whose researchers discovered them, has used its control, both through patents and material rights to the cell lines, to impose limits and conditions on use by other academics.[117]

Earlier examples relate to mice. The OncoMouse (see previous discussion and Chapter 2) was patented by Harvard and licensed exclusively to DuPont. (DuPont had provided unrestricted funds to the laboratory of Phil Leder, the Harvard professor who developed the Onco technology, in return for the right of first refusal on any patentable results.) The Cre-lox mouse was developed by DuPont and patented by the company. Those who wished to use the mice faced extremely restrictive terms.[118] There was widespread discontent within the academic community regarding DuPont's practices, especially given the community's long tradition of sharing mice. In 1998, and after pressure from the academic community, Harold Varmus, the director of NIH (and a Nobel laureate) announced a Cre-lox memorandum of understanding (MOU) among DuPont, Jackson Labs, and NIH that greatly increased openness regarding the use of Cre-lox mice by academic researchers. A year later, an OncoMouse MOU was signed.

As discussed in Chapter 2, the MOUs had a profound effect in increasing research based on the mice. Moreover, as in the establishment of biological research centers, the MOU had a democratizing effect. Post-MOU citations to the original mouse articles grew at a faster rate both from authors and from institutions that had not cited the original papers prior to the MOU.[119] The logic for the finding is that—prior to the MOUs—accessibility to mice was considerably more restricted. Researchers at institutions where a colleague had either engineered a mouse or already had access to a mouse were likely to share the benefits, while researchers at institutions that did not have a mouse found access more difficult. The lesson: it is not patents per se that impede research, but the way that patents are managed.[120]

Space

Research also requires space. Not just any space, but special space that is suited for the specific purposes of the researcher. Some of this space can be quite expensive. At a minimum, laboratories generally require access to water and electricity. But oftentimes labs require considerably more than this. Scientists doing research in solid state or nanotechnology, for example, need "clean" rooms to avoid contamination. Some research requires special exhaust systems; other research requires exceedingly cool facilities. Some research requires exceedingly stable facilities so that experiments will not be affected by vibrations. The specifications for lab space designed for the study of viruses can be particularly exacting in order to minimize the threat of acquiring infections from agents manipulated in the labs.

Space is often allocated at the time the faculty member is recruited. In the biomedical sciences, a new faculty member at an elite research institution often gets a lab with eight work stations (desks plus bench space for lab personnel) and approximately 1,500 square feet with an additional 500 "common" square feet that is shared among labs. On other campuses, "starter" labs in the biomedical sciences are considerably smaller, on the magnitude of 600 square feet, and accommodate only four to six lab personnel. In some other fields, the amount of space allocated to labs is generally lower than it is in the biomedical sciences, depending on the type of research the faculty does. Astronomers and experimental particle physicists, for example, generally require considerably less lab space on campus than do physicists working in the fields of optics or solid state.

The amount of space a principal investigator has affects the size of the team and thus the researcher's productivity. David Quéré, for example, was only able to double the size of his group after he got a second lab, this one at the École Polytechnique. The allocation and reallocation of space can be highly contentious. An associate provost once recounted his university's efforts to come up with alternative ways to reclaim lab space from research-inactive faculty after mandatory retirement was abolished.

There is also the question of whether space is allocated fairly. It was an issue of space that energized Nancy Hopkins to confront the MIT administration regarding gender disparities in the early 1990s. Hopkins, who was switching fields at the time, requested an increase of 200 square feet of lab space above the 1,500 square feet she already had. She "noticed male junior faculty were given 2,000 square feet when they began." Yet her request for 200 additional square feet was initially denied.[121]

Approximately 180 million square feet are devoted to research in science and engineering at U.S. academic institutions. Over 45 percent of this is for research in the biological, medical, and health sciences. Engineering and the physical sciences each have claim to about 17 percent of the space, and agricultural sciences to another 16 percent. The remainder is shared by computer sciences and "other sciences."

The amount of research space by field for the period 1988 to 2007 is shown in Figure 5.1. As can be seen, the amount of research space in the biological, biomedical, and health sciences has grown dramatically over time, especially since the mid-1990s, while space for most other fields has only increased modestly. Indeed, the only field to have come even close to rivaling the rate of growth of that in the biological, biomedical, and health sciences was engineering. Much of the growth for the former was spurred by the doubling of the NIH budget, a process that began in 1998 and continued until 2003. In response to what they perceived as increased opportunities for funding, many campuses went on a building binge. Elias Zerhuni, the former dean of the Johns Hopkins School of Medicine and the former director of NIH, described this as an era in which deans routinely boasted to other deans regarding the number of cranes that they had on their campus constructing new buildings.

It was not only universities that went on a building binge. Biomedical research institutes and hospitals also went on an NIH-induced binge. Including research space at such facilities in the calculation raises the share of space devoted to the biological, biomedical, and health sciences to approximately 50 percent in 2005, the latest year for which data on institutes and hospitals are available. The comparable figure in 1988 was 43 percent.[122]

Surveys conducted by the Association of American Medical Colleges (AAMC) provide detail concerning the dramatic increase in research facilities at medical schools.[123] Before the NIH's budget began its doubling,

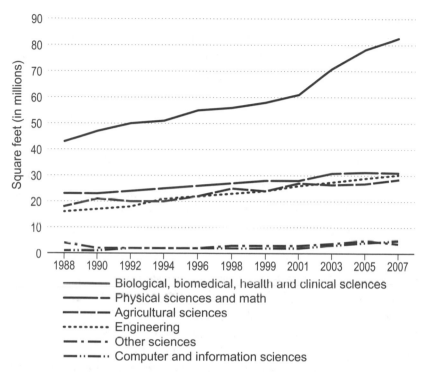

Figure 5.1. Net assignable square feet for research by field, at academic institutions, 1988–2007. *Note:* "Physical sciences" includes earth, atmospheric, and ocean sciences, astronomy, chemistry, and physics. *Source:* National Science Foundation (2007d); National Science Board (2010).

medical schools reportedly were spending approximately $348 million annually on the construction and renovation of buildings for research. That jumped to $760 million a year during the period of NIH expansion and was projected to be $1.1 billion annually from 2003 to 2007. (All figures are in 1990-adjusted dollars.) In many instances, campuses did not have the funds to construct the buildings but floated bonds to do so, assuming that much of the debt would be recovered through increased grant activity engendered by better facilities housing more research-active faculty. The AAMC survey (seventy medical institutions responded) found that the average annual debt service for buildings in 2003 was $3.5 million; it grew to $6.9 million in 2008.

The brakes were applied to the NIH budget beginning in 2004, and in constant dollars the budget shrank by about 4.4 percent between 2004 and 2009.[124] Success rates for NIH grants declined, and universities found that revenues from grants did not live up to their expectations. This has put considerable pressure on U.S. universities as they scramble to service the debt associated with these buildings and also provide "bridge" funding to faculty whose grants have not been renewed. We return to this and related issues in Chapter 6.

Policy Issues

The importance of equipment and materials for the production of research raises several policy concerns and research questions. First, although increased access to materials can have a democratizing effect, the increased importance of equipment and the high costs of equipment can increase the disparity between the haves and have-nots. This pertains not only to the disparity within the public sector of research universities and institutes but also to the disparity between the private sector and the public sector. Industry has the financial resources to stay on the cutting edge; the public sector increasingly does not. As one scientist wrote, "I have worked in some of the best-funded [academic] laboratories in the world, and even these laboratories do not have access to fancy next-generation machines in a way that large biopharmaceutical companies do. I strongly believe that this is changing the nature of the public/private divide and the extent to which academic science manages to stay at the technological frontier." Other scientists have expressed a similar view. An interesting research question is the degree to which this is happening and how it relates to the productivity divide between the two sectors and the ability of academe to attract researchers interested in pursuing fundamental research.

Second, despite the important role that equipment plays in research, little is known about the degree of competition in the market for equipment. Casual empiricism suggests that the market is highly concentrated. Illumina, for example, currently controls about two-thirds of the sequencing market.[125] It is important to know the extent of concentration in these markets, because highly concentrated industries price products considerably above the marginal cost of producing the product. Resources are only efficiently utilized if price reflects the marginal cost of producing another unit, but clearly this is not the case in the equipment market, where price often depends upon "how hard you bargain." How much loss of efficiency is there in equipment markets, and are the associated monopoly profits necessary to entice suppliers into markets where technology changes so quickly?[126]

Third, similar concerns arise regarding the market for extremely large equipment. Much of the equipment for a telescope or a collider is one of a kind. How is such equipment supplied, and how is it priced?

Fourth, large research projects, such as the HGP and the PSI, require a considerable amount of resources. In a similar vein, extremely large pieces of equipment come with price tags of billions of dollars and tie up resources for years to come. Whether these are good investments is a question that we will return to in Chapter 6.

Fifth, how much does a scientist's success depend upon having a monopoly on new types of equipment or securing a monopoly on a time slot on a scarce resource such as a telescope or on a submergence vehicle such as Alvin?

Sixth, there is reason to be concerned that universities may have borrowed themselves into deep financial trouble, building biomedical research facilities that they can only pay for by cutting programs in other fields of science as well as the humanities and social sciences. The effects of the NIH doubling may be felt on university campuses for years to come.

Conclusion

The overwhelming importance of equipment and materials to the production of research—and the associated costs—means that in most fields access to resources is a necessary condition for doing research. It is not enough to want to do research—one must also have access to the inputs with which to do research. At U.S. universities, equipment, materials, and funding for graduate and postdoc stipends are generally provided by the dean at the time of hire in the form of start-up packages. Thereafter, equipment, materials, some buy-off for faculty time, and the stipends that graduate students and postdocs receive become the responsibility of the scientist. Scientists whose work requires access to "big" machines off campus must also obtain grants to procure time (e.g., beamtime) and to pay for time at the research facility.

This means that in a variety of fields funding is a necessary condition for doing "independent" research that is initiated and conceived by the scientist. Scientists working in these fields in the United States take on many of the characteristics of entrepreneurs. As graduate students and postdocs they must work hard to establish their "credit-worthiness" through the research they do in other people's labs. If successful in the endeavor, and if a position exists, they will subsequently be provided with a lab at a research university. They then have several years to leverage this capital into funding. If they succeed, they face the onerous job of continually seeking support for their lab; if they fail, the probability is low that they will be offered a start-up package by another university. The reliance on the individual scientist to generate resources is not nearly as common in many other countries, where researchers are hired into government-funded and government-run laboratories such as CNRS in France. Nevertheless, fits and starts in funding for such programs translate into the possibility that certain cohorts of scientists enter the labor market when conditions are favorable for research while other cohorts do not. In the next chapter, we examine funding for research in the United States as well as in other countries. In Chapter 7 we examine the labor market for scientists and engineers.

CHAPTER SIX

Funding for Research

STANFORD UNIVERSITY receives approximately $759 million a year in support of research, the University of Virginia about $306 million, and Northwestern University about $428 million. In the case of Stanford, this represents 23 percent of university revenues; it represents 25 percent for Virginia and 27 percent for Northwestern.[1] Where does the money come from? What criteria are used for allocating it? More generally, why support research at universities?

Recall that scientific research has properties of what economists call a public good. Once made public, others cannot easily be excluded from using it. Neither is knowledge depleted once it is shared. As noted in the earlier discussion, the market is not well suited for producing goods with such characteristics. Unlike the baker, whose customers must pay if they wish to eat his cake, or the symphony orchestra which can sell tickets to its concerts, thereby excluding those who do not pay from attending, the researcher has nothing to sell once her findings have been made public. Thus, she has no means of appropriating the benefits.[2] It is particularly difficult to appropriate the benefits arising from basic research, which at best is years away from contributing to products that the market may or may not value. Equally, if not more important, it is virtually impossible to appropriate the benefits that arise from the contribution that basic research makes to future fundamental research.[3]

Society, however, is more ingenious than the market (to use a phrase of Kenneth Arrow's), and the priority system has evolved in science to create a reward system that encourages the production and sharing of knowl-edge. Scientists, as discussed in Chapter 2, are motivated to do research by a desire to establish priority of discovery. The only way they can do so is to share their findings with others.

Priority thus addresses the appropriability problem. It does not, how-ever, address the resource question. Research costs money—lots of money. The typical lab at a public university, for example, composed of eight re-searchers—a faculty principal investigator (PI), three postdoctoral fellows (postdocs), and four graduate students—plus an administrator has annual personnel costs of just over $400,000 after fringe benefits but before indi-rect costs. (That is $53,000 for each postdoc, $35,000 for each graduate student, $53,300 for the administrator, and, at 50 percent of the faculty member's salary, $55,850 for the PI.)[4] Add in 500 mice and $18,000 a year in supplies for each researcher, and lab costs come to about $550,000 a year before one has even opened the equipment catalogue, which could easily set the lab back another $50,000 to $100,000. Big science costs magnitudes more.[5]

Other forms of intellectual property, such as patents and copyrights, ad-dress the appropriability and resource problem by awarding monopoly rights to the inventor. From society's point of view, however, the monopoly solution can be problematic in that, despite the requirement of disclosure, patents can restrict others from building on the knowledge that has been produced, thereby creating hurdles to cumulativeness.[6]

Consider, for example, the case of gene sequencing discussed in Chapter 5. The initial Human Genome Project (HGP) was financed by the governments of six countries. But in 1998 Craig Venter and the company he helped to found, Celera, entered the race to sequence the human genome. When the announcement was made in June 2000 that a working draft of the genome had been compiled, it was joint—issued by the HGP and Celera. When the genome was published in February 2001, it was published simultaneously by the two groups. So far so good. But while the government-funded HGP project made data available with few restrictions, Celera used copyright law to limit access to the genes the firm had sequenced. Intellectual prop-erty restrictions were removed from the Celera-sequenced genes when they were resequenced by the HPG. The work of Heidi Williams shows that this made a difference. Using indicators such as patents, numbers of papers published, and commercially available diagnostic tests, Williams found that Celera's policy led to a reduction in subsequent research and product devel-opment on the order of 30 percent.[7]

The rationale for public investment in research and development (R&D) is thus twofold: to provide the needed resources for basic research and to invest in research that provides for openness.[8] The two, of course, are re-lated. Those who engage in basic research have incentives to disclose be-cause priority is the primary extrinsic reward they receive from doing research. However, as research has increasingly moved to what one can think of as Pasteur's Quadrant—producing knowledge that is both funda-mental and useful—the two have become more distinct in terms of the rationale for public support of research.

The public's rationale for supporting scientific research also rests on the importance of R&D to specific outcomes deemed socially desirable and not directly provided by the market, such as national defense and better health. The late British science policy scholar Keith Pavitt was fond of say-ing that America's fear of Communism and cancer played a leading role in shaping its science policy.[9]

The relationship between research and economic growth provides an-other rationale for government support of science and has been a particu-lar rallying cry for more resources in recent years. In the summer of 2006, for example, the state of Texas decided to invest $2.5 billion in science teaching and research in the University of Texas system. A primary focus of the initiative was to build up the research capacity at campuses in San Antonio, El Paso, and Arlington in an attempt to turn these cities into the next Austin, Texas, if not the next Silicon Valley. The National Academy of Sciences report, *Rising above the Gathering Storm,* received considerable attention when it was issued later the same year. The message: the U.S. competitive position in the world has begun to erode and will continue to decline unless more U.S. citizens are recruited into careers in science and engineering and the United States steps up its investment in research.

This chapter examines sources and mechanisms for supporting research conducted in the public sector, especially at universities. The chapter be-gins with an overview of sources of funds and then focuses on mechanisms for the distribution of the funds. It continues with a discussion of the ben-efits versus the costs of different mechanisms and presents a case study of funding for biomedical research in the United States. It concludes with a discussion of policy issues related to the funding of public science, such as whether there is a right amount to invest and whether the national research portfolio is well balanced.

Several themes emerge from the discussion. One is the tendency of most systems of support to experience stop-and-go periods. This has ef-ficiency implications; it can also have implications for careers. Scientists who have the bad fortune to enter the labor market during a "stop" period can feel the adverse effects for years. Another theme is the loss of efficiency that accompanies various mechanisms for funding science. By way of example, an investigator-initiated mechanism provides maximum freedom of intellectual inquiry and consequently may have the greatest intellectual payoff. But it also comes at considerable cost, as it requires time both on the proposing and reviewing end. It may also discourage risk taking.

Sources of Funds

Federal Funding

In 2009, U.S. universities spent almost $55 billion on research. The largest contributor to research by far was the federal government (59.3 percent), followed by universities themselves (20.4 percent).[10] Considerably less came from state and local governments (6.6 percent), industry (5.8 percent), and other sources (7.9 percent) such as private foundations.

The composition and amount of funding for university research has changed considerably during the past fifty-five years, as can readily be seen in Figure 6.1. (Dates correspond to fiscal years and begin in October of the previous year and end in September of the corresponding year.)

Several trends emerge. First, the amount contributed by the federal government has gone through considerable fits and starts beginning in the mid-1950s. Prior to *Sputnik,* the federal government, in 2009 dollars, was spending less than a billion a year on university research and was contributing about 55 percent of the amount universities and colleges spent on research. The role of the federal government changed dramatically in response to the launch of *Sputnik:* in constant dollars, the amount the federal government spent on research at colleges and universities grew by a factor of six from 1955 to 1967. The proportion of funds coming from the federal government also dramatically increased, going from 54.2 percent to 73.5 percent. Funding for research was sufficiently plentiful to lead scientists to parody a well-known advertisement at the time for Grant's whisky: "While you're up [in Washington] get me a grant!"

The increase had profound effects on the practice of science at U.S. universities. Universities expanded, and new universities were created. Not only were there more federal funds for research, the coming of age of baby boomers meant that increasing numbers of students headed to college. To meet the rising college enrollments and research demands, new universities were established, programs added, and faculties greatly expanded. Thus, for example, between the late 1950s and the early 1970s, the number of doctorate-granting institutions in the United States grew from 171 to 307. Over the same period, the number of doctoral programs in physics went from 112 to

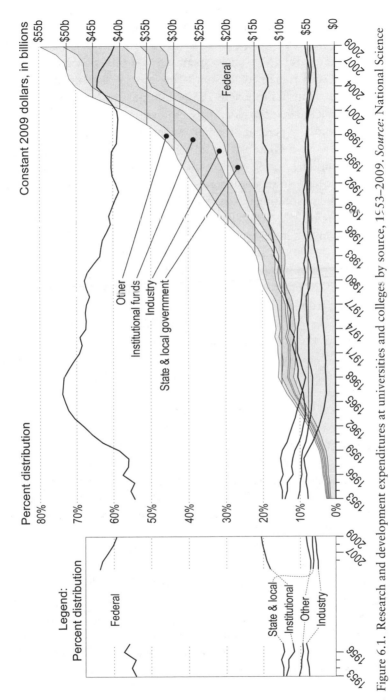

Figure 6.1. Research and development expenditures at universities and colleges by source, 1953–2009. *Source:* National Science Foundation (2010a and 2010b).

194, and the number in the earth sciences more than doubled, going from 59 to 121; in the life sciences, the increase went from 122 to 224.[11]

The Vietnam War brought the nation's love affair with science to a halt. The amount the federal government spent in real dollars on university research declined and remained flat until the late 1970s. University jobs, which had been plentiful in the late 1950s and most of the 1960s, became scarce; the percentage that the federal government contributed to university research fell from 73 percent to approximately 66 percent. The roller coaster continued in the late 1970s and 1980s: increases, followed by a plateau in real funding during the recession of the early 1980s, followed by increases. The fits and starts dissipated considerably during the next fifteen years, as federal contributions to university research continued to increase. Despite the increase, the federal share of university R&D continued to decline and, beginning in 1989, hovered at or slightly below 60 percent.

This changed in 1998 with the commitment to double the National Institutes of Health (NIH) budget in a period of five years. The federal government's contribution to university research grew dramatically in the next five years, going from $19.1 to $28.4 billion (constant 2009 dollars);[12] the federal share increased from 58.4 percent to 63.9 percent. Although many had assumed that the doubling would be followed by a period of "normal growth" in the NIH's budget, this was not to be the case. The years after the doubling were followed by real decreases in the amount of funds allocated for the NIH as well as to several other federal agencies that support university research. The federal contribution to university research, until the 2009 stimulus package arrived, remained flat.

The start button was pushed again with the passage of the American Recovery and Reinvestment Act of 2009, which provided $21 billion for science and engineering research and infrastructure support, much of which was targeted for universities. The act was revolutionary in the sense that it was the first time that funding for research had been specifically provided as a countercyclical measure. Moreover, heretofore research funding had been procyclical. Witness the declines in federal funding for university research during the recessions of 1973 and 1981.[13]

The majority of stimulus funds were directed to individual research projects. Some of the stimulus funds, however, were used to support large-scale projects put on hold the previous year when the funding stream slowed down. The stimulus package, for example, added $400 million to the major facilities account of the National Science Foundation (NSF), which supports such large-scale projects as telescopes and supercomputers. NSF chose to use the funds to support the Alaska Region Research Vessel, the Advanced Technology Solar Telescope, and the Ocean Observatories

Initiative.[14] Although the scientific community welcomed the increase, they almost immediately began to wring their hands over what would happen when the stimulus money was expended and funding went back to its pre-2009 level.

Support from Industry

Universities have a long tradition of receiving support for research from industry. In the 1950s, for example, the earliest years for which data were collected, universities were receiving about a twelfth of their research funds from industry. But by the 1960s and throughout the 1970s, industry support for university R&D had fallen and stood between 3 and 4 percent. The decline in industry's share was partly due to increased support from the federal government, but it was also because the amount of funds industry invested in university R&D grew only modestly during the period.

The ups and downs in federal funding for universities in the late 1960s and 1970s led universities to seek alternative sources for funding research. Industry was a likely candidate, and the importance of industrial support for university research grew substantially during the 1980s and 1990s. There were several contributing factors. First, growth in university patenting and licensing (see Chapter 3) meant that faculty had more opportunities to work with industry. Second, the increasing number of freshly minted PhD students who went to work in industry provided growing opportunities for faculty to work with colleagues in industry (see Chapter 9). Third, faculty became increasingly involved with industry as a result of faculty start-ups, examples of which are noted in Chapter 3.

Industrial support for university research in the United States reached its peak in the late 1990s when industry contributed approximately 7.4 percent of all university research funding. Since then, the proportion of university research supported by industry has declined, and the amount, in constant dollar terms, remained fairly flat until 2006—a victim of the 2001 recession as well as the large number of corporate mergers in the new century, which resulted in the consolidation of R&D efforts among companies that merged.[15] Beginning in 2006, however, industry support for R&D began to modestly increase. It is too early to know how the financial meltdown of 2008 affected the amount of support that industry provides, but it would be a miracle if it increased significantly.

Industry support means that it is not uncommon for faculty to receive funding from industry for specific projects or for the further development of proofs of concept licensed by the firm. For example, the research of

Philip Leder, who developed the genetically modified OncoMouse as a model for studying cancer, was supported by a grant from DuPont. He is not alone. By the mid-1990s, more than 25 percent of life science faculty reported receiving support from industry through grants and contracts.[16]

A worrisome consequence of industry support is the control that industry may exert over publications and intellectual property coming out of the research. Leder's agreement with DuPont allowed DuPont to have an exclusive license on the ensuing mouse that he invented and Harvard patented. The consequences for follow-on research were substantial, as Murray and colleagues have shown (see the discussion in Chapter 2). But Leder is not an isolated case. A survey of biomedical faculty by Blumenthal and his colleagues found that those with industrial support were four times as likely as their colleagues to state that trade secrets resulted from their research and five times as likely to state that they had restrictive publication arrangements with the sponsor.[17]

An even more worrisome threat to open science arises when universities form research alliances with a firm. Examples include Monsanto's research alliance with Washington University School of Medicine, established in 1982, and the Massachusetts Institute of Technology's 1997 collaborative agreement with Merck. In the case of Washington University, the alliance initially provided $6 million in research funds for faculty to engage in "exploratory and specialty" research. In exchange, the university agreed to a 30-day publishing delay while Monsanto patent attorneys reviewed research.[18] Merck's 1997 collaborative agreement with MIT provided for up to $15 million in funding over a five-year period. In exchange, Merck received certain patent and license rights to developments resulting from the collaboration.[19]

By far the most controversial of these agreements was struck between the Department of Plant and Microbial Biology at the University of California–Berkeley and Novartis in 1998. In exchange for up to $25 million in research support over a five-year period and access to Novartis's gene-sequencing technology and DNA database on plant genomics, Novartis was given first rights to negotiate licenses to patents on a proportion of the discoveries made in the department. It is no wonder that the agreement was highly controversial. Although industrial grants to individual researchers or research teams had occurred in the past, as well as strategic alliances with universities, this was the first time that an entire department had been funded by one firm. Such alliances clearly dampen the speed with which knowledge is disseminated. They may also, by directing research in specific directions, threaten a fundamental tenet of the research culture of the university: the ability of faculty to choose their own research topics.

Nonprofit Foundations

Nonprofit foundations provide another source of funds for university research. Indeed, long before either the federal government or industry had become a ready source of funds for university research, the Carnegie Foundation, the Rockefeller Foundation, and the Guggenheim Foundation were supporting scientific research. The Sarah Mellon Scaife Foundation provided the funds to renovate Jonas Salk's laboratory when he moved to the University of Pittsburgh in 1948.[20] In 1951, James Watson was able to go to the Cavendish Laboratory at the University of Cambridge thanks to a fellowship from the National Foundation for Infantile Paralysis.[21] Although the federal government does not track funding from nonprofit foundations as a separate category, funding from nonprofits represents a large component of the "other source" shown in Figure 6.1. The figure suggests that the amount of support coming to universities from nonprofits has increased in recent years.

Some nonprofit foundations support a wide range of initiatives, research (including university research) being but one of them. Currently, the largest of these is the Bill and Melinda Gates Foundation. Its net worth, which was over $29 billion in 2006, was significantly increased that year when Warren Buffet signed a letter of intent pledging $31 billion to the foundation in Berkshire Hathaway shares.[22]

Many nonprofit foundations focus on a specific area, such as global warming or infantile paralysis. Other examples that readily come to mind are the Cystic Fibrosis Foundation, the American Cancer Foundation, the American Heart Association, and the Ellison Medical Foundation (with its focus on aging).[23] In addition to creating public awareness for their cause and lobbying Congress for funds, such foundations also support university research. Some nonprofits have a quite narrow focus, such as the Kirsch Foundation, which currently is focused almost exclusively on finding a treatment for Waldenström macroglobulinemia (WM), which affects about 1,500 people a year in the United States.

There are even foundations that are devoted to establishing a new area or field. The Whitaker Foundation, for example, devoted its entire resources to transforming biomedical engineering from a barely recognized discipline into a firmly established field. During its 30 years of existence, the foundation gave away more than $800 million to help create departments of bioengineering at universities and provide support for graduate student training and faculty research.[24]

Although it is difficult to find an exact accounting, casual empiricism suggests that target-specific foundations have been on the rise in recent years as a growing population of individuals find themselves in possession of great wealth and either face a health threat or have a loved one who does. Examples include the Prostate Cancer Foundation, funded by Michael Milken after he was diagnosed with prostate cancer, the Michael J. Fox Foundation for Parkinson's Research, established by the movie actor after he was diagnosed with Parkinson's disease, and the Kirsch Foundation, which changed its focus from social issues to Waldenström macroglobulinemia after the disease was diagnosed in its cofounder, Stephen Kirsch.[25]

Perhaps no other nonprofit organization has had as powerful an impact on academic research in the United States as the Howard Hughes Medical Institute (HHMI). Established in 1953 by the late aviator and highly eccentric engineer, industrialist, and movie producer Howard Hughes, the institute acquired a stronger footing when it sold the Hughes Aircraft Company to General Motors in 1985, thus establishing the institute's endowment at $5 billion.[26] At the close of the 2010 fiscal year, the endowment was valued at close to $14.8 billion (down from $18.7 billion in fiscal 2007).[27] By law, HHMI is required to distribute 3.5 percent of its assets each year. It has done so by supporting between 300 and 350 HHMI investigators at research universities, funding a number of training programs, and establishing the "farm"—the Janelia Farm Research Campus, in Ashburn, Virginia, which opened in 2006 with the goal of bringing twenty-five interdisciplinary teams together to study neural circuits and imaging.[28]

HHMI's largest outlay by far is in support of investigators. In 2010, for example, HHMI supported approximately 350 investigators from more than seventy universities and other research organizations and spent more than $700 million doing so.[29] The institute prides itself on "supporting people, not projects."[30] The selection process is relatively straightforward. Candidates nominate themselves, supplying a curriculum vitae, a 250-word account of their major achievements, and a 3,000-word summary of their ongoing and planned research. Applicants also supply five selected publications and a paragraph describing each. Initial applications are reviewed by a panel of experts and winnowed down to a group of semifinalists, for whom three reference letters are requested. Final selection is then decided by a panel after reviewing all material. Renewal of the five-year appointment is based on peer review that "centers on an evaluation of the originality and creativity of the investigator's work relative to others in the field as well as on the investigator's plans for future research."[31]

Nonprofit-foundation support for research can suffer from the same ups and downs related to the business cycle as does government funding and industrial support. Foundations that rely on donations can be particularly hard hit during a recession. Moreover, foundations that fund grants out of their endowment can experience severe problems when the stock market takes a deep dive, as it did in 2001 and again in 2008. During the 2001 downturn, for example, the HHMI endowment plummeted by $3 billion in two years. The downturn came at a particularly bad time— just when the Foundation had started to build the $500 million Janila Farm facility. In order to continue with construction, the foundation chose to cover the shortfall in part by cutting investigator grants by 10 percent for one year.[32]

An unwise investment strategy can also take its toll. Foundations that invested almost exclusively with Bernard Madoff, for example, found their balance sheet at close to zero in 2008. The Picower Foundation, which reported assets of almost $1 billion in 2007, announced in late 2008 that it would "cease all grant making effective immediately." Investigators supported by the foundation received e-mails from Barbara Picower, a cofounder of the foundation, informing them that their funding was terminated.[33]

Self-Funding

Universities have also used their own resources (labeled "institutional funds" in figure 6.1) to support research and to smooth out the peaks and valleys of federal funding. Although in the mid-1950s universities contributed about 14 percent, their share declined considerably during the 1960s when the federal government's contribution was growing at a fast pace. By 1963, only about 8 percent of research expenditures were "self-funded" by universities. This did not last: as the federal budget for research deteriorated, universities directed increasing amounts of their own funds to research. By 2009, slightly more than 20 percent of the funds for research, or approximately $11 billion, were coming from universities themselves.

At least two other factors have contributed to universities picking up a larger share of research funding.[34] First, there is the issue of indirect cost recovery. Historically, external funding agencies have funded much of the infrastructure of universities, as well as the cost of administering research, by paying indirect costs on grants. This means that a university marks up its direct-cost request for research (for example, for graduate students, postdocs, equipment, and faculty salaries) by a multiple, known as the indirect rate. Government auditors, however, began to take a much harder look at the rate after a much publicized case involving Stanford University in the early 1990s, and caps were put on expenses that universities could claim in a number of areas. The end result was that the average indirect rate at private research and doctoral universities, which was over 60 percent in 1983, fell to about 55 percent in 1997 and has remained fairly constant since.[35] Rates at public institutions average about 10 percentage points lower.[36] A 2000 Rand report suggests that "universities recover between 70 to 90 percent of the facilities and administrative expenses associated with federal projects."[37]

A second reason universities are picking up a larger share of the cost for research relates to start-up packages. As already discussed, in recent years it has become the norm for universities to provide start-up packages for new hires. Universities can easily spend $10 million a year on such packages. Not only do they play an important role in recruiting senior faculty; they also provide the time and the resources for newly minted faculty to develop the preliminary results necessary for bringing in their own research funds.

Where do the funds that universities spend on research come from? No one has done a precise accounting, but it is safe to say that some of the funds are diverted from the instructional budget as universities increasingly replace tenure-track faculty in the classroom with cheaper part-time, adjunct, and non-tenure-track faculty. Some funds come from endowments, which—until the 2008 recession—had performed well at most private and public institutions and spectacularly well at Ivy League institutions.[38] Some funds come from licensing revenues generated by technology-transfer programs.

Do students pay for the increased costs of research that universities are contributing? This is a question that Ron Ehrenberg, Michael Rizzo, and George Jakubson have investigated for 228 research and doctoral universities for the twenty-year period spanning the late 1970s to the late 1990s.[39] Their goal was to determine whether increases in internal funding for faculty research are associated with increased student-faculty ratios and increased tuition payments. Their findings suggest that students bear some of the cost, especially at private institutions, where the student-faculty ratio grows as internal funding for research grows, and where tuition levels increase as internal funding for research grows. The first effect is smaller at public institutions, and the tuition effect is nondiscernable for public institutions. They also found that institutions that increase the size of their graduate student enrollments compensate by increasing tuition. This is true for both public and private institutions.

Other Countries

Trends in the support of university research in nine European countries and in Japan are given in Table 6.1. The classification scheme builds on that developed by the Organization for Economic Co-operation and De-velopment (OECD), splitting sources into seven categories. Government funds are subdivided into direct government funds (DGF), such as contracts and grants, and general university funds (GUF), which come in the form of block grants, distributed either incrementally or on a formula basis. Addi-tional categories include funds from business, from abroad (including contracts for research with foreign companies), from private nonprofits, and higher education's own funds. Data are not reported for certain cate-gories for Denmark, Germany, and Italy, and these categories are excluded in calculating shares. Note also that for certain countries data are only available for a different year than that for other countries. These quirks of the data are noted on the table.

The country patterns mirror, in many ways, those of the United States. That is to say that in most countries there has been a decrease in the share of research funds coming from the government—and an increase in re-search funds coming from business, nonprofits, and higher education it-self. With the exception of France, the decrease in government support has come from a decline in general university funds.[40]

But there are substantial differences by countries. In the United King-dom, for example, the amount of funds coming from government grants and contracts has grown considerably. The same pattern is observed for Ireland and the Netherlands and for the earlier period for Denmark and Spain. The growth in business support for university research has occurred primarily in Germany, although business support for research has also in-creased in the Netherlands and Japan and in Belgium and the United King-dom for the earlier period. All countries (with the exception of Japan) ex-perienced a substantial increase in funding coming from abroad from 1983 to 1995, some of which was from foreign companies. With the ex-ception of Ireland and Spain, the increase from business persisted to 2007.

The increasing role of nonprofits has been particularly important in the United Kingdom. Nonprofits also play an increasing role in Denmark and Ireland and, for the earlier period, in the Netherlands. The largest non-profit in Europe is the Wellcome Trust, which in 2008 had assets of ap-proximately £15.1 billion and gave away (2007–2008) approximately £620 million to support research, both within the United Kingdom and internationally. Like other foundations, it was hit hard by the financial crisis, losing an estimated £2 billion; accordingly, it cut its support for re-search in 2009.[41] Specific nonprofits play a minor but growing role in sup-port of research in other countries as well. For example, L'Association Française contre les Myopathies (AFM) raises approximately 100€ million a year through a telethon and spends approximately 60 percent of it on research on rare neuromuscular diseases.[42] And in Italy, bank foundations,

Table 6.1. Funding for research in higher education by country, source, and year, percentage

	Belgium	Denmark	France	Germany	Ireland	Italy	Japan	Netherlands	Spain	UK
All government										
1983	86.2	95.0	97.6	95.0	82.2	99.3	54.8	96.2	98.8[d]	85.3
1995	76.2[a]	89.5	90.6	90.7	62.0	93.3	52.3	85.7	70.4	67.9
2007	66.3	79.7	89.8	82.2[b]	83.3	90.8	51.6	86.7[c]	73.1	69.2
Direct government funds										
1983	39.4	11.3	46.3	—	13.6	—	14.0	6.4	19.3[d]	20.5
1995	26.7[a]	22.6	46.0	20.2	20.0	—	10.4	6.3	30.1	30.1
2007	40.3	18.8	33.8	23.6[b]	45.5	12.7	13.2	15.9[c]	26.2	35.0
General university funds										
1983	46.8	83.7	51.2	—	68.6	—	40.8	89.8	79.5[d]	64.8
1995	49.5[a]	66.8	44.6	70.5	42.0	—	42.0	79.3	40.3	37.7
2007	26.0	60.8	56.0	58.6[b]	37.9	64.7	38.4	70.8[c]	46.9	34.3
Business										
1983	9.3	0.9	1.3	5.0	7.2	0.5	1.2	0.6	1.2[d]	3.1
1995	15.4[a]	1.8	3.3	8.2	6.9	4.7	2.4	4.0	8.3	6.3
2007	11.1	2.2	1.6	14.1[b]	2.3	1.3	3.0	7.1[c]	9.0	4.5

Nonprofit organizations										
1983	0.0	2.7	0.1	—	2.1	—	0.1	2.6	0.0[d]	5.6
1995	0.0[a]	4.5	0.5	—	2.5	—	0.1	6.5	0.5	14.0
2007	2.4	11.1	0.3	—	6.1	1.1	1.0	2.7[c]	1.2	13.5
Higher education institutions										
1983	2.9	—	1.0	—	1.0	0.0	44.0	0.3	0.0[d]	3.8
1995	3.6[a]	—	4.0	—	4.5	—	45.1	0.3	13.7	4.2
2007	12.9	1.0	6.1	—	1.5	4.0	44.3	0.0[c]	12.4	4.3
Abroad										
1983	1.6	1.4	0.1	0.0	7.6	0.2	0.0	0.3	0.1[d]	2.2
1995	4.8[a]	4.2	1.6	1.1	24.0	2.0	0.0	3.5	7.0	7.6
2007	7.2	6.0	2.2	3.7[b]	6.8	2.7	0.1	3.4[c]	4.3	8.4

Source: Organisation for Economic Co-operation and Development (2008), stats.oecd.org; gross domestic expenditure on R&D by sector of performance and source of funds.

—: Data are not available.

a. Year is 1991, not 1995.
b. Year is 2005, not 2007.
c. Year is 2001, not 2007.
d. Year is 1984, not 1983.

established by law during the restructuring of the mutual savings banks in 1990, regularly support research at Italian universities.

The People's Republic of China is not included in Table 6.1 because of lack of early data, not because of its place in world science. Over the past ten to fifteen years, China has become a major force in world R&D. Indeed, by 2007 (the latest year for which good data are available), China was spending approximately $100 billion a year on R&D, or 10 percent of the world R&D total, leading it to rank third behind the United States (33 percent of world R&D) and Japan (13 percent of world R&D).[43] China's increasing commitment to R&D can readily be seen by tracking the percentage of its gross domestic product (GDP) that it devotes to R&D. In 1998, it was 0.7 percent. By 2007, it had more than doubled to 1.49 percent. The United States, by comparison, spends 2.68, and Japan 3.44.[44]

Universities receive about 11 percent of the $100 billion China spends on R&D; research institutes receive 26 percent.[45] A third of the research funds going to universities come from industry—an impressive figure compared with that for other countries (see Table 6.1). The high percentage reflects the common practice for Chinese universities to have joint research programs with firms.[46] This close relationship is one reason that Chinese universities perform a smaller percentage of basic research (38 percent) compared with that performed by academic institutions in the United States (56 percent).[47]

In recent years, the Chinese government has singled out a select group of universities, known as the 985 institutions, in an effort to direct resources to institutions the government sees as having the greatest potential for success in the international academic community. Special treatment means the universities have been able to hire more competitively on the international academic market; they also have been able to attract star visiting professors by creating positions known as *jiangzuo,* or lecture chairs. These special chairs are designed to provide financial support to young and middle-aged leading scholars in targeted disciplines working abroad to return for short stays in China (usually three months).[48] It is not necessarily the salary that attracts these visitors back to China but the opportunity to return to China, to work in new research facilities, and to develop their own research agenda.

A case in point is Tian Xu, a professor of genetics at Yale University (and a Howard Hughes Investigator) who has been coming back to Fudan University in Shanghai since 2002. What really brought him back was the opportunity to run a genetics program, the scale of which is unimaginable in the United States, and to work with young Chinese scientists. To wit: Xu has facilities to house thousands of mice (45,000 cages, to be precise) in two separate buildings—something that would be absolutely impossible for a faculty member to have in the United States, not only because of the annual cage costs ($11 million plus) but because it would simply not be possible to get that much research space at a U.S. university.[49] Compare Nancy Hopkins's struggles to increase her laboratory space by a mere 200 square feet, as recounted in Chapter 5.

China's strategy is not without its critics. An editorial in *Science,* signed jointly by the deans of the Schools of Life Sciences at Tsinghua University and Peking University (the two most highly rated universities in China), charges that "rampant problems in research funding—some attributable to the system and others cultural—are slowing down China's potential pace of innovation."[50]

The deans are particularly critical of the way in which grants are awarded. They concede that scientific merit plays a key role in success with regard to winning small research grants, such as those awarded from China's National Natural Science Foundation. But in the case of megaprojects, "the guidelines for grants are so narrowly described" that the "intended recipients are obvious." They elaborate: "To obtain major grants in China, it is an open secret that doing good research is not as important as schmoozing with powerful bureaucrats and their favorite experts." Not precisely earmarking, but close! The two also lament, "A significant proportion of researchers in China spend too much time on building connections and not enough time attending seminars, discussing science, doing research, or training students (instead, using them as laborers in their laboratories)." Part of their concern is specific to China. But, as we saw in Chapter 4, scientists in the United States, too, find themselves diverted from their research by other demands on their time—in their case, applying for and administering grants.

Focus of Research

Not all science is created equal when it comes to funding. Moreover, what is favored during one period may lose favor in another, and the research focus often depends on who's paying. When state funding was the major source of resource support, for example, universities directed their research to topics of interest to the state. Wisconsin focused on dairy products, Iowa on corn, Colorado and other western states on mining, North Carolina and Kentucky on tobacco, Illinois and Indiana on railroad technology, and Oklahoma and Texas on oil exploration and refining.[51]

Defense-related funding from the federal government altered the focus of university research beginning with World War II. It also contributed to the expansion of several universities, including the Massachusetts Institute of Technology and the California Institute of Technology. Other universities were quick to learn from their sister institutions and used postwar defense contracts to propel themselves into the all-star league. Stanford was an early example of this; more recently, the Georgia Institute of Technology and Carnegie Mellon have benefited from defense-related research.[52]

In recent years, the tremendous growth in biomedical research funds has contributed to the growth of universities with a heavy focus on medical research, such as the University of California–San Francisco, Johns Hopkins University, and Emory University. It has also played a role in the strategic plans of universities. By way of example, membership in the American Association of Universities (AAU) is viewed as highly prestigious within the university community. The organization currently has but sixty-one members, and membership is by invitation only. A key criterion is research performance, one metric of which is money. The dominant role that funding for medical research plays means that those outside the AAU club have a much greater chance of admittance if they have a strong program in the biomedical sciences. Such logic was a factor leading the University of Georgia to adopt plans in 2007 to develop a medical school.[53]

The share of federal funds for university research by field is given in Figure 6.2 for the period 1973–2009. The figure makes abundantly clear that funding for the life sciences dominates all others and that its share grew, even before the NIH doubling. In sharp contrast, the share of funds going to the physical, environmental, and social sciences has declined throughout almost the entire period. Mathematics and computer sciences, however, have been able to increase their share over the period, especially in the middle years. The fortunes of engineering have been somewhat erratic: up considerably in the early years, flat during much of the middle years, followed by a sharp decline which only recently has been reversed.

The U.S. love affair with funding for the life sciences—especially the biomedical sciences—is not difficult to understand. It is far easier for Congress to support research that the public perceives as benefiting their well-being. Moreover, a large number of interest groups constantly remind Congress of the importance of medical research for "their" disease. The age distribution of Congress does not hurt. The average age of members of the House of Representatives in 2009 was 56.0; the average age of senators was 61.7.[54] Both chambers are considerably older than they were at their "youngest" in 1981, when the average age in the House was 48.4 and the average age in the Senate was 52.5.[55] Certain senators are particularly focused on biomedical research. Until he lost his seat, Senator Arlen Specter (born in 1930), for example, had been a long-term champion of NIH funding; he

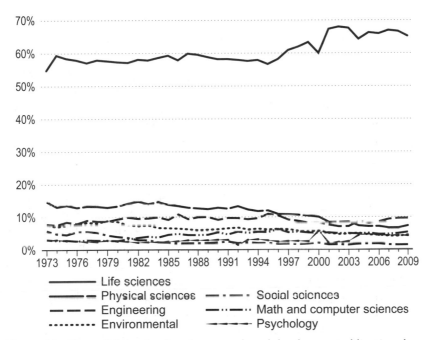

Figure 6.2. Share of federal university research and development obligations by field, 1973–2009. *Source:* National Science Foundation (2004, 2007b, 2010b).

almost single-handedly increased the amount that the NIH got out of the 2009 stimulus funds from a "modest" $3.9 billion to $10.4 billion. He is also a two-time survivor of two forms of cancer and had cardiac bypass surgery in 1998.[56]

The focus of research has also changed over time in other countries, but lack of consistent data makes this difficult to document in a systematic way. Suffice it to say that certain countries, among them Japan, Australia, and Sweden, have experienced increases in the proportion of funds going to biomedical research while others, such as Spain and Germany, have not.[57]

Mechanisms for Allocating Funds

University research has traditionally been funded out of resources the university has received from the state or, in the case of private institutions, from tuition and private donations. In many European countries, research universities have a long tradition of receiving general funds from the state in the form of block grants, a portion of which has been used in support of research (see Table 6.1). In some countries, research in the public sector has been primarily conducted at government-run institutes that operate independent of the university system, such as the Centre National de la Recherche Scientifique (CNRS) and Institut National de la Santé et de la Recherche Médicale (INSERM) in France and Max Planck in Germany. Institute researchers have also taught and held an appointment at a university. But the funds for research have come primarily through the institute. By 2008, approximately 80 percent of research in France was done in these so-called mixed units, many of the labs for which are physically located at a university or other host institution.[58] Such funding arrangements have meant that the responsibility of raising resources for research has not rested with the faculty member. Neither, in many cases, has it rested with the institution. Moreover, there has generally been no tradition of evaluating the research outcomes produced as a result of funding. More to the point, funds were often provided independent of results.

More recently, as documented in Table 6.1, researchers in Europe increasingly are supported by competitive grants. The Engineering and Physical Science Research Council (EPSRC) in the United Kingdom, for example, funds research in engineering and physical sciences; the Research Council in Norway funds all types of research in the university, as does the Flemish Science Foundation in Belgium.

Since the 1950s, resources for research at U.S. universities, including the funds to buy out part of their own time from teaching responsibilities, increasingly have become the responsibility of faculty members.[59] The university, as we have noted, provides start-up funds, but thereafter faculty are responsible for raising their own funds through the submission of proposals to funding agencies, be they nonprofit, federal, or in some cases state or local. The majority of these funds come from just four federal agencies: the National Institutes of Health (NIH), the National Science Foundation (NSF), the Department of Defense (DOD), and the Department of Energy (DOE), in that order. In addition, the U.S. Department of Agriculture (USDA) and National Aeronautics and Space Administration (NASA) provide over a $1 billion a year in support of university research.

The NIH and NSF evaluate proposals through a process of peer review, with some agency variation. The DOD and DOE are more likely to base their funding decisions on in-house evaluation, as are the USDA and NASA. Given the size of the NIH and NSF relative to other federal agencies, this means that approximately six out of ten federal dollars coming to U.S. universities in support of research are distributed through the process of peer review.[60]

Peer Review

Peer review begins at the NIH when a proposal is assigned to a "study section." Study section members review proposals in advance of the meeting and provide a score on each of five criteria (1 = excellent; 9 = poor): significance, investigator, innovation, approach, and environment. Members also provide a preliminary overall impact score, designed to synthesize the other scores. Preliminary impact scores are used to determine which proposals will be discussed when the study section meets. Applicants whose proposals are triaged receive the reviewers' scores and comments. All discussed applications receive a final impact score from each member of the study section, and an average impact score is calculated from these. Scores and accompanying written reviews are forwarded to the specific institute (there are twenty-seven institutes and centers at the NIH) and are reviewed by the institute's national advisory committee. Percentile cutoffs are important in determining who gets funded, although the NIH does fund PIs whose proposals fall below what is referred to as the payline. NIGMS at NIH, in the interest of distributing the wealth, singles out for special scrutiny proposals that would place the PI over $750,000 excluding indirect costs. Investigators whose proposals are turned down have the right to resubmit one additional time, and most do.

The current NIH review process was implemented in January 2010 and is a substantial modification of the process that existed for many years. That earlier process allowed for longer proposals (25 pages rather than the 12-page limit now) and did not provide a score for proposals that were triaged. Investigators whose proposals were turned down had the right to resubmit two additional times. Resubmission was particularly challenging for individuals whose proposals were not scored.

Historically, the NIH review process has put considerable weight on past accomplishments, which are enumerated on a standardized NIH biosketch form.[61] Results from the previous grant (if there was one) also play an important role in evaluation. The presence of demonstrated expertise and strong preliminary data play an especially key role in the review process: "No crystal, no grant." A major reason that universities provide start-up funds is to permit the newly hired faculty member time to continue the process of collecting preliminary data for an NIH proposal. The "lineage" of the scientist is often noted, in terms of where the scientist trained and in whose lab the scientist did his or her postdoc work. Researchers must also demonstrate that they have adequate space at their university in which to conduct the research.

Preliminary analysis of proposals reviewed by the National Institute of General Medical Sciences (NIGMS) under the new system suggests that the two criteria that are most highly correlated with the overall impact score are approach (Pearson correlation coefficient of 0.74) and significance (0.63). The criteria with the lowest correlation are investigator (0.49) and environment (0.37).[62]

Anywhere between 10 and 40 percent of the NIH applications are funded. Success obviously depends on the number of applications, the cost of the proposals being considered, and the availability of funding. It is also institute specific. For example, in 2010, the highest success rate was for applications reviewed by the National Institute of Deafness and Other Communication Disorders (30.2 percent); the lowest was for proposals reviewed by the National Institute of Aging (14.5 percent). The largest institute, Cancer, had a success rate of 17.1 percent.[63] In 2001, during the doubling, six institutes had success rates above 35 percent, and many more had success rates above 30 percent.[64]

The R01 grant, the "bread and butter" for university investigators, typically lasts for three to five years, and researchers can apply to renew their grant. This is the norm, not the exception. It is greatly encouraged by the fact that renewals do better in the review process than new proposals. It is not unknown for researchers to be supported on the same grant for over forty years. Harold Scheraga (Cornell University) has had the same NIH grant to study protein folding for fifty-two years.[65] In rare instances, a university can nominate a new investigator to take the place of an investigator who is stepping down, and the same grant is passed on to a new generation.

The NSF peer-review process follows a slightly different procedure. Investigators submit proposals to programs, which are generally organized around fields of study. Programs vary as to whether they use mail reviews exclusively or panel reviews supplemented by mail reviews to evaluate proposals.[66] Reviewers rank proposals on a five-point scale, from Excellent to Poor. Reviewing is voluntary: of the 60,400 requests made in fiscal year 2008, the NSF got almost 37,000 reviews (61 percent).[67] Unlike the case of the NIH, program officers have considerable discretion in making funding decisions, especially with regard to proposals that fall between "clearly fund" and "clearly do not fund." There is not a tradition of continuing a grant at NSF as there is at NIH, although researchers can and do submit proposals for follow-on research.

NSF has the appearance of putting less emphasis on reputation than does NIH and limits the number of publications the researcher can list to a maximum of ten (NIH used to not have a limit; it now "recommends" that one limit the number to fifteen). Anywhere between 20 and 37 percent of NSF research proposals are funded.[68] As in the case of NIH, the rate depends upon the number of applications and the availability of funding. It also depends on NSF policies with regard to size of award and length of award. In an effort to "increase productivity by minimizing the time PIs spent writing multiple proposals and managing administrative tasks," the NSF tried to extend the length of the average grant and increase the size of the grant. Between 2000 and 2005, the average size of an award increased by 41 percent; the average length of an award stayed approximately the same, at almost exactly three years. Success rates plummeted as more proposals chased fewer grants.[69] Not only was there an increase in the number of applicants, there was also an increase in the number of proposals per applicant. Both effects were no doubt due in part to the increased dollar value of an award, although the increase was also likely due to increased ease of submitting through the NSF fast-track system and the pressure universities brought to bear on faculty to engage in grantsmanship.[70]

Peer review also plays a role in allocating resources for university research outside the United States. It is used, for example, by all research councils in the United Kingdom as well as by the Wellcome Trust. It is the basis for decisions made by the Flemish Science Foundation and by the Norwegian Research Council. The European Union, which has long supported research through the Framework Program, now in its eighth form (Eighth Framework Program), has always used peer review to distribute resources. In an effort to encourage "cutting-edge" basic research, the European Research Council was established in 2007.[71] Again, decisions are based on peer review. Likewise, the Fund for Investing in Fundamental Research, which was established in Italy in 2005, makes decisions by peer review, as does the Agence Nationale de la Recherche in France, which made its first grants in 2005.[72]

Other Mechanisms

There are at least three other mechanisms, in addition to block grants in the form of unrestricted funds and peer review, for allocating research funds.

Assessment. This approach distributes government funds through an assessment of the strength of the department. The method has become increasingly important in recent years outside the United States. For example, in the United Kingdom, the Research Assessment Exercise, which in 2009 distributed £1.57 billion in support of university research, includes quality of publications as one of the measures for evaluating departments.[73] Publications also play a role in the distribution of research funds to Norwegian universities, as they do in Denmark, Australia, and New Zealand.

In Flanders, 30 percent of university research funds are distributed based on bibliometric measures.

Earmarks, the Money Schools Love to Hate. In 1978, the president of Tufts University, Jean Mayer, hired two lobbyists to press the university's case to obtain funds from the Department of Agriculture to build a nutrition center at the university. Their efforts were successful: Tufts received $32 million toward the building of a nutrition center, which, not surprisingly, is today known as the Jean Mayer USDA Human Nutrition Research Center on Aging.[74] "Once the genie was out of the bottle, nothing could put it back."[75] Money for earmarks for university research has grown in leaps and bounds ever since. In 2008, earmarks equaled $4.5 billion or 14 percent of all federal funding for research.[76]

Politicians often justify earmarks on the rationale that the peer-review system concentrates research funding among a few elite universities. Without earmarks, research at second-string institutions would never get a chance to develop. The proclivity of peer review to be risk averse is also sometimes used as a rationale for providing funds to universities through earmarks.

Occasionally, universities and colleges get earmarks without asking for them. Marywood College, for example, once received earmarked funds from the Department of Defense that they had not asked for, thanks to John Murtha (D-Pennsylvania).[77] But most universities that receive funds hire lobbyists to make their case in Washington, D.C. Moreover, it is not only the second string who lobby. Despite the public disdain that most elite universities hold for earmarks, they too engage in the practice. In fiscal year 2003, 90 percent of AAU institutions accepted at least one earmark and received a total of $336 million in earmarks.[78] Despite their efforts, earmarking redistributes funds away from top research universities toward lower-ranked institutions.

Not all lobbying meets with equal success. It helps considerably to be from a state having a member on the Senate Appropriations Committee. Universities with representation on the committee receive, for example, $56 for every $1 spent on lobbying, almost four times more than universities without representation received for every $1 spent lobbying. Membership on the House Appropriations Committee is not nearly as lucrative.[79]

Set-asides are another way Congress affects the allocation of resources for research. In this case, funds are provided for pet projects, often projects in which a state may have a considerable advantage. For example, buried in the federal spending measure adopted in the spring of 2009 was a $3 million directive for the NSF "to establish a mathematical institute devoted to the identification and development of mathematical talent." The directive was backed by Harry Reid, the Senate Majority Leader (D-Nevada). Not surprisingly, the University of Nevada at Reno supports the Davidson Academy, a public school for exceptionally gifted students.[80] But not all set-asides meet with success. The NSF folded the $3 million into a competitive grants program to fund a network of seven mathematics institutes at universities around the country. When the winners were announced in August 2010, the Davidson Institute was not one of them.[81]

Congressional representation also affects NIH allocations and (indirectly) the distribution of grants. Powerful congressmen, for example, can provide guidance on the allocation and disbursement of appropriated funds, direct reallocations among various NIH institutes, and support funds for specific diseases. Having an additional member on the appropriate subcommittee of the House Appropriations Committee that deals with the NIH budget has been shown to increase NIH funding to public universities in the member's state by 8.8 percent.[82]

Prizes. In recent years, there has also been considerable interest in stimulating R&D by offering inducement prizes. The idea is not new: the British government, for example, created a prize in 1714 for a method to solve the longitude problem. More recently, the Ansari X Prize was established in 1996 for the first private manned flight to the cusp of space. The $10 million was awarded eight years later to Burt Rutan. In 2006, the X Prize Foundation announced that it will pay $10 million to the first privately financed group to sequence 100 human genomes in ten days at a cost of less than $10,000 per genome. The winner will get another $1 million to decode the genomes of 100 additional people selected by the X Prize Foundation.[83] There are also prizes for dogs and cats: the Michelson Prize in Reproductive Biology, for example, will be awarded to the first entity to provide a nonsurgical sterilant that is "safe, effective and practical for use in cats and dogs."[84]

Prizes have been particularly embraced by the private sector and philanthropists. A 2010 McKinsey study reported that more than sixty prizes of at least $100,000 each debuted between 2000 and 2007, representing almost $250 million in prize money.[85]

The public sector (at least in the twentieth and twenty-first centuries) was a Johnny-come-lately to the use of prizes as a means to foster innovation. But since 2009, prize fever has struck Washington. President Obama in September of that year called on agencies to increase their use of prizes as part of his Strategy for American Innovation. In March 2010, the Office of Management and Budget issued a memorandum to agency heads affirming the administration's commitment to prizes and providing a policy and legal framework to guide agencies in their use of prizes. In September 2010, the White House and the General Services Administration launched the website Challenge.gov, where interested parties can readily find information about various incentive prizes sponsored by government agencies. In its first three months, the site featured forty-seven challenges from twenty-seven agencies. The prize frenzy got a further boost when prize authority was adopted as a component of the reauthorization of the America Competes Act, signed by the president in January 2011.[86]

The Pros and Cons of Different Allocation Systems

Earmarks

Earmarked projects are virtually never peer reviewed, and it is therefore impossible to know what was given up in order to fund them. This, in and of itself, makes them the bête noire of the research community. But they do have some pluses. For example, once established, earmarked projects often receive a steady stream of funding for years to come. Stability can encourage a long-term horizon and, theoretically, increases risk taking.

Prizes

Prizes have much to recommend them: they invite alternative approaches to a problem, not being committed to a specific methodology. They are awarded only in instances of success; the incentive to exaggerate is eliminated. In addition, prizes attract participation from groups and individuals who might otherwise not participate. A recent contest to foster "apps for healthy kids," for example, attracted a number of student entrants. The winner, a game called "Trainer," was developed by students at the University of Southern California.[87] Close to 200 individuals entered Harvard's 2010 Challenge contest to spur research on type 1 diabetes. One of the twelve winners, a diabetes patient, proposed an easier way for patients to measure whether they are successfully controlling their diabetes; another winner (a Harvard undergraduate) proposed that studying diabetes from a chemical perspective could yield new insights.[88]

But there are some serious downsides. Like the priority system, prizes encourage multiples. They are not well suited for research that has unknown outcomes—the desired outcome must be known and carefully specified. There is also the temptation for the awarding agency to raise the bar after a solution is proposed. There is also the problem of determining the size of the award. Ideally, one wants it to be sufficiently large to attract entrants, but not so large as to overcompensate the winners.

The greatest problem with using prizes as a way to encourage academic research is that funding is only awarded after completion: entrants are on their own to find the funds needed to compete. This means that prizes are a suitable mechanism for stimulating academic research that requires substantial resources only if the work complements research supported by other means or if partnerships can be built with industry.[89] Scientists at Carnegie Mellon University and the University of Arizona are doing precisely the latter. They are collaborating with Raytheon to compete in the $30 million Google Lunar X Prize, which will award $20 million to the first team to "safely land a robot on the surface of the Moon, travel 500 meters over the lunar surface, and send images and data back to the Earth." The second team to do so will receive $5 million, and another $5 million will be awarded "in bonus prizes."[90]

Block Grants and Assessments

Both direct government funding through a system of block grants and funding through peer review have benefits. Both also have downsides, or costs. The block grant approach to funding ensures that scientists can follow a research agenda with an uncertain outcome over a substantial period of time. It also exempts scientists from devoting long hours to seeking resources, or reviewers from spending hours evaluating proposals. These are not trivial benefits.

But block grants with no strings attached have costs. There is no built-in incentive for faculty to remain productive throughout their research career when neither funding nor salary depend on performance. Moreover, the research agenda is often set by the director of the laboratory or by full professors in the university. As a result, younger faculty may be constrained from following leads they consider promising and must wait for their senior colleagues to retire prior to leading a research effort.[91]

Perhaps most important, the no-strings-attached approach fails to meet the criterion of accountability. In recent years, this has proven to be the Achilles' heel of such a system, as the public, especially in Europe where the system had flourished, has demanded to know what they are getting for their investment in research—in terms of both the quality of the research and its contribution to economic development. Like it or not, a number of countries in Europe (mentioned above), as well as Australia and New Zealand, have moved away from using unrestricted funds in supporting research to a system that allocates university resources on the basis of past performance or through peer review. In France, the call for reform has been a bit different and a bit later in coming, but the rationale is lack of quality.[92]

Allocating resources on the basis of past performance invites universities to game the system. In the United Kingdom, for example, there have been numerous instances of just-in-time hires, where universities hire highly cited researchers just before the cutoff for the next evaluation period in order to boost their performance score.[93] In some instances, universities have hired faculty who retained a position at another university. This has proved to be a common practice in China, where performance affects resource allocation, and where a number of highly cited U.S.-Chinese faculty, as noted earlier, have been granted *jiangzuo,* or lecture-chair positions that require them to spend at most three months a year in China.[94] It is not only to enhance the research environment that Chinese universities are luring these professors back. It is also to enhance their resource base.

The criteria used for evaluation can also affect the quality of the research. For example, the formula used in Australia initially focused on publication counts in Institute of Scientific Information (ISI) journals (now Thomson Reuters Web of Knowledge). Not surprisingly, the largest increase in publications was in journals in the bottom-quality quartile, with the exception of medical and health sciences, where the largest growth was in the bottom two quartiles.[95]

Peer Review

The peer-review system also has its benefits. It provides for freedom of intellectual inquiry and encourages scientists to remain productive throughout their careers. To the extent that success in the grants system is not completely determined by past success, the system provides some opportunity for last year's losers to become this year's winners. Peer review arguably promotes quality and the sharing of information. The system also, as noted in Chapter 3, encourages entrepreneurship among scientists. Getting money from a venture capitalist is not that different from getting money from a funding agency—both require making a strong pitch.

Just as some of the benefits of a competitive grants system are costs of the unrestricted grant approach, so too some of the benefits of the latter are costs of the former. First is the question of time: grant applications and administration divert scientists from spending time doing research. A 2006 survey found that faculty scientists in the United States serving as PIs on federally sponsored grants spend 42 percent of their research time filling out forms and in meetings, tasks split almost evenly between pre-grant (22 percent) and post-grant work (20 percent).[96]

Reviewing the proposals of others also takes time. According to Antonio Scarpa, director of the NIH Center for Scientific Review, the now defunct 25-page R01 grant took as much as thirty hours to evaluate, including seven hours for each of the three assigned reviewers.[97] If senior faculty are involved, that comes to about $1,700 per proposal.[98] It is not surprising that in recent years concern has been raised at both the NIH and NSF that it is increasingly difficult to attract experienced reviewers and that the quality of the reviews has declined.[99] Nor is it surprising that the NIH cut the length of proposals by almost 50 percent beginning in 2010.[100]

A competitive funding system can also discourage risk taking. Grants are often scored on their "doability," selected because they are "almost certain to 'work.' "[101] To quote the Nobel laureate Roger Kornberg, "If the work that you propose to do isn't virtually certain of success, then it won't be funded."[102] There is a perception among older scientists that peer review, at least at NIH, used to be a different game, with reviewers focused on "ideas, not preliminary data."[103] The problem is compounded when funding is difficult to come by. The recently released *ARISE* report (*Advancing Research in Science and Engineering*) from the American Academy of Arts and Sciences concluded that in tight times "reviewers and program officers have a natural tendency to give highest priority to projects they deem most likely to produce short-term, low-risk, and measureable results."[104]

The underlying incentive system encourages risk aversion on the part of the PI: failure is not rewarded. Future funding clearly depends on obtaining successful outcomes during the current grant period. The system particularly discourages risk taking when one's own salary is at stake, as is often the case for researchers at medical institutions and always the case for summer support. The rubric for today's faculty has gone from publish or perish to "funding or famine," to use a phrase coined by Stephen Quake, a professor of bioengineering at Stanford University.[105] The most painful of appeals come from scientists whose labs will have to close and whose careers as an independent investigator will come to an end if their grant is not renewed.

The way funding is structured, at least that at NIH, also discourages scientists from taking up new research agendas during the course of their career. Because renewals have a much higher chance of receiving a thumbs up, researchers stay with a known course and specialize in a line of research over their career. An established scientist once told me of the disdain he held for his colleagues who kept the same grant going for years, seeing it as a sign of lack of creativity. He is clearly in the minority: the current system encourages such behavior. He also has greater flexibility in choosing his research agenda: he is an HHMI Investigator.

Neither has the competitive grants system proved to be friendly to the young. In recent years, for example, the number of new investigators funded by NIH has remained almost constant while the number of experienced investigators has increased (see discussion to follow).[106] And success, when it comes, increasingly comes to older scientists. The average age at which scientists receive first independent funding increased from 37.2 to 41.8 between 1985 and 2008.[107] At least three factors have contributed to this outcome. First, the need for preliminary results biases funding decisions toward more established researchers and delays the submission of grants by investigators who are just starting out. Second, more than 70 percent of new investigators must resubmit their proposals before receiving funding; thirty years ago, over 85 percent of all new investigators received funding on their first submission. Resubmission can easily add an additional year to the process. Third, people increasingly are older at the time that they get a faculty position.[108]

The grants system comes up particularly short when the odds of receiving funding are extremely low. It is inefficient in terms of the time and resources expended in submitting and evaluating proposals that have an extremely low probability of being funded. It lowers morale.[109]

There is also the problem that the grants system provides incentives to secure as much funding as possible for one's work, irrespective of whether an increase in funding leads to a proportionate increase in productivity. Money can become an end, not a means, and the amount of funding a measure of productivity.[110]

Granting agencies are aware of many of these problems. NIH, for example, has repeatedly made efforts to increase the number of young investigators it funds. A recent initiative, for example, created "Kangaroo grants" to help investigators transition from postdoc positions to new faculty positions. Reviewers of R01 proposals are made aware of whether the proposal comes from a new investigator, and the payline is generally lowered for new investigators. Moreover, new investigators now routinely receive an additional one year of funding without asking for it. One of Elias Zerhouni's last actions before stepping down as the Director of the NIH in the fall of 2008 was to make room for new investigators by declaring it formal NIH policy to "support new investigators at success rates comparable to those for established investigators submitting new applications."[111] The institutes got the message: in 2009, NIH supported 1,798 new investigators, a considerable increase over the 1,361 supported in 2006.[112] One consequence of the action was a substantial increase in the number of grants funded below the payline.[113]

In an effort to increase risk taking, the NIH created Pioneer and Eureka Awards. The former are "designed to support individual scientists of exceptional creativity who propose pioneering—and possibly transforming—approaches to major challenges in biomedical and behavioral research."[114] The latter are designed "to help investigators test novel, often unconventional hypotheses or tackle major methodological or technical challenges."[115] Laudable as these efforts are, the numbers are miniscule. In 2009, for example, the NIH made eighteen Pioneer awards, the most ever. But the odds are less than 1 percent: over 2,300 applications were received for the multimillion five-year award.[116]

NSF also undertook a new, foundationwide initiative to encourage "transformative research" in 2007. Among other things, the agency expanded its merit-review criteria to explicitly include "review of the extent to which a proposal also suggests and explores potentially transformative concepts."[117]

The NIH Doubling: A Cautionary Tale

It is tempting to assume that money is the answer to many of the problems that plague peer review and, more generally, the university research enterprise. Additional funds should translate into higher success rates, which in turn should encourage increased risk taking. More money should also mean more jobs and grants for young researchers.

But anyone who thinks so should be careful what they wish for. The doubling of the NIH budget between 1998 and 2002 ushered in a host of problems. By the time it was over, success rates were no higher than they had been before the doubling. By 2009, and in part because of the real decreases that the NIH experienced in the intervening years, success rates were considerably lower than they had been before the doubling.

Faculty were spending more time submitting and reviewing grants. Although early in this century 60 percent of all funded R01 proposals were awarded the first time they were submitted, by the end of the decade only 30 percent were awarded the first time.[118] More than one-third were not approved until their last and final review. This not only took time and delayed careers, but the perception was that these "last chance" proposals were favored over others, creating a system that, according to Elias Zerhouni, awarded "persistence over brilliance sometimes."[119] Moreover, and jumping ahead to Chapter 7, there is little evidence that the increase translated into permanent jobs for new PhDs, as had been the case in the 1950s and 1960s when government support for research expanded.

It is also not clear that the doubling resulted in the United States being relatively more productive, at least as measured by publications. Frederick Sacks's study of U.S. publications in biomedical fields for the period during the doubling found no "upward jump" in U.S. publications relative to those from laboratories outside the United States where funding did not double.[120]

A major cause of this seeming paradox was the response of universities to the doubling. Some universities saw the doubling as an opportunity to move into a new "league" and establish a program of "excellence." Others saw it as an opportunity to augment the strength they already had. For still others, expansion of their existing programs was simply necessary if they were to remain a player in biomedical research. Regardless, the end result was that the majority of research universities went on an unprecedented building binge. Recruiting senior faculty—with their large grants and capacity to generate still larger grants—required space—lots of it. Deborah Powell, dean of the Medical School at the University of Minnesota, put it bluntly: "The problem in recruiting senior professors is that they want lots of space . . . Getting a group of four or five neuroscientists means that you have to look at thousands of square feet of space and lots of money."[121]

Universities used philanthropic, local, and state resources as well as debt to finance the expansion. They hired additional faculty and research scientists, many in soft-money positions. Universities also encouraged faculty who had heretofore not applied for grants from the NIH to "go where the money is." And they encouraged those who had grants to get more: not one grant or two grants but three became the expectation at many research institutions. New buildings with larger laboratories required more resources to support them.

Not surprisingly, the number of applications for new and competing research projects grew. In 1998, the NIH received slightly over 24,240 applications for R01 awards; by the end of the doubling in 2003, it received 29,573. By 2009, long after the doubling had ended, it received 27,365.[122] Success rates, which were over 30 percent at the beginning of the doubling, fell to 20 percent by 2006. By 2009, they had "rebounded" to 22.2 percent.[123]

One reason for the decline in success rates was the substantial growth in budgets accompanying the proposed research: in 1998, the average annual budget of the typical grant was $247,000; by 2009, it had grown to $388,000.[124] Several factors contributed to the increase: first, more faculty were on soft-money positions and thus writing off a larger proportion of their salary on grants.[125] Second, the cost of equipment and supplies grew considerably during the period. Mice and magnetic resonance imaging equipment are expensive: the Biomedical Research and Development Price Index increased by 29 percent between 2000 and 2007; the Consumer

Price Index, by comparison, rose by 20 percent.[126] Third, tuition for graduate students (which is included in grants) was increasing. The increase provided a way for universities to get more federal funds.[127]

Another factor contributing to the decline in success rates was that NIH had less money with which to support R01 grants. Not only did the NIH budget decline in real terms after the doubling, but commitments made during the doubling to fund grants of four- to five-years' duration meant that fewer resources were available as the doubling ended. In 2003, the NIH had $2.6 billion for competing R01 grants; at the nadir in 2006, it had $2.2 to spend on R01s.[128]

The NIH also chose to devote a smaller percentage of its budget to R01 grants, opting instead to put funds into large project grants as well as a portion of the budget into the Roadmap initiative created by Director Zerhouni in 2002 in an effort to provide more flexibility and address major opportunities and gaps in biomedical research. In 2001, 53 percent of the funding for new awards went to R01 grants; by 2006, R01s received only 45.1 percent of the funding for new awards. The percentage had slightly increased by 2010 and stood at 47.4 percent.[129]

Some of the new grants during the doubling went to researchers who had heretofore not received NIH funds. But the vast majority of new grants went to established researchers: the percentage of investigators who had more than one R01 grant grew by one-third during the doubling, going from 22 percent to 29 percent.[130] The number of first-time investigators grew by less than 10 percent during the doubling.[131]

Young researchers were at a disadvantage competing against more seasoned researchers who had better preliminary data and more grantsmanship expertise; at every submission stage, the success rates of new investigators were lower than those for established researchers submitting a proposal for a new line of research.[132] As seen in Figure 6.3, the increased number of grants for experienced investigators and minimal growth in grants for first-time investigators resulted in a dramatic change in the age distribution of PIs. In 1998, less than a third of awardees were over 50 years old: almost 25 percent were under 40. By 2010, almost 46 percent were over 50, and less than 18 percent were under 40. More than 28 percent were over 55 years old.

One response of the biomedical community was to lobby (unsuccessfully) for more funds. There was even a move to generate another "storm," given the perception that the earlier "Gathering Storm" report had proved helpful to the physical sciences, its primary focus. (The report was written soon after the NIH doubling).[133] Thus, some believed that a similar report focusing on the biomedical sciences might be the way to attract Congress's attention.

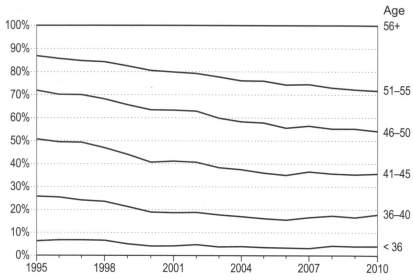

Figure 6.3. National Institutes of Health competing R01 equivalent awardees by age, 1995–2010. *Source:* Provided by Office of Extramural Research, National Institutes of Health.

The Stimulus Bill

No one expected that help would come in the form of a stimulus bill brokered in the middle of the night. But when the biomedical community woke up on the morning of February 4, 2009, they found themselves to be the recipients of more than $10 billion in stimulus funds to be spent *in two years*. If it was difficult to have a smooth landing after the doubling, what would happen after an infusion of $10 billion, scheduled to disappear after two years?

NIH chose to spend a third of the funds by extending the payline, funding (but only for two years) proposals that had fallen below the initial cutoff. They spent another third on administrative supplements designed to accelerate the tempo of research (more people and more equipment). But it was the Challenge grants—which represented less than 10 percent of the expenditures—that got by far the most attention. Almost as soon as they were announced in early March 2009, universities, hungry for indirect costs and with faculty whose grants had not been renewed, put on a full court press.[134] In less than ten weeks, more than 20,000 proposals were submitted for the award, which could fund up to $1M of direct costs over the two-year period. The University of Minnesota, which submitted 224, accounted for 1 percent of these, as did the University of California–

Irvine, which submitted approximately the same number.[135] Deans at some universities reportedly told faculty members that they would be judged on the number of applications they submitted.[136]

In the end, the NIH funded 840 Challenge grants; the success rate was slightly less than 4 percent.[137] But this is not necessarily the end of the proposals—they can be resubmitted as R01 proposals. Resubmission is facilitated by the fact that the format of the Challenge grant (12 pages) is a perfect fit for the format of the newly streamlined R01 proposal.[138] The stimulus funds may have helped many researchers and universities through a difficult period, but they were not a "fix."[139] If success rates were low in 2009, they will assuredly be lower in the foreseeable future.

Policy Issues

The United States spends between 0.3 and 0.4 percent of its gross domestic product on R&D at universities and medical schools. This represented almost $55 billion in 2009 or approximately $170 for every man, woman, and child. Over 30 billion of this comes from the Federal government and two-thirds of this goes toward research in the life sciences, especially the biomedical sciences. Moreover, the percentage going to the life sciences increased during the first years of this century.

To an economist, facts such as these raise questions of efficiency. Is the 0.3 to 0.4 percent enough? Too much? Is the two-thirds allocation to the life sciences "right"? Before one can even hope to answer such questions, it is helpful to know what economists mean when they use the term *efficiency*. Not surprisingly, it has a specific meaning. To wit, resources are said to be efficiently deployed if one cannot increase the size of the proverbial pie by reallocating those resources.

How does one tell if resources are efficiently allocated? The straightforward way, and ignoring risk, is to compare the rate of return between investment opportunities: if the rate of return on investment X is 20 percent and that on investment Y is 10 percent, clearly one should invest more in X, taking the resources from Y. Marginal returns will eventually decline on X and increase on Y as the reallocation occurs.

This seems quite straightforward. Compute the rate of return resulting from investments in research at public institutions and compare it with alternative rates of return. Or compute the rate of return for putting another dollar into biomedical research and compare it with the rate of return for putting another dollar into research in physics. It sounds easy, but the devil is in the details—and the lack of data with which to measure the details.[140]

For example, how narrowly or broadly does one define the benefits? Take the atomic clock. The idea of using atomic vibration to measure time was first suggested more than 130 years ago by Lord Kelvin in 1879; the practical method for doing so was developed in the 1930s by Isidor Rabi.[141] The clock has contributed to numerous new products and innovations, including GPS. Or take fundamental research in physics, which has led to a number of new products and processes, including nuclear magnetic resonance. Where does one draw the line?

When does one draw the line? Often, as Chapter 9 details, the benefits from research are years away. This means that society may often have to wait years for the benefits to show up in the economy. There is also the issue that many of the benefits arising from public research are not traded in the market and thus are hard to value. How does one put a value on the images transmitted from the Hubble telescope? Or the satisfaction derived if and when the mysteries of dark matter are unraveled?

Partly as a result of these challenges, studies of public rates of return on investments in R&D have been rather narrowly focused, looking either at rates of return to specific types of research or rates of return to the research that led to the development of specific products. There has been, for example, considerable research regarding the rate of return to research on corn as well as to agricultural R&D more broadly defined. A recent study, for example, finds the rate of return to research sponsored by the U.S. Department of Agriculture to be 18.7 percent.[142] The study also reports rates of return for agricultural research funded by specific states. When the estimates include spillovers to other states, they average 32.1 percent, with a minimum of 9.9 percent and a maximum of 69.2 percent. There have also been numerous studies regarding rates of return to investments in medical research. One study, for example, estimates that investments by the National Institutes of Health on factors related to cardiovascular disease, coupled with visits by people at risk of cardiovascular disease to their doctor, have had a return of about 30 to 1.[143]

A study, now quite dated, that was prepared for the NSF traced the key scientific events that led to five major innovations (magnetic ferrites, video tape recorders, oral contraceptives, electron microscopes, and matrix isolation). Of particular significance is the finding that in all five cases non-mission scientific research (defined to be research "motivated by the search for knowledge and scientific understanding without special regard for its application") played a key role and that the number of non-mission events peaked significantly between the twentieth and thirtieth year prior to an innovation. The study also finds that a disproportionate amount of the non-mission research (76 percent to be precise) was performed at universities and colleges.[144]

Case studies such as these are valuable, especially given the long lags between public investments in R&D and economic outcomes, which make estimation difficult. However, it is important to recognize that they suffer from a winner's bias, focusing on areas where public R&D has made a difference, rather than sampling across the spectrum, thereby including successes and failures as well as areas where public R&D has not made a difference.[145]

An alternative way to study rates of return to public investments is to survey firms, inquiring about the role that public research plays in the development of new products and processes. Using such an approach, Mansfield found that 11 percent of the new products and 9 percent of new processes introduced in the seventy-six firms he interviewed could not have been developed (without substantial delay) in the absence of recent academic research. He uses this data to estimate social rates of return of the magnitude of 28 percent.[146]

Taken together, studies such as these suggest that the return to past investment in public research has been substantial. Whether returns will continue to be substantial in the future is, of course, uncertain. To quote Mansfield, "Because such studies are retrospective, they shed little light on current resource allocation decisions, since these decisions depend on the benefits and costs of proposed projects, not those completed in the past."[147]

The answer to the efficiency question regarding the right amount for the United States to spend on research in the public sector is thus difficult to answer. But one is on safer ground if the question is rephrased to ask whether the amount being spent should be increased. We may never know the right amount, but—given the fairly healthy returns to previous investments in research—the right amount is likely to be greater than the .3 to .4 percent of the GDP that is currently being spent.

What about the balance in the U.S. R&D portfolio? Is the heavy and until recently increasing focus on biomedical research warranted from an efficiency point of view? No one has made the calculations to determine whether the return to a marginal dollar spent on biomedical research is greater than a marginal dollar spent on research in, say, solid-state physics. But one can make a credible case, as I do in Chapter 10, that the current situation may not be efficient. Rather, it reflects the public's interest in health and the strength of various lobbying organizations in supporting medical research. It also reflects the reality that funds for research in some areas of science are tied to the mission of federal agencies, and certain of these agencies in recent years have found their budgets either cut or growing at a lower rate than those of other agencies. The end of the Cold War, for example, resulted in cuts in the amount allocated to the Department of Defense and, consequently, to defense-supported research at universities.

There are other efficiency issues, such as whether large grants are more effective than small grants and whether the selection process and structure of a grants program, such as that employed by HHMI, is more effective than that employed by NIH. With regard to the size of grants, there is some evidence that productivity, as measured by the number of publications, has a low correlation with the amount of funds received in grants. A study by the National Institute of General Medical Sciences (NIGMS), for example, found the correlation coefficient between the number of publications by NIGMS investigators and the total direct costs of their grants to be only 0.14.[148] This, of course, is but one study, and it does not address the question of whether it is more efficient to fund large projects involving multiple PIs or fund more individual projects. In NIH terms, it does not address whether R01s are more effective than P01s. Nor does it even begin to address the efficiency concern as to whether large pieces of equipment that come with price tags of billions of dollars and tie up resources for years to come are good investments.

With regard to the latter, a recent study suggests that the HHMI system encourages creativity and, by implication, greater risk taking than does the NIH system. In an effort to control for selection issues, the authors compared the productivity of researchers funded by HHMI with that of researchers funded by NIH but who had been awarded early-career prizes from one of several foundations. They found that HHMI investigators produce high-impact papers at a much higher rate than the control group. They also found evidence that the direction of HHMI investigators' research changes compared with that of the control group. At least three factors may account for why the HHMI system appears to do better than the NIH system: it evaluates people, not projects; it funds individuals for a longer period of time than does the typical NIH grant; and it is reasonably forgiving of "failure," at least the first time the individual comes up for review.[149] The Wellcome Trust appears convinced: in 2009, it announced that it would begin to evaluate people rather than projects in making its awards and make the awards for a longer period of time.[150]

Conclusion

Research is an expensive business. Because of the characteristics of basic research and the motives of individuals who are drawn to doing it, and partly by historic accident, in many countries scientific research is performed in the university. It is paid for by a coalition of forces, with the government, regardless of country, picking up the largest part of the tab.

Other contributors include industry, private foundations, and universities themselves. In recent years, the trend has been for universities to pick up an increasing portion of the expenses and for the proportion supported by the government to decline. But these patterns vary by country.

Increasingly, the criterion for the support of university research is performance: no output, no funding. Although this may seem to be a straightforward proposition, it has not always been so, especially in Europe. Moreover, increasingly it has become the responsibility of faculty to generate the resources to support their research, either indirectly by building reputation or directly by submitting grants. The United States is the extreme case: the university's direct support of a faculty member's research virtually disappears after two to three years. In addition, faculty are increasingly expected to raise the funds to pay for their own salary. This is especially the case at medical institutions, and not only for non-tenure-track research faculty but also for faculty holding tenure.

At the same time, the resources to support research, as measured by success rates in getting grants, have become more scarce. This is in part because funds for research, especially in recent years, have been virtually flat, but it is also because the size of the university research enterprise and the expectations of universities have expanded.

Such a system has led faculty, and the government agencies that support faculty, to be risk averse. "Sure bets" are preferred over research agendas with uncertain outcomes. It is not just the peer-review system that fosters risk aversion. The Defense Advanced Research Projects Agency (DARPA), which once boasted that "it took on impossible problems and wasn't interested in the merely difficult," has increasingly shifted to funding research that is more near term and less risky.[151] Playing it safe may generate research, but it is, to quote Donald Ingber, "not science in its truest sense because science is the process by which we define the unknown."[152]

The system, at least in the United States, has particularly failed young investigators. It is no wonder: they have fewer preliminary results and less grant expertise than their grey-haired colleagues. But failure to adequately support young faculty is a recipe for more problems now and down the road. Exceptional contributions are more likely to be made by the young.[153] Future discoveries, as well as the education of future generations of scientists, depend on building up a base of new investigators. Moreover, supporting early-career scientists makes careers in science and engineering more appealing to younger people who are in the process of choosing careers.[154]

Many of the problems that confront the funding of science are scale related. A system that worked relatively well when the research community was small does not work nearly as well when the enterprise grows by a factor of twelve, as the U.S. enterprise has done in the past fifty years. As the system becomes larger, there is a need to codify the rules and allocation mechanisms. This can discourage risk taking. A larger system also makes it more difficult for scientists to engage in intensive peer review. The process used by the HHMI to appoint investigators might prove difficult to replicate on a significantly larger scale.

Other problems with the current system of support for university research relate to its proclivity to experience periods of stop and go. Stop-and-go funding is harmful for careers; it also makes it difficult for agencies to engage in long-term planning. The NIH assumed that its budget would grow at a "normal" rate after the doubling. Universities assumed that the manna would continue. Although the NIH might have behaved differently had it known that its budget would decline in real terms, it is not clear that universities would have. There was too much at stake: if the university did not expand, it would be left behind. The situation was a bit like going to a football game. The first person who stands up can see better, as can the second and third. But by the time everyone stands up, no one sees better. And everyone is colder for having stood up.

CHAPTER SEVEN

The Market for Scientists and Engineers

WHEN THE PRICE OF GAS began to increase in the mid-2000s, the demand for hybrid cars increased. The result was waiting lists of two to three months, and customers who paid more than the sticker price for the car. The same thing occurred in 2008 when gas prices went above $4 a gallon: a shortage of hybrids existed. In both instances, it was relatively short lived. Within a matter of months, the number of hybrids produced increased, the shortage ameliorated, and the premium that people were willing to pay fell.[1] The market responded quickly. Within a relatively short period of time, more hybrid cars could be produced.

Substitute engineers for hybrids, and we get a different outcome. Historically, when the demand for engineers increases (due, for example, to an increase in the defense budget), the market adjusts slowly. It takes, after all, at least four to five years to educate an additional PhD engineer. Or when demand decreases, it takes four to five years before the number of PhDs awarded begins to decrease.

Consider what happened in mathematics in the 1990s by way of example. After the supply of math PhDs increased between 1989 and 1996 (partly as a result of the breakup of the Soviet Union), the real nine-month teaching and research salaries of new PhDs declined by 8 percent. Unemployment rates increased, as did the percentage of jobs held by new PhDs that were temporary.[2] The number of full-time non-tenure-eligible faculty in traditional math departments increased by 37 percent, while the number of tenure-track faculty fell by 27 percent.[3] Not surprisingly, between the fall of 1994 and 1996, the number of applicants to math graduate programs decreased by 30 percent in response to the dismal job prospects for new PhDs.[4]

Past chapters have alluded to factors that affect the market for scientists and engineers. This chapter addresses them directly. It begins with a discussion of factors affecting the supply of new PhDs. Next, because of the growing importance of postdoctoral training in the United States, the focus shifts to this market. The chapter then turns to a discussion of the market for academics. The chapter closes with a case study of what happened to the market for PhDs in the biomedical sciences during the NIH doubling. Because of the extraordinarily important role that the foreign born play in U.S. academic science and engineering, Chapter 8 is devoted exclusively to the foreign born.

The Market for PhD Education

Approximately 550,000 PhDs trained in science and engineering are in the U.S. labor force; 39 percent of them work in academe and 41 percent in industry. The other 20 percent work in government, "other," or are unemployed.[5] Each year, U.S. universities produce another 24,000-plus PhDs in science and engineering. Other PhDs, who received their doctoral training outside the United States, come to the United States—many as postdoctoral researchers (postdocs). Indeed, almost half of the 36,500 science and engineering postdocs working at U.S. universities in 2008 came to the United States with a PhD in hand.

The number of PhDs awarded to citizens and noncitizens is given in Figure 7.1 for the period 1966–2008.[6] Awards for citizens are further divided by gender. Three broad trends clearly stand out: a decline in the number of citizen-men receiving PhDs, especially during the period 1970 to the late 1980s and 1998 to 2002; a gradual increase in the number of citizen-women getting degrees; and a substantial increase in the number of noncitizens (both permanent residents and those on temporary visas) obtaining a PhD in the United States, although there have been brief periods when the number of noncitizens declined. When the data are further differentiated by race (not shown), we find that the decline is largely among white men.[7] The number of Asian citizens receiving PhDs has increased slightly over time, as has the number of African Americans and Hispanics.

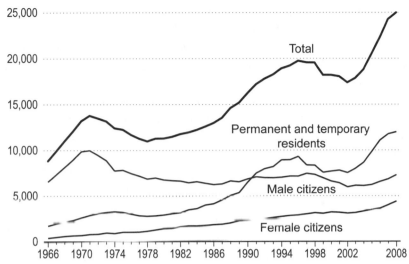

Figure 7.1. Science and engineering PhDs by citizenship and gender, 1966–2008. *Source:* National Science Foundation 2010c and National Science Foundation 2011c. For purposes of consistency over time, "medical/health sciences" and "other life sciences" are excluded from totals.

Relative Earnings

There is considerable evidence that the number of individuals choosing to follow a course of study in science and engineering is responsive to market signals. This is not to say that everyone contemplating a career in science and engineering bases their decision on market signals. Clearly some individuals have a sufficient taste for science that they would choose such a career regardless of relative earnings. But there are a number of individuals who, at the margin, contemplate careers in other fields. For them, money matters.

The fraction of college graduates with a degree in engineering, for example, closely tracks the career prospects of engineers four years earlier— when the students were freshmen—as measured by the present value of earnings in engineering relative to other occupations. It is also highly correlated with relative wages in engineering at the time the students entered college—an easier measure than relative present value—for students to compute.[8] Or consider the choice of majors at Harvard College. In the four-year period before the financial collapse of 2008, the average number of declared economics and applied math majors (both considered excellent preparation for a career on Wall Street) outnumbered the combined total of majors in biology, biochemistry, chemistry, math, neurobiology, and physics

(812 to 780).[9] Careers in science and engineering looked relatively unattractive in the long run. Even in the immediate short run the relatively low payoff made them unattractive. The $3,000 offered for a summer research stipend in a faculty member's lab was a mere drop in the bucket compared with the $15,000 that financial firms offered for summer interns.[10]

This was, of course, before the financial crisis and economic downturn of 2008, when Wall Street jobs disappeared in a matter of seconds and law firms laid off not only associates but full partners. Not surprisingly, applications to graduate school increased. Doctoral institutions reported an average increase of 10 percent in the number of applications from U.S. citizens and permanent residents; they enrolled on average 8 percent more domestic graduate students in the fall of 2009 than they had in 2008.[11]

Salaries for PhDs in science and engineering have been relatively low for a substantial period of time. One way to see this is to examine the earnings of PhD scientists and engineers relative to the "average" educated person. Figure 7.2 does this, showing the earnings of science and engineering PhDs relative to the earnings of individuals whose highest degree is a bachelor's. The top panel shows mean earnings for PhDs within ten years of receiving the doctorate relative to mean earnings of bachelor's degrees who are aged 25 to 34 for the period 1973–2006. The bottom panel shows earnings for PhDs who have been out ten to twenty-nine years relative to those for bachelor's degrees aged 35 to 54.[12]

Early-career engineers with a PhD earn about 1.6 times what those with a bachelor's degree earn; PhDs in the physical sciences earn about 1.4 times the benchmark; those in the life sciences generally earn less than 1.3 times the benchmark. (The spike in relative earnings in 1991 is due to the heavy impact that the 1991 recession had on the earnings of the benchmark group. At the same time, salaries of early career scientists and engineers increased.) There was a downward trend in the early-career PhD premiums throughout the 1990s, especially in the life sciences where PhDs earned only 5 percent more than a bachelor's degree in 1999. The dot-com build-up and the doubling of the NIH budget contributed to an increase in the earnings of PhDs in the early years of the millennium, while at the same time the recession of 2001 took a toll on the earnings of the baccalaureate group. The result was an increase in relative earnings. By 2006, relative earnings had declined again in all fields.

The conclusion: Seven-plus years of training less than doubles one's pay. In the case of the life sciences, the premium is never more than 50 percent and generally 30 percent or less. But this is for the early years. What happens as the career progresses? Does the educational premium increase with experience? The answer (see Figure 7.2 bottom panel) is generally no

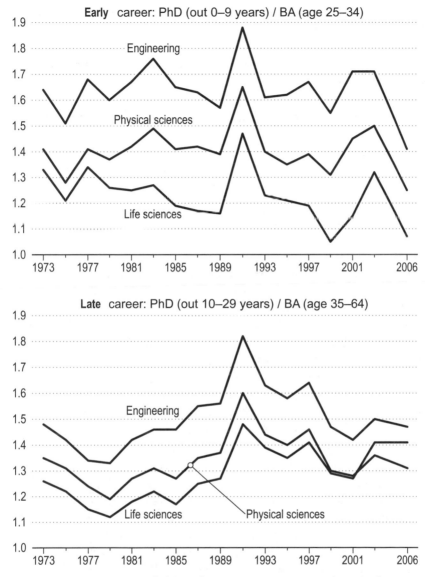

Figure 7.2. Mean earnings of PhDs relative to mean earnings of terminal baccalaureate recipients, by field, 1973–2006, early career and late career. *Note:* All data are in 2009 dollars and are for full-time and part-time workers. Analysis is restricted to men. PhD salary is adjusted for twelve months. *Source:* National Science Foundation (2011b) and Current Population Survey (2010). The use of NSF data does not imply NSF endorsement of the research methods or conclusions contained in this book.

and reflects the observation in Chapter 3 that the earnings profile, at least for academic scientists and engineers, is generally less steep than that in many other occupations.

It is not just relative salaries that affect the attractiveness of a career in science. The amount of time required for training and the value of that time also enters in. Consider an individual trying to decide between whether to pursue a PhD or to get an MBA. Even if there were not a salary differential, there is a huge differential in the amount of time it takes to train. The MBA degree takes two years; the typical degree in science and engineering takes seven-plus years. Moreover, the time it takes to get an MBA has remained constant for a number of years, while the time it takes to get a PhD has not. In the early 1980s, the median time to degree was between 6.2 and 6.7 years, depending upon field; by the mid-1990s, it had increased by more than a full year in the life sciences, being just shy of 8 years, and by approximately half a year in the physical sciences and in engineering. In recent years, the median time to degree has declined a bit and now stands at 7.1 in the life sciences, 6.8 in the physical sciences, and 6.9 in engineering.[13]

There is a cost associated with these extra years of training. Suppose our hypothetical individual (I'll call the individual a "he") were choosing between an MBA and a PhD in the biological sciences in 2004 and that if he did neither he could earn $42,300 the first year out of college.[14] Thus, both possibilities require "foregoing" $42,300 in earnings the first year in school, and $42,300 plus a presumed salary increase the second year. But things change dramatically when the MBA graduates in 2006, and receives a starting salary of $95,400, while the PhD student is still in graduate school.[15] The disparity persists. The PhD candidate continues to "forego" earnings; the MBA recipient begins to progress through his career. It is not unreasonable to assume that five years later, when the PhD candidate graduates in 2011, the MBA will be earning $120,000,[16] while the PhD's first job at a research university will pay approximately $70,500.[17] The disparity is even greater if the PhD student takes a postdoc position for another several years, receiving approximately $40,000 a year.

The present value computations presented in Table 7.1 are fairly straightforward. The present value of the MBA is approximately $3.2 million dollars. The present value of the PhD is just over $2 million.[18] Little wonder that the propensity to get an MBA has increased over time (for both men and women) while the propensity to get a PhD, especially for men, has declined for many years![19] It is even less of a wonder when one realizes that MBAs who graduate from the very top programs and go into finance can expect to earn four to five times as much as the typical MBA in our example, while PhDs who are hired as faculty at top programs can expect to earn only about three times as much as the typical PhD.[20]

Table 7.1. Projected lifetime earnings of MBA versus PhD in biological sciences holding a position at a research university (Present value, U.S. dollars)

MBA degree	PhD completed in 7 years	PhD completed in 8 years	PhD completed in 7 years and 3 year postdoc	PhD 7 years, support in graduate school
3,230,642	2,011,385	1,902,261	1,957,962	2,171,811

Note: See text for explanation and source.

A three-year postdoctoral position drives the difference up by another $53,000. Another year in graduate school contributes another $109,000 to the differential. And these estimates are on the conservative side. For those in math and statistics, the differential would be greater, given the relatively low pay that PhDs in this field receive; for those going to a lower tier institution, the differential would also be greater. If the stock options and bonuses that many MBAs receive are included, the differential would be significantly larger. Indeed, a 2001 study estimated that the present value of expected lifetime earnings of bioscientists is approximately $2 million less than the present value of the lifetime earnings plus stock options and bonuses of an MBA.[21]

Of course, the typical graduate student receives some type of support while in graduate school, which covers tuition and provides for a stipend. The MBA student does not. Neither does the law student nor the student enrolled in medical school. The most common type of support is a research assistantship, which pays between $16,000 and $30,000, depending upon the department and the discipline. A smaller and more select group of individuals are supported on fellowships, which pay in the same range but allow the student more freedom in the choice of a faculty member to work with. Once one takes the stipend into consideration, the cost of training decreases (see the last column), but the PhD is still expensive relative to a career in business.[22]

Table 7.1 makes it quite clear that reasons other than money enter into the decision to pursue a career in science and engineering. If it were only money, virtually no one would choose such a career. But it also makes clear that a number of variables can make the career financially less appealing: increased time to degree and increased propensity to take a postdoc are two factors that have certainly done precisely this in recent years, especially to American males.

On the other hand, increases in graduate stipends make the career more appealing. This is not surprising: stipends arrive early in the career, and, given the "power" of discounting, their early arrival greatly augments their value.[23]

Students understand this. Recent work shows that the number and quality of U.S. citizens choosing to apply for a National Science Foundation (NSF) Graduate Research Fellowship respond quite strongly to the value of the award. Research also shows that the number of applicants responds positively to an increase in the number of awards, and thereby an increase in the probability of receiving an award, while the "quality" of those applying is only modestly reduced.[24] Furthermore, although the link is difficult to prove, the evidence strongly suggests that even though the NSF program is small, bestowing only 1,000 fellowships a year, an increase in the value of the NSF award increases the number of domestic students going to graduate school, perhaps because other stipend-granting agencies as well as universities base their stipend rate on that of the NSF.

Students also understand that a decrease in the value of foregone opportunities makes graduate school more attractive. Thus, for example, when the unemployment rate increases and it becomes increasingly more difficult for recent graduates to find a job, it is not uncommon to find more people—especially men—heading to graduate school. A study spanning the years 1950 to 2006 found that the number of men getting PhDs in science and engineering was positively related to the unemployment rate that existed six years prior to their receiving their degree.[25] The collapse of the dot-com bubble undoubtedly contributed to the recent uptick in men going to get PhDs in engineering and the physical sciences that started right after the bubble burst.

People may also choose to go to graduate school if the alternative is perceived to be extremely undesirable. Such was the case during the Vietnam War, when the availability of a student deferment (2-S) from military service encouraged many men to go to graduate school rather than risk the draft. The effect was striking: the propensity for men to get a PhD grew by more than 60 percent in the short span of a few years; it then dramatically fell with the end of Vietnam draft deferments in 1967–1968.[26] The increase and dramatic decline can be seen in Figure 7.1, which reflects entry conditions occurring five to six years before the degree was awarded.

Job Availability

The computations of Table 7.1 make the strong assumption that individuals *will* get a full-time job in their field of training after investing seven-plus years in training. But this is not always the case. The physics market was severely stressed in the 1970s and again in the early 2000s; the math market was severely stressed in the 1990s; the market for chemists has fallen on difficult times in recent years because of merger and acquisition activity in industry; and the market for those trained in the biomedical sciences has been seriously depressed for a number of years, as we will see in the case study.[27] Multiple postdocs, underemployment, a position as a staff scientist, or working outside one's field have been outcomes that numerous highly trained individuals have experienced. Some have even experienced unemployment, despite their high level of skill. Indeed, the percentage of new doctorate recipients in math who were unemployed in 1994 and 1995 exceeded 10 percent—at a time when the overall unemployment in the economy was less than 6.5 percent.[28] One consequence of the 2001 recession was that unemployment rates though low—doubled among doctoral scientists in the life sciences and computer and information sciences and increased by more than 50 percent in the physical sciences, engineering and math, and statistics.[29]

Sometimes it is hard for those in secure positions to comprehend the math of Table 7.1. But graduate students get it, especially graduate students who, having foregone a considerable amount to get a degree, face bleak job prospects at the time of graduation. During the height of the physics employment crisis in the 1970s, the economist Richard Freeman gave a talk to the physics department at the University of Chicago. "When I finished the presentation, the chairman shook his head, frowning deeply . . . 'You've got us all wrong,' the chairman said gravely. 'You don't understand what motivates people to study physics. We study for love of knowledge, not for salaries and jobs.' But . . . I was prepared to give . . . arguments about market incentives operating on some people on the margin, when the students—facing the worst employment prospects for graduating physicists in decades—answered for me with a resounding chorus of boos and hisses. Case closed."[30]

PhD programs have historically focused on training a workforce that would replicate the career of those doing the training. It was assumed that after at most two years in a postdoc position, the newly minted researcher would get a job in academe. Some, of course, would go to the "dark side," taking a job in industry. And in certain fields, such as engineering and chemistry, the dark side was not that dark. Faculty at prestigious institutions such as MIT and Stanford had a long tradition of sending graduates to industry, after all. But for many fields, an academic job was the expected norm.

Thus, over 55 percent of the PhDs who received their degrees in the biological sciences in the late 1960s, when the academic market was flourishing,

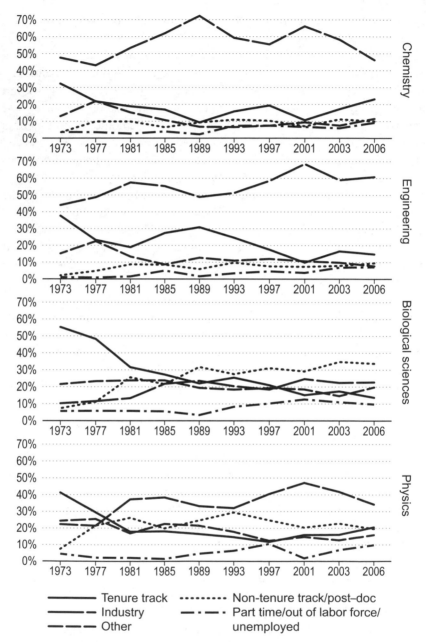

Figure 7.3. Job position by field, five- and six-year cohort, 1973–2006.
Source: National Science Foundation (2011b). The use of NSF data does not imply NSF endorsement of the research methods or conclusions contained in this book.

had settled into a tenure-track faculty position by 1973—five-to-six years after getting their PhD. The rate in physics was 41 percent, in chemistry it was 32 percent, and in engineering it was 38 percent. But by the early 1980s, market conditions had changed considerably for recently minted PhDs. Only 32 percent in the biological sciences had a tenure-track position, 18 percent in physics, 19 percent in chemistry, and 19 percent in engineering.

Where did they go? Some went into positions in industry. As Figure 7.3 shows, the percentage of new PhDs working in that sector increased considerably during this period for all fields. But in physics and the biological sciences, many ended up in the types of positions that virtually did not exist a decade earlier—non-tenure-track positions in academe and protracted postdoctoral positions. Moreover, in certain fields, a number began to work part time, withdrew from the labor force, or were unemployed.

This overall trend has continued during the last twenty-five years, although it has shown considerable fluctuations in response to market forces. By 2006, the last year for which data are available, less than 25 percent of the early career scientists in chemistry and physics held a tenure-track position; in the biological sciences and engineering, the figure stood at 15 percent or lower. By way of contrast, over a third of the recent PhDs in the biological sciences held a non-tenure-track position or a postdoc position; almost 20 percent in physics held either a non-tenure-track position or a postdoc position. Moreover, with the exception of engineers, close to 10 percent of those who had been out five-to-six years were working part time, unemployed, or were out of the labor force.

A 2006 editorial in *Nature Immunology* asked, "Is the 'conventional' career path of student to postdoctoral fellow to tenure becoming the 'alternative' career path?" The answer, given these data, is a clear "yes"—and not only in the biomedical sciences.[31]

Information Flows and Demographics

Information, or the lack thereof, also affects the supply of individuals going to graduate school. In the United States, information, especially with regard to the job outcomes of recent graduates, has typically not been readily available from graduate programs. The point was made abundantly clear when, in the late 1990s, the economist Paul Romer asked a research assistant to initiate application to the top ten graduate departments of mathematics, physics, chemistry, biology, computer science, and electrical engineering in the U.S, as measured by *U.S. News and World Report*. The student also began to apply to the top ten business and law schools. Not one of the sixty science and engineering programs provided any information about the distribution of salaries for graduates, either in the initial information packet or in response to a follow-up inquiry from him. But seven of the ten business schools included salary information in the application packet; one of the three nonrespondents directed the research assistant to a webpage with salary information. Four out of the ten law schools gave salary information in the application packet, and three more directed him to the information in response to a second request.[32]

The spread of information technology has not improved the amount of information that departments make available concerning the job outcomes of their graduates. A 2008 study of the webpages of fifteen top programs in the fields of electrical engineering, chemistry, and biomedical sciences found that only two of the forty-five programs listed actual information on placements. Four others provided some information on placements but did not list specific information regarding the placements. By way of contrast, seven of the fifteen programs in economics provided a list of students and where they were placed, year by year.[33]

Why are departments reluctant to provide information concerning the placements of their graduates? A cynic would point out that the research enterprise has come to rely on the 120,000 graduate students supported on research assistantships and fellowships to staff their labs (see discussion in Chapter 4). They are cheap—and temporary. Placement data could discourage potential applicants and put faculty research in jeopardy by killing the geese that incubate the golden eggs.

The culture of the university also stresses careers in academe, rather than careers in industry. Most graduate students with academic ambitions, especially in the biomedical and physical sciences, take a postdoc position, after receiving their PhD. In this sense, they have a job, albeit a temporary position, after they graduate. The ready availability of postdoc positions also conveniently lets the department off the hook. They have, after all, placed the student. The MIT program in biology can safely state on its webpage that the "majority [of PhD recipients] . . . go on to a postdoctoral position in an academic setting."[34]

The fact that faculty know little about careers outside of academe also affects the lack of information that is provided. When graduate students in the Yale molecular biophysics and biochemistry program wanted to learn about careers outside of academe, it was the students—not the faculty—who eventually created a seminar series to hear from alumni working outside of academe.[35]

Yet slightly more than 40 percent of all PhDs in science and engineering work in industry today, compared with fewer than 25 percent thirty years ago.[36] In some fields it is significantly higher. More than half of all PhDs in chemistry and engineering have worked in industry for a number of years. The percentage in physics and astronomy who are working in industry has grown by 50 percent in recent years. The percentage of those in math and computer science working in industry has tripled, and today stands at about one-third of all those with degrees in the two fields. The percent of life scientists working in industry has also grown dramatically. Despite this growth, fewer than 30 percent of those with a degree in the life sciences work in industry.[37]

Students, of course, get much of their information from other students, rather than from faculty. This may be one reason that liberal arts colleges have a relative edge in sending students to graduate school.[38] Undergraduates at the Swarthmores and Carltons of the United States do not have the "opportunity" to interact with graduate students and postdocs in the lab. They do not learn of their travails—sufficiently real to have spawned the comic strip *Piled Higher and Deeper (PHD)* that centers on the life (or lack thereof) of a "group of overworked, underpaid, procrastinating graduate students and their terrifying advisers."[39] Neither are the faculty at liberal arts colleges likely to spend long hours applying for research grants. Instead, the students are in an environment that stresses learning rather than "producing" science.[40]

The number of individuals receiving PhDs also depends on underlying demographics and college graduation patterns. For example, the large increase in the number of women receiving PhDs is due in large part to the increase in the number of women graduating from college, not to a change in the propensity of those going to college to get a PhD.[41] The same is true for underrepresented minorities. Indeed, the most effective way to increase the supply of underrepresented minorities receiving PhDs is to increase the number receiving bachelor's degrees. This is not a trivial observation: a policy maker would achieve larger increases by building the base of students eligible to go to graduate school than by investing, as many institutions do, in changing the propensity of those who graduate to go to graduate school.

By way of summary, the supply of individuals receiving PhDs is responsive to relative salaries, the availability of financial support, and underlying demographics. But preferences also matter. Rewards are intrinsic as well as extrinsic. The probability of enjoying these intrinsic rewards, however, depends upon the availability of research positions. It is also clear that it is often difficult for students to get good information from graduate programs concerning career outcomes of recent graduates.

Shortages?

Predictions of shortages of scientists and engineers occur with some frequency, despite evidence to the contrary. Such pronouncements have a long history, dating back to at least the late 1950s. Although predictions during the 1950s were perhaps on target, especially given the large sums that the government invested in R&D after the launch of *Sputnik,* predictions of shortages since have often strayed considerably from the underlying reality.[42]

Several predictions deserve special mention. First, in 1989, the National Science Foundation predicted "looming shortfalls" of scientists and engineers in the next two decades.[43] The same year, William Bowen and Julie Sosa published a book entitled *Prospects for Faculty in the Arts and Sciences: A Study of Factors Affecting Demand and Supply, 1987–2012.* The authors predicted faculty shortages in the ensuing period, basing their prediction on the assumption that an aging faculty, hired when higher education was expanding in the late 1950s and early 1960s, would be retiring at the same time that the baby boomers' children were headed to college.

By 1992, it was abundantly clear that the shortage had failed to materialize. The House Committee on Science, Space, and Technology's Subcommittee on Investigations and Oversight conducted a formal investigation, leading to considerable embarrassment at NSF. The next director of NSF apologized to Congress, acknowledging that "there was really no basis to predict a shortage."[44] Moreover, by 1992 the economic, legal, and political climate facing higher education had changed substantially. Universities faced budget problems as a result of economic recession. The elimination of mandatory retirement meant that faculty retired at a much slower rate than predicted. There was political pressure to downsize the federal budget. Mergers and acquisitions led to a dampening in demand from industry. And the demise of the Cold War led to cuts or plateaus in federal funding, especially federal funding for defense.

But getting egg on their face did not stop the forecast pundits. In June 2003, the National Science Board, the governing body of the NSF, released a draft task-force report for public comment that spoke of the "unfolding crisis" in science and engineering, stating, "Current trends of supply and demand for [science and engineering] skills in the workplace indicate problems that may seriously threaten our long-term prosperity, national security, and quality of life."[45]

Predictions of shortages are not limited to the United States. A 2003 European Commission Communication, "Investing in Research: An Action

Plan for Europe," for example, concluded that "Increased investment in research will raise the demand for researchers: about 1.2 million additional research personnel, including 700,000 additional researchers, are deemed necessary to attain the objectives, on top of the expected replacement of the aging workforce in research."[46]

Several issues arise when it comes to predicting shortages. First, to the extent that the shortage is real, the prediction of a shortage may lead to an under response on the part of students, given evidence that students have rational expectations and base their decisions partly on the expectation that others will respond, thereby putting downward pressure on wages.[47] Second, the models underpinning the projections are subject to substantial error, in part because political events that have a profound effect on scientific labor markets—such as the fall of the Berlin Wall, the doubling of the NIH budget, and the events of 9/11—are extremely difficult to predict.[48]

Third, shortages are often predicted by groups who have a vested interest in attracting more students to graduate school and into careers in science and engineering. To quote Michael Teitelbaum, "On the issue [of shortages] where one stands depends upon where one sits."[49] Most of the assertions come from four groups: universities and professional associations, government agencies, firms that hire scientists and engineers, and immigration lawyers. All have a considerable amount to gain by an increase in supply: universities, for example, in terms of students (and lab workers); companies in terms of the lower wages associated with an increase in supply.[50]

Blue ribbon commissions charged with addressing scientific labor market concerns have not disappeared.[51] However, their strategy changed in the first decade of this century. Although the message is still one of "we need more," the term *shortage* is not used. Instead, the underlying theme of these reports is that the United States is losing its dominance in science and engineering, in large part because the science and engineering enterprise has been expanding in Europe and Asia. A prime example is the *Rising above the Gathering Storm* report, released by the National Academy of Sciences in 2006,[52] which expressed deep concern that "scientific and technology building blocks critical to our economic leadership are eroding at a time when many other nations are gathering strength."[53] Areas of special concern included the number of individuals majoring in science, engineering, and math in college and pursuing graduate degrees. The message: without more scientists and engineers, the United States will lose its dominance in science and engineering.

A strength of *Rising above the Gathering Storm* was that it did not put all of its emphasis on supply-side initiatives—as is often the case—but instead also stressed measures that would enhance the demand for innovation and, by extension, the demand for S&E workers.[54] The importance of this should not be minimized. Initiatives that lead to an increase in the number of scientists and engineers without sufficient growth in demand (from industry, government and academe), can create a newly trained workforce with high hopes and poor job prospects—a perfect recipe for discouraging the next generation from entering careers in science and engineering.

The Market for Postdoctoral Training

Regardless of where one sits, there is almost unanimous agreement that there is not a shortage of postdoctoral fellows. Although it is difficult to get a precise handle on the exact number of postdocs working at U.S. universities, it is clear that it exceeds 36,000 and that it has grown considerably over time.[55] Problems with counting occur in part because postdocs work for individual faculty members, and this makes it more difficult to collect data. It is also difficult to determine who exactly is a postdoc because it is not uncommon for individuals who are essentially postdocs to be called by another title, such as research scientist. Thus all estimates must be taken with a grain of salt.

With this in mind, turn to Figure 7.4, which shows the number of postdocs by field working in the United States in academic graduate departments over the period 1980 to 2008.[56] The figure documents the considerable growth that has occurred since 1980 in the number of postdocs as well as changes that have occurred in the composition of the postdoc pool by field. In terms of size, the academic postdoc pool has almost tripled during the period, going from just over 13,000 to over 36,000.

Growth has been stimulated in part by the increased availability of research funds for hiring postdocs. It has also been stimulated by the cost advantage, discussed in Chapter 4, that can arise from staffing labs with postdocs rather than graduate students. The cost advantage is particularly relevant at private institutions, where tuition for graduate students can exceed $30,000 and is paid for in part from the principal investigator's grant.[57]

Almost 60 percent of academic postdocs work in the life sciences. The increase in the number of postdocs in the life sciences was especially notable during the period that the NIH budget doubled. But the number of postdocs in engineering has increased at a far greater clip over time, as has the number in the geosciences.

Postdocs are also increasingly likely to be temporary residents. In 1980, about four in ten postdocs were temporary residents; by 2008, almost six

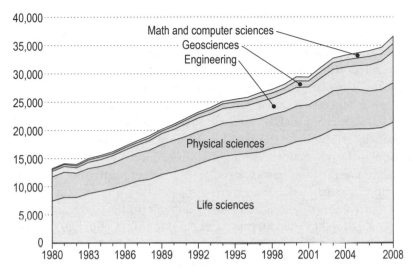

Figure 7.4. Number of science and engineering postdocs by field, 1980–2008. *Source:* National Science Foundation (2011d). Multidisciplinary studies was introduced as a subfield in 2007; the forty-nine recipients in 2007 and the seventy in 2008 receiving multidisciplinary degrees are distributed across the five fields affected by the change. Degrees in neurosciences, which was introduced as a subfield in 2007, are counted in the life sciences for 2007 and 2008.

out of ten were temporary residents. Again, the dramatic increase came during the period of the NIH doubling. Many of the postdocs who are temporary residents received their PhD training outside the United States. Indeed, the best estimate is that almost five out of ten academic postdocs in the United States earned a doctorate in another country and that four out of five postdocs with temporary visas earned their doctorate outside the United States.[58]

The response of noncitizens to employment opportunities arising from the NIH doubling is one reason why scientific labor markets respond more quickly to changes in demand than they did in the past. Twenty-five years ago, when the United States was producing the majority of PhDs, an increase in demand could only be met (or be primarily met) by growing the supply in the United States.[59] This takes time. But the expansion of PhD programs in other countries has created a supply of PhDs who are ready and willing to come to the United States to work, assuming they can get a visa. The postdoc market has proven particularly responsive to changes in demand. It not only provides an opportunity to come to the United States, with a starting salary of approximately $37,500, but there is also the distinct possibility that, once here, the trainee can stay.

Individuals on postdoctoral appointments are generally selected by the person in whose lab they will work. In the case of academe, where the vast majority of postdocs are located, this means that the principal investigator selects the postdoc. Although established investigators can choose among applicants who contact them, beginning investigators, who have yet to establish a reputation, must rely on the Internet and on advertisements posted in science journals to fill their postdoc positions.[60]

Financial support for the postdoc position is provided either through the principal investigator's grant (or start-up funds) or through a fellowship that the postdoc has received. Postdocs supported on fellowships have more independence than those supported by a faculty member because they come with funds and project in hand (or have a project in mind and get a fellowship soon after arriving) and in theory could go to another lab. They are also in the minority and are more likely to work with high-profile, established investigators.[61] Most of the postdocs in Susan Lindquist's laboratory at the Whitehead Institute, for example, come with their own funding.[62] Ninety percent of those in Roberto Kolter's lab at the Harvard Medical School have their own funding.

Postdoc stipends range anywhere from the high $30,000s to the low $50,000s, depending on the department, field, and years of seniority of the postdoc. As noted in Chapter 4, the NIH provides guidelines for those supported on NIH grants. In 2010, the suggested minimum was $37,740.[63] Some institutions pay considerably more. The starting pay at the Whitehead Institute, which was voted the "best place for postdocs to work" in 2009 by readers of *The Scientist,* was $47,000.[64] The Institute also provides health, dental, and retirement benefits to postdocs, something many programs do not do.[65]

The probability that an individual takes a postdoc position depends partly on the job market for recent PhDs. To quote the American Institute of Physics, "The proportion of new PhDs accepting postdoctoral positions has been a better job market indicator than the unemployment rate for physics PhDs, which is traditionally low and does not fluctuate a great deal."[66] Only 12 percent of newly minted engineers had definite plans to take a postdoc position in 2001 at the peak of the high-tech market; 54 percent had a definite job offer. The remaining 34 percent had no definite plans at the time of graduation. Five years later, when the market for engineers had cooled down considerably, it was a different story: 18 percent had definite plans to take a postdoc upon receiving their degree, only 42 percent had definite job plans, and 40 percent had yet to make definite plans.

More generally, the proportion of new PhDs with definite plans to take a postdoc increases when the size of the graduating class increases, consistent with the idea that job market prospects are depressed due to an increase in supply. The proportion taking postdoc positions is also inversely related to the availability of jobs in academe, as proxied by the percentage change in "fund revenue" for private and public academic institutions.[67]

The postdoc position is often described as a holding tank, where individuals sit until the market improves. One in eight respondents in a national survey reported that they had taken their most recent postdoc position because other jobs were not available. Those who report "bad jobs" as the reason for having taken the most recent postdoc position hold the position for a significantly longer period of time than those who do not report "bad jobs" as the reason.[68]

Although it is difficult to prove, it is assumed that the ones most likely to wait it out as a postdoc are those who most aspire to an academic position and have difficulty finding one. It is not uncommon for individuals to remain in a series of postdoc positions for five, six, or even seven years. Some stay even longer. Julia Pinsonneault, for example, was a postdoc for eleven years before finally taking a research scientist position at Ohio State.[69] Postdoc work allows one to build up a curriculum vitae and hedge one's options.[70] It also puts food on the table, although a salary in the $50,000's is a far cry from what one could have gotten if one had entered a different career.[71]

Postdoc working conditions and job prospects for an independent research career have been sufficiently bleak in recent years to lead postdocs to form the National Postdoctoral Association in 2003.[72] One outcome has been a gentleman's agreement among many research universities that the postdoc position would last for no more than five years.

On some campuses, including Stanford, Yale, Johns Hopkins, the University of Illinois, and the University of Chicago, postdocs are either unionized or have formed a local association. Issues often include the availability of fringe benefits, university privileges (such as use of the library!), and job prospects. The largest successful organizing campaign to date took place in 2008, when the California Public Employment Relations Board officially recognized the PRO/UAW (the Postdoctoral Researchers Organize/ International Union, United Automobile, Aerospace and Agricultural Implement Workers of America) as representing postdocs on the ten-campus California system.[73] The first five-year contract was signed in August 2010. It gave postdocs a slight raise and committed to raise rates to conform with the NIH guidelines. Postdocs agreed to a no-strike provision in the contract.[74] It is noteworthy that in a world in which only about one in seven full-time workers in the United States are represented by a union, that after only ten years of organizing effort, the California agreement brings the number of postdocs represented by a union to more than one in 10.[75]

The Academic Market

The academic market is a buyer's market, and has been for a number of years—given the strong preference of many new PhDs and postdocs to take a job at a university. For example, 59 percent of physicists who received their PhDs in 2005 and 2006 had the long-term employment goal of working at a university or college.[76] A recent survey of postdoctoral fellows found that 72.7 percent of those looking for a job were "very interested" in working at a research university.[77] A survey of postdocs at the University of Texas Medical Center found that 79 percent want a job in academe after the postdoc is over.[78] An earlier survey of U.S. doctoral students in the fields of chemistry, electrical engineering, computer science, microbiology, and physics found that 55 percent of the respondents aspired to a career in academe, either doing research or teaching.[79] Whether the percentage is 55 or 79, one must conclude that there is considerable disparity between the aspirations of young scientists and engineers and the reality that, depending upon field, at most 25 percent will get a permanent position in academe.

Several factors explain the softness of the academic market—especially that for tenure-track positions—in the United States. First, the pool of trained individuals available for positions has grown dramatically over time, as can be seen from Figure 7.1. Moreover, it is not just U.S. PhD production that has grown dramatically; the pool of individuals trained outside the country who come to the United States to take a postdoctorate position and want to stay has grown dramatically as well.

Second, salaries of tenure-track faculty are considerably higher than those of non-tenure-track faculty. This leads universities to substitute other, cheaper labor for tenure-track faculty. Undergraduate classes are staffed increasingly by part-time faculty or by faculty holding non-tenure-track positions, which come with higher teaching loads and little opportunity for doing research. By 2001, more that 35 percent of the full-time faculty at public research universities and over 40 percent at private research universities held non-tenure-track positions.[80]

Third, public institutions have experienced a decline in the proportion of funding coming from the state, as states faced increasing demands for funds for prisons and health care.[81] From 1970 to 2005, state support, adjusted for inflation and enrollment, fell by 11 percent.[82] A number of state institutions currently receive less than 10 percent of their budget from the state, including the University of Washington (4 percent of its $2.9 billion budget), Pennsylvania State University (9.4 percent of its $3.4 billion budget),[83] and the University of Michigan (6.3 percent of its $5.1 billion budget).[84] These problems were exacerbated with the financial crisis of 2008, when most states faced substantial declines in tax revenues.

Fourth, the high cost of start-up packages makes universities very selective in hiring. It is better to hire one highly productive scientist than two whose combined productivity may be slightly higher but for whom the combined costs are much higher.

The situation is somewhat different at medical schools, where tenure has become decoupled from the guarantee of a salary (or as one medical administrator put it, it is out of fashion to link tenure to salary). To be more specific, only 62 of the 119 medical schools that offer tenure to basic science faculty equate tenure with a specified financial guarantee; at only eight schools is the guarantee equal to "total institutional support." At the other 54, some type of limit is put on the guarantee. At 42 of the 119 institutions, tenure comes with absolutely "no financial guarantee."[85]

Similarities and Differences between the United States and Other Countries

It is not only in the United States that the academic market for scientists and engineers has been soft in recent years. The academic job prospects of young Italian PhDs have also been bleak for a number of years. The age of faculty reflects these problems. In 2003, the average age was 45 for faculty in research positions (*Ricercatore Universitario*—equivalent to the rank of assistant professor in the United States), 51 years for those in associate-professor positions (*Professore Associato*), and 58 years for those in full-professor positions (*Professore Ordinario*).[86] The Italian academic market is also subject to "stop and go." For example, a "no new permanent position" policy was in place from 2002 to the end of 2004 and again from 2008 to mid-2009.[87]

The academic labor market has also been soft in Germany. The number of professors at German universities peaked in 1993 at about 23,000 and has, with few exceptions, declined steadily since then.[88] In 2004, for example, the total stood at just slightly over 21,000. The decline is not due to a decrease in the number of students. During the same period, the number of high school graduates increased significantly while the ratio of university professors per 100 high school graduates went from 11.26 to 9.43.[89] Moreover, the decline has come at the same time that the number of individuals who have earned a Habilitation, a requirement for obtaining an appointment as a professor at most German universities, has grown considerably.[90] The result has been a dramatic increase in the number of applicants for job openings. One estimate, for example, is that the ratio of new applications to job openings rose from roughly 1.5 to 2.5 over a recent fourteen-year period.[91]

A similar situation exists in South Korea, where universities, particularly private universities, under pressure to reduce expenditures on teaching personnel, are increasingly relying on part-time instructors. In 2006, for example, the number of full-time instructors in four-year colleges and universities was approximately 43,000, while the number of part-time instructors was more than 50,000.[92]

But there are other respects in which the U.S. academic situation differs considerably from that of other countries. One dimension relates to tenure, another to the degree to which universities are staffed by their own graduates, a third to how salary is determined, and a fourth to the process of selection.

The U.S. university system is characterized by a tenure system, which usually determines within a period of seven years whether the individual has a permanent job or is forced to seek employment elsewhere.[93] Those without tenure can be treated as second-class citizens. Mathematicians at Harvard are said to wait to learn the names of junior colleagues until after they have been promoted. (The practice is reminiscent of the medieval one of parents only naming their children after they survive infancy.) By way of contrast, in many other countries, job security comes at the moment of hire into an entry position.[94] This is true in France, where the entry position *Maître de Conférence* is accompanied with job security. It is also true in Italy and until quite recently in Belgium. In Norway, job security, if not instant, is assured within several months of hiring.

Academic systems also differ in terms of the amount of inbreeding practiced. While the hiring of one's own PhDs is relatively rare in the United States, the practice is common in Europe. By way of example, over 59 percent of university professors in Spain work at the university where they received their doctoral training.[95] The percentage would be higher if the Spanish university system had not expanded, creating new universities without a history of PhD programs. Inbreeding is widespread in Italy, France, and Belgium as well. It is less common in the United Kingdom. In Germany, by law, promotion requires institutional mobility.

The way in which academic salaries are determined also varies considerably across countries. In the United States, it is the norm for faculty of the same rank to earn widely different salaries both within institutions and across institutions, as the data summarized in Chapter 3 demonstrate. Mobility, or the threat of mobility, plays a key role in determining salary. Indeed, one of the primary ways by which faculty receive salary increases in the United States is to court an offer from another university, and thereby receive a counter offer from their own university.

Many stay after receiving a counter offer, but some leave. In recent years, for example, when private institutions had more resources than public universities, a number of highly productive faculty moved from state-supported institutions to the "privates." During the past ten years, the University of Wisconsin lost a number of faculty to private universities. Although it is still too early to know how the financial crisis will affect hiring, one suspects that it will be the public universities that are hit the hardest. The University of California system, for example, implemented in July 2009 a furlough policy that effectively cut salaries by 10 percent.[96] The response, as a faculty member from Berkeley recently said, is that "phones are ringing." Other states, including Florida, Arizona, and Georgia, have also furloughed faculty.

In many other countries, faculty are civil servants; they receive a salary based on years in service and rank. This is true, for example, in Belgium, France, and Italy.[97] In such a system, the threat of movement has virtually no effect on salary at one's employing institution. The only way to earn substantially more is to leave the country (going, for example, to the United Kingdom or the United States) or to supplement one's salary with an additional position or by consulting.

Finally, the way in which faculty are selected is idiosyncratic to a country. In the United States, academic departments have considerable autonomy in making hires. The department negotiates for a position with the dean, forms a search committee, and interviews candidates for the position. Candidates usually are drawn from departments having equal or higher status. The decision regarding whom to recommend to the dean is made by the department. After the offer is formally made, salary negotiations begin in earnest.

In other countries, the recruitment and hiring process can involve national committees. Undoubtedly the most complex is that of Italy, where the selection process is dominated by a committee selected by the discipline, rather than by the university or department. To be a bit more specific, assuming there to be no government ban on new hires, the university launches a call for applications (*concorso*). The university then establishes a selection committee, all of whose members belong to the discipline in which the position is being offered; only one of the members is selected by the university. In a practice reminiscent of guilds, all other members are elected by the discipline at the national level. The committee is then charged with selecting the best possible candidate, based primarily on publication record. In principle, if the university is unhappy with the selection, it can refuse to hire the candidate and launch a new search. In practice, there is considerable behind-the-scenes maneuvering to steer the process toward selecting a candidate suitable for the university. The process is used not only for initial hires but also for promotions. Thus, an Italian "assistant professor" can only be promoted at his institution if and when a *concorso* is launched for an "associate" position in the department. This means that Italian faculty spend considerable energy lobbying for the creation of new positions.[98]

The French recruitment system is also centralized and discipline centered. The process begins with the central government issuing a list of vacancies, by discipline and institution, for the ranks of Maître de Conférence and Professor.[99] Only qualified candidates can apply: applicants must first obtain a certificate from the *Conseil National des Universités*—whose members are either elected or designated by the Ministry of Education. Once obtained, the *qualification* is valid for four years. Applications are then examined at the university level by a disciplinary committee, elected every four years and made up of faculty members as well as members invited from other institutions and disciplines.[100] Hiring decisions are made at the university level.[101]

In theory, both the French and the Italian systems should discourage "inbreeding," since selection is made by a national committee. In practice, however, considerable inbreeding exists in both countries because lower-tiered universities have strong incentives to lobby for home-grown faculty who will be supportive of their institution. Having neither a carrot nor a stick in terms of control over salary or teaching load, the university, if it were to hire the "best" candidate, could find itself stuck with a prominent researcher who spends as little time as possible at the university.[102]

Cohort Effects

A distinguishing characteristic of the market for scientists and engineers is the presence of cohort effects. Careers of scientists and engineers are affected by events occurring at the time their cohort graduates.[103] To put it succinctly, there can be a right time for getting a PhD and a not-so-right time. Careers can be affected for years to come. Some scientists graduate when jobs are plentiful; they have a choice among jobs and have little difficulty obtaining funding for their research. Their careers flourish. Many scientists who are now at the end of their career or have recently retired graduated when university jobs were readily available and success rates on grants exceeded 40 percent. They were able to take a chance on risky research. They had options; their careers blossomed. Likewise, in the 1990s, computer scientists were "hot." It was a seller's market, just as was the market for those working in the field of bioinformatics in the late 1990s and early 2000s.[104]

Others graduate when jobs are considerably less plentiful. They move from postdoc to postdoc position, or non-tenure track to non-tenure track position, hoping to eventually land a tenure-track position and become a principal investigator. They often settle for a job as a staff scientist, working either for an exceptionally talented (and lucky) member of their cohort or for a member of a cohort who graduated when jobs were plentiful. Such was the experience of individuals who graduated in 1969, after federal funding for scientific research was severely curtailed (see Chapter 6). Such was the experience of mathematicians in the early 1990s—especially those trained in areas closely related to the expertise of recently hired Soviet émigré mathematicians.[105] Such was the experience of biomedical PhDs in recent years. And such will be the experience of PhDs who graduated during the recent financial crisis, given that, according to one survey, 43 percent of colleges and universities imposed partial faculty-hiring freezes and 5 percent completely stopped hiring altogether.[106] The ecology postdoc who entered the PhD market in the fall of 2008 with fifteen papers in top-tier journals and $400,000 in grant funding and did not get a "single sniff" in response to her initial job applications is emblematic: Cohort matters![107]

Cohort matters because a scientist's productivity is related to where he or she works and the conditions of employment. Location in a prestigious department or research institute as an independent researcher fosters productivity.[108] Although the relationship between location and research productivity is due in part to selective hiring, there is much to suggest that organizational context has its own effect. Place matters. And matters and matters. A study of economists, for example, found that holding innate ability constant, placement at a higher-ranked institution leads to higher productivity for years to come. Initial placement depends in part on the state of the job market when one graduates. The study concluded that "initial career placement matters a great deal in determining the careers of economists."[109] If initial career placement matters for economists, it certainly matters for scientists, whose research is considerably more dependent on access to equipment and materials.[110]

Several factors—explored earlier in Chapter 2—explain why location is important. First, and as already suggested, top research institutions provide better resources for research. Their start-up packages are "richer," and their lab space is larger. Second, scientists working at top research institutions have lively colleagues to interact with and excellent graduate students to staff their labs. They also have lower teaching loads. Third, although it is difficult to measure, reputation matters. A proposal from the California Institute of Technology, other things equal, is likely to get a more favorable rating than a proposal from the Illinois Institute of Technology. This, in turn, can jump-start the process of cumulative advantage, or what Robert Merton so aptly named the Matthew Effect: "the accruing of greater increments of recognition for particular scientific contributions to scientists of considerable repute and the withholding of such recognition from scientists who have not yet made their mark."[111]

Case Study

In 1996 the National Research Council formed a committee to study trends in early careers of life scientists. The committee was initially co-chaired by Shirley Tilghman, who at the time was a professor of genetics at Princeton University and later became president of Princeton University, and Henry Riecken, who was the Boyer Professor Emeritus of Behavioral Sciences at the School of Medicine of the University of Pennsylvania. The impetus for the study was that the number of PhDs in the life sciences had grown substantially in recent years but the job market opportunities for young life scientists had not kept pace. Increasingly, young life scientists had found themselves in a holding pattern, waiting for a permanent position.[112]

There were a number of disturbing trends. Time to degree had increased, the percentage of life scientists holding postdoc positions had grown, and the duration of the postdoc position had also increased. Moreover, the likelihood that a young life scientist would hold a tenure-track position, especially at a research university, had declined. Furthermore, young faculty were experiencing increasing difficulty getting NIH grants funded and were getting funded for the first time at later and later ages.

To be a bit more specific, during the ten-year period of 1985 to 1995, the number of PhDs awarded in the biomedical sciences in the United States had increased by almost 40 percent, and stood at 6,000 by 1995.[113] Median time to degree, which was just over seven years in 1995, had increased to eight years. Sixty percent of all new PhDs took a postdoc position, up from around 55 percent a decade earlier. Over 30 percent of PhDs who had been out of graduate school for three to four years held a postdoc position, up from 25 percent a decade earlier. And the percentage who held a postdoc position for five to six years had grown by approximately 50 percent.[114]

Obtaining a tenure-track position had become increasingly unlikely. In 1985, the odds were about one in three that someone who had received her PhD five to six years before (1979–1980) held a tenure-track position at a PhD-granting institution. By 1995, the odds were approximately one in five that a recent PhD held a tenure-track position. It was not just the odds that had declined; the actual number of young faculty holding tenure-track positions at PhD-granting institutions had declined. The big growth was in "other" positions, a category that included postdocs, staff scientists, and other non-tenure-track positions as well as those who were working part time.

After documenting and studying these trends, the committee made five recommendations: (1) restraint in the growth of the number of graduate students in the life sciences, (2) dissemination of accurate information on career prospects of young life scientists, (3) improvement of the educational experience of graduate students, (4) enhancement of opportunities for independence of postdoctoral fellows, and (5) alternative careers for individuals in the life sciences. The committee's intent regarding recommendation five was to convey the conviction that "the PhD degree [should] remain a research-intensive degree, with the current primary purpose of training future independent scientists."[115] In other words, the committee did not endorse the idea of training PhDs in the life sciences who would then pursue alternative careers.

The committee expanded on the third recommendation in the text of the report, encouraging federal agencies to place greater emphasis on training grants and individual fellowships for supporting predoctoral training—as opposed to indirectly supporting training through the funding of graduate research assistantships on research projects. Their rationale was that it is pedagogically superior to support students on training grants because the quality of the training is peer reviewed when the training grant is up for renewal, while the quality of training provided to research assistants is not considered in the review of research projects. In addition, training grants minimize potential conflicts of interest that can arise between the trainer and the trainee since the graduate student is not "indentured" to a faculty member. Despite the apparent advantages of training grants, the number of students supported on training grants had remained fairly constant for a number of years, while the number supported on research assistantships had grown dramatically.[116] The training-grant recommendation was sufficiently controversial to lead Henry Riecken to resign as co-chair and write an "alternative opinion" in which he expressed the view that "the recommendation is unsupported, outside the study charge, and inconsistent with the committee's overall study findings."[117]

It will come as little surprise that the life science community did not rush to embrace the recommendations. Graduate programs continued to grow, little effort was made to disseminate information on career outcomes, and there was virtually no reallocation of funds between training grants and research assistantships. Primarily at the initiative of the Alfred P. Sloan Foundation, a number of professional master's programs were started in the life sciences in the late 1990s. The hope was that such programs could prepare individuals for nonresearch positions in industry.[118]

Then, in 1998, the NIH budget began its five-year doubling. Many hoped that it would provide salvation for the young. This was not to be the case, although conditions initially improved marginally. The probability that a PhD trained in the biomedical sciences and who had been out of graduate school five to six years held a tenure-track position, which had declined from around 19 percent in 1995 to 9.9 percent in 2001, rebounded to 15.0 percent by 2003. By 2006, the latest year for which there is reliable data, the figure stood at 12.0 percent.[119] The percentage of individuals remaining in a postdoc position for six-plus years, which had declined between 1995 and 2001, increased slightly. There was considerable growth in non-tenure-track positions, especially at medical schools, although the percentage of early career scientists holding these positions had declined by 2006. There was virtually no change at all in the percentage of recently trained PhDs working in industry. So much for the idea that a growing biotech sector is providing jobs for an increasingly large percentage of the newly trained workforce.

In short, the pickup in academic jobs was relatively modest for the young, and the indications are that it rapidly petered out. There were some other disquieting trends. Approximately one out of ten recent PhDs was either working part time, unemployed or out of the labor force.[120] The age at which new faculty with PhDs were hired at medical schools increased by two years between 1992 and 2004, reaching 39.[121] The young had a hard time competing for funding. The number of awards to first-time investigators, which had initially increased, declined.[122] The "spread" between the success rate on grant applications from established investigators and that for new investigators grew. In 1996, the difference was about 2.6 percentage points. By 2003, it was over 6 percentage points.[123] Career trajectories of young life scientists were sufficiently bleak to prompt the journal *Nature* to run an editorial titled "Indentured Labour," which argued that "too many graduate schools may be preparing too many students, so that too few young scientists have a real prospect of making a career in academic science."[124]

Once again, and in response to problems the young faced getting positions and funding for research, a National Research Council committee was established, chaired this time by the Nobel laureate and then president of the Howard Hughes Medical Institute, Thomas Cech. The committee issued its report in 2005, *Bridges to Independence*. Its key recommendation was that the NIH establish a new grants program, informally known as a "Kangaroo" grant, in which individuals in a postdoc position receive research funding that can be taken with them when (and if) they get a faculty position. A key objective was to provide incentives for universities to hire young investigators.[125]

The postgraduation experiences of the class that entered the molecular biophysics and biochemistry program at Yale University in 1991 exemplify the situation.[126] Only one of the initial thirty had received tenure by the fall of 2008. She had been a postdoc in Susan Lindquist's lab and was an associate professor at Brown. Another held a tenure-track position but had not yet received tenure. Four others held research positions at universities; one had an adjunct teaching position at a university. Only the tenured professor had received NIH funding, despite the stated mission of the Yale program "to prepare students for careers as independent investigators in molecular and structural biology."

The low numbers in academe do not prove that it was the academic market or problems with the NIH that led eleven to follow careers in industry or four to follow alternative careers, such as becoming a patent lawyer, entering the information technology industry, or starting a home-care business for seniors. Other factors entered into their decision. Several had doubts by the time they graduated that an academic career was right for them. For some, becoming an academic was never an option. For example, according to Deborah Kinch, associate director for regulatory affairs at Biogen in 2008, "I never bought into the concept of being a professor . . . Being a grad student is the last bastion of indentured servitude, and being a faculty member is pretty much the same thing, at least until you get tenure. Earning the same low salary and foraging for every grant—that was the last thing I wanted to do."[127]

Others from the Yale class sought careers that were more compatible with marriage. Several made a special effort to investigate careers in industry and found them to be particularly appealing. They are not alone in seeking a nonfaculty path. Several indicators suggest that graduate students today are less interested in obtaining research and teaching positions than they were in the past.[128] Clearly, stressful experiences as graduate students and postdocs and the paucity of tenure-track jobs contribute to this view.

One could discount the Yale study on the basis that a number of those who left science—or did not take up positions in academe—had shown signs of straying from the traditional path in graduate school. The evidence that problems exist is perhaps even more striking when one studies the over 400 National Institute of General Medical Sciences NIH Kirschstein

National Research Service Award (NRSA) postdoctoral fellows awarded during 1992–1994. Kirschstein fellows are supposedly the very best, selected for their research promise. This particular group of Kirschstein fellows also had the good fortune of launching their careers when the NIH budget was doubling.[129]

What happened to their careers? By 2010, slightly more than a quarter of the former Kirchstein fellows had tenure at a university; 30 percent were working in industry. What about the others? A handful (about 6 percent) were working at a college; 4 percent were research group leaders at institutes. Another 20 percent were working as a researcher in someone else's lab and a startling 14 percent could either not be located after extensive Google searches or had not published a paper since 1999. This was not exactly what one would expect from "the best" who came of professional age during the doubling of the NIH budget. If times were tough for them, times will be much tougher for those who have graduated since or will graduate in the near future.

Policy Issues

In many fields of science, such as chemistry, physics, and math, the market has been soft in recent years—not just the academic market but in certain fields the market in industry as well. Job prospects have been particularly dismal in the biomedical sciences. But still students continue to enroll in PhD programs. Many are foreign born, but some are U.S. born. Why? Why, given such bleak job prospects, do people continue to come to graduate school?

It is undoubtedly a combination of things. Dangle stipends that cover tuition and the prospect of a research career in front of students who find puzzle solving rewarding and who have been a star in their undergraduate pond, and it is not surprising that they come, discounting the all-too-muted signals that all is not well in the research community. Overconfidence likely also enters in: they perceive themselves to be considerably above average—others may not make it, but they will.

Active recruitment on the part of faculty also plays a major role. Faculty need students (and postdocs) to staff their labs. Faculty can be persuasive: they stress the positives such as stipends and the opportunity to do pathbreaking research—downplaying or failing to mention the negatives, such as the low probability of having an independent research career.

From the point of view of the faculty member and the university, the system works. Graduate students and postdocs constitute perhaps as much as 50 percent of the workforce in the biomedical sciences.[130] They bring fresh points of view; they are temporary. To quote a 2011 National Research Council report tasked with evaluating training programs at the NIH, the "body of graduate students and postdoctoral fellows [supported by NIH training grants] provides the dynamism, the creativity and the sheer numbers that drive the biomedical research endeavor."[131] Although it notes problems that trainees encounter in finding jobs, the report goes on to describe the system as "incredibly successful in pushing the boundaries of scientific discovery."[132]

Faculty members rationalize the system that provides them a workforce by arguing that the system is "fair"—that students know the outcomes but continue to come in spite of this. Faculty also point out that alternative careers exist—that being a research scientist is not for everyone. Some institutions, such as the University of California–San Francisco, actively support ways for students to explore alternative careers. But at what cost? The same NRC report went so far as to recommend that "one highly needed and valuable outcome is for biomedical and behavioral sciences trainees to teach in middle school and high school science." Turn them into teachers! That's a socially acceptable way to deal with the excesses that the current system creates.[133]

This raises serious efficiency concerns to economists. Yes, there is an apparent shortage of math and science teachers in the United States. But surely there is a more efficient way to increase the supply than by transforming people who have invested seven years of training in graduate school and another three to four as a postdoc into teachers.

It raises the more general question of whether the U.S. model that couples research to training is efficient. Is it a good use of resources? Or would the United States get more from its resources if it were to loosen the link between research and training, and conduct at least some research in a non-training environment. We return to these issues in the final chapter of the book.

Conclusion

The market for scientists and engineers differs in many respects from that of other markets. The gestation period is extremely long, the cost of getting the degree is exceptionally high, and the job prospects at the time of graduation are difficult to predict. Moreover, aspirants often lack reliable information concerning the job outcomes of recent graduates. Somehow, in this era of information technology and social networking, the young still make career decisions, especially with regard to science and engineering careers, in the dark. This undoubtedly is due in part to their "love" of the subject. Love, after all, is blind. But it is also because faculty do not readily provide information. Either they don't know, or, if they know, they do not want to tell.

The global nature of the market for scientists and engineers is another way in which this market differs from many other markets. We turn to a discussion of this in Chapter 8.

CHAPTER EIGHT

The Foreign Born

FULLY A THIRD OF THE FACULTY in electrical engineering at the Georgia Institute of Technology received their undergraduate degree outside the United States. A third of Stanford's physics department received their doctorate training abroad. Forty-four percent of the PhDs awarded by U.S. institutions in science and engineering (S&E) are to foreign students on temporary visas. The percentage awarded to foreign students is approximately 48 percent when students with green cards are included. The presence of the foreign born is even higher among postdocs, almost 60 percent of whom are temporary residents.[1] In terms of country of origin, 7.5 percent of S&E PhDs working in the United States in 2003 were born in the People's Republic of China.[2] Although some Chinese work outside academe, a substantial number are at universities, as faculty, staff scientists, or postdocs.

Clearly the foreign born play a large and, as we will see, growing role in U.S. S&E. Their presence at universities is especially noteworthy: as faculty they teach classes and conduct research, as graduate students they take classes and work with faculty on research projects, as postdocs they staff research labs. In a book about academic science, it is absolutely crucial to examine their presence and role in some detail. That is what this chapter sets out to do. It begins with a description of the presence of the foreign born at U.S. universities and continues with a discussion of whether the foreign born crowd out U.S. citizens from graduate school slots and faculty positions; that is, whether the foreign born take positions away from U.S. citizens. The chapter closes with a discussion of the contributions that the foreign born make to U.S. science and whether a case can be made that the foreign born are disproportionately productive.

The Presence of the Foreign Born

A crash course on visa classifications is in order before proceeding. The term *temporary resident* refers to anyone who has been granted entry to the United States for a limited time on a temporary visa. Most foreign-born graduate students are in the United States under such a provision. The criteria for granting a student visa place considerable weight on the student's ability to support herself financially while studying in the United States.[3] Most foreign-born postdoctoral fellows working in the United States are also in the country as temporary residents.[4] By way of contrast, *permanent residents*—those with green cards—are just what the title implies: they can remain permanently in the United States. Students and postdocs in the United States who are permanent residents usually got their residency status because someone in their family (a spouse or parent) obtained permanent residency status, but there are exceptions. For example, in 1982 and in response to events at Tienanmen Square, Congress enacted the Chinese Student Protection Act, which bestowed permanent residency status on Chinese students in the United States at the time.[5] A sizeable number of the foreign born eventually become naturalized citizens. In 2003, for example, approximately 15 percent of the S&E doctorates in tenure-track positions in the United States had become citizens through the process of naturalization.[6]

The visa measure used to study the presence of foreign scientists and engineers in the United States depends upon the question at hand; it also depends upon the richness of available data. Temporary residency is the measure to use if one wants to know the number of scientists and engineers who lack the right to permanently stay in the United States. If one wants to know the number who are foreign born, then it is appropriate to include permanent residents in the count as well as naturalized citizens, if the information is available.[7]

Faculty

A quick look at almost any department's webpage provides convincing evidence of the large role that the foreign born play on the faculty of S&E

departments in the United States. For example, almost 25 percent of the chemistry faculty at research-intensive universities in the United States received their undergraduate education outside the United States. The most likely source country is China, followed by the United Kingdom, Canada, and India.[8] Their background is somewhat similar to that of foreign electrical engineers at Georgia Tech: among the forty-two electrical engineers who went to undergraduate school outside the United States, half come from three countries: India (nine), China (seven), and Taiwan (five).[9]

A 2007 study of ninety-five U.S. universities identified 6,199 individuals from mainland China on the faculty. The University of Michigan–Ann Arbor headed the list with 139 Chinese faculty (2.6 percent); the University of Pittsburgh was next with 133 (3.1 percent), and the University of Missouri Kansas City followed with 131 (7.0 percent). When institutions are ranked in terms of the proportion rather than the number of Chinese, Stevens Institute heads the list with 27 percent. Georgia Tech is a distant second, with 7.6 percent, and Cornell University is fifth with 6.2 percent.[10] Although not all of the Chinese faculty members identified are in S&E departments, the vast majority are.

The widespread presence of foreign faculty at U.S. universities has a considerable impact on some countries. Israel is a prime example. For every 100 physicists on the faculty of a department of physics in Israel, there are ten Israeli physicists on the faculty of a top-forty U.S. department; there are twelve Israeli chemists in top-forty U.S. departments for every 100 Israeli chemists at Israeli universities, and thirty-three Israeli computer scientists for every 100 Israeli computer scientists at Israeli universities.[11] Recently, there has also been an exodus of Russian physicists and mathematicians to U.S. universities, although it is difficult to get a specific count. The impact of migration on many other countries does not result so much from a loss of trained scientists who come to the United States to work but rather, as we will see, from the inability to attract students who train in the United States back to their home country after graduation.

Changes in U.S. visa policy in recent years undoubtedly have facilitated the hiring of foreign faculty. Universities used to compete with firms for the limited number of available H-1B visas, but since October 2001, and as a result of the American Competitiveness in the Twenty-First Century Act, the cap on H-1B visas is no longer applicable to universities, government research labs, and certain nonprofits.[12] Many faculty and postdocs now initially take a position at a university on an H-1B visa.

As pervasive as the foreign born are on U.S. faculties, it is tricky to get an accurate count across departments and over time. The most comprehensive data are for faculty who receive their doctoral training in the United

Table 8.1. Percentage of foreign-born faculty at U.S. universities and colleges by field and year

	1979	1997	2006
Engineering	17.5	28.4	34.9
Life sciences	10.0	12.1	15.5
Biological sciences	8.9	10.5	15.2
Earth/environmental	10.3	12.4	14.7
Physical sciences	10.7	17.8	18.1
Chemistry	9.5	11.6	14.6
Math/computer science	10.4	24.5	31.4
Physics and astronomy	12.4	17.7	23.3
All fields	11.7	16.3	21.8

Source: Survey of Doctorate Recipients, National Science Foundation (2011b). The use of NSF data does not imply NSF endorsement of the research methods or conclusions contained in this book.

Notes: The sample is restricted to those who worked full time and received their PhD in the United States. Those holding postdoc positions are excluded. "Foreign born" refers to permanent and temporary residents and those who indicated that they had applied for citizenship at the time the doctorate was received.

States. When the analysis is restricted to this group, one readily sees that the proportion of foreign born (defined by visa status at the time the PhD was received) almost doubled between 1979 and 2006 (see Table 8.1), going from just under 12 percent in 1979 to about 22 percent in 2006, the latest date for which data are available.[13] The field with the highest proportion of foreign-born faculty is engineering, with over one-third, closely followed by math/computer science. Chemistry and earth/environmental sciences are the least foreign-intensive fields.

These data, however, fail to include the not insignificant number of faculty who come to the United States with PhD in hand. For example, one-third of all faculty hires in physics in 2005 received their PhDs outside the United States; 21 percent of basic-science faculty at U.S. medical schools received their MD or PhD equivalent degrees outside the United States.[14] Ten percent of the chemistry faculty at research universities received their PhDs abroad.[15]

There is one database that includes faculty who received their PhD training abroad. The proportion of the foreign born is, of course, higher when they are included in the analysis. To wit: based on these data, 35 percent of all faculty at four-year colleges, universities, and medical schools in 2003 were foreign born.[16] But the percentage is larger not only because it includes those who received their PhD abroad but also because it uses a more inclusive measure of the foreign born, classifying faculty as foreign if they were born outside the country rather than by their citizenship status at the time the degree was awarded, as does Table 8.1. The conclusion: although it is difficult to pin down the exact proportion of faculty who were not citizens when they got their PhD, the widespread presence of the foreign born on faculty at U.S. universities and medical colleges is indisputable. At a very minimum, and based on a back of the envelope calculation, at least 26.5 percent of faculty in S&E were not citizens at the time they received their PhD degree—whether the degree was received in the United States or outside the United States.[17]

Graduate Students

There have been ups and downs in the number of U.S. students getting PhDs in S&E over the past 40 years (see Chapter 7). But since 1970 the pattern for the foreign born has been one of consistent growth over time, except for a decline in the late-1990s that is partially accounted for by an increase in the number of individuals who chose not to declare their citizenship status, and the decline in temporary residents receiving PhDs in 2008, which reflects visa restrictions after 9/11.[18] This is readily seen from Figure 8.1, which charts foreign-born PhD recipients from 1966 to 2008 by visa status. The considerable dip in the number of temporary residents

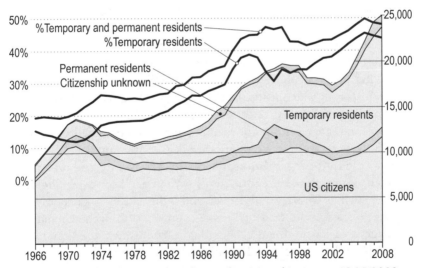

Figure 8.1. Science and engineering degrees by citizenship status, 1966–2008. *Source:* Natational Science foundation (2011c). For purposes of consistency over time, the S&E fields exclude "medical/health sciences" and "other life sciences."

in the early 1990s is due to the passage of the Chinese Student Protection Act in 1992, and is reflected in the large increase in permanent-resident degree recipients during this period.[19]

Figure 8.1 tells a remarkable story. In the late 1960s to the early 1970s, only one in five PhD recipients was foreign. By 2008 almost one in two was foreign. The proportion going to the foreign born grew most dramatically in the late 1980s and early 1990s.

Tightened visa restrictions associated with the U.S. response to the 9/11 attack created considerable concern that the flow of graduate students coming to the United States would decline. And initially such policies did take a toll, as can be seen in the slight decline in the percentage of PhDs being awarded to temporary residents in 2008. However, in recent years the number of first-time, full-time graduate students on temporary visas has rebounded to pre-2001 levels, allaying the concern. In the biomedical sciences, there was never a drop, reflecting the large pool of funds available for the support of graduate students in the biomedical sciences (discussed in Chapters 6 and 7) and the hot nature of the field.[20]

Fields vary considerably in terms of how foreign they are. Engineering has the longest tradition of attracting foreign-born students. Since the late 1970s, the number of engineering PhD degrees going to foreigners has exceeded the number going to U.S. citizens; in 2008, the percentage stood at 61.5 percent. Math and computer science programs are also heavily populated by students from abroad; slightly over 57 percent of the degrees in the field went to foreign students in 2008; in the physical sciences, 44.4 percent were awarded to foreign students in 2008. The field least populated by the foreign born is the life sciences, but even in this field by 2008 fully one-third of the PhD recipients are foreign born.[21]

U.S. PhD programs have become increasingly international because of trends both within and outside the United States. As discussed in Chapter 7, low salaries of PhDs relative to those in other occupations, the long time to degree, and stagnant pay for faculty have all contributed to making a PhD relatively less attractive than other degrees to Americans, especially American men. At the same time, the demand from foreign-born students expanded as a result of the enormous growth in bachelor-degree holders in countries such as China, South Korea, and India as well as changes in government policies in the students' home countries and in the United States that made it easier to attend U.S. graduate schools. Another key factor is that faculty with research funding need students to staff their laboratories, and the foreign-born provide a ready source. The stipend associated with a graduate research assistantship may not be a princely sum, but it has a relatively higher value to the foreign born than it has to U.S. citizens.

Foreign students may also be less selective than U.S. students in choosing programs.

The data bear this out. Foreign students are considerably more likely to be a research assistant than are citizen-students (49 percent versus 21 percent). The difference reflects the larger range of alternatives and resources available to citizens, including employer support for attendance at graduate school. It also reflects that citizens are more likely to be supported on fellowships than are foreigners (22 percent versus 13 percent) and on grants/stipends (15 percent versus 6 percent).[22]

Almost half of the noncitizens receiving a PhD in the United States come from just three countries: China, India, and South Korea.[23] Tsinghua University in Beijing sent more students to graduate school in the United States than any other institution in the world. Peking University, which is just down the road from Tsinghua in Beijing, holds second place. Seoul National University takes fourth place. Third place belongs to the University of California–Berkeley and fifth place belongs to Cornell University.[24]

China has not always held the dominant position.[25] In the 1970s, the largest number of foreign PhD recipients came from India (13.3 percent) and Taiwan (13.2 percent).[26] The next largest number came from the United Kingdom (4.5 percent) and South Korea (4.1 percent). There were also a number of Iranian students studying in the United States. Indeed, 3.0 percent of all PhDs awarded in the 1970s went to Iranians, but the number coming to study in the United States declined precipitously after the fall of the Shah in 1979.[27]

Political events and the availability of assistantships and fellowships are not the only factors affecting enrollment patterns. The number of South Korean students coming to study in the United States has depended in part on the availability of faculty jobs at South Korean universities for U.S.-trained scientists and engineers. Although historically South Korea outsourced the graduate training of future faculty to the United States, the job prospects for new PhDs at South Korean universities had diminished considerably by the late 1980s. As a result, more graduate students opted to stay in South Korea for their PhD training so as not to lose contact with faculty who could help them obtain a faculty position.[28] Changes in currency value also affect the likelihood that students will study in the United States. The depreciation of the bhat, for example, during the East Asian financial crisis was accompanied by a decline in the number of students from Thailand studying in the United States.

Political events play a large role in determining whether students come to the United States to study, as is abundantly clear from events in China in the past 30 years and, more recently, events in Russia. The establishment of diplomatic relations between China and the United States in 1979 and the lifting of restrictions on Chinese students' studying abroad, first partially in 1981 and then totally in 1984, opened up the possibility for Chinese students to study in the United States. Not only did the opportunity become available, but just as importantly, the demand for studying in the United States was there because of the large number of Chinese undergraduates who were able to go to college after the Cultural Revolution ended in 1976. As a result, in a very short time in the mid-1980s, the number of Chinese students at U.S. universities rose dramatically.[29] The number has continued to increase during the last twenty-five years; in 2007, the last year for which data are available, the United States awarded 4,629 PhDs in S&E to students from China.[30]

When Chinese students first started coming to the United States in large numbers, they headed disproportionately to lower-tier graduate programs. Indeed, more than 50 percent of Chinese degree recipients who entered PhD programs between 1981 and 1984 in chemistry, physics, and the life sciences got their PhDs from programs rated outside the top-fifty.[31] But this has changed considerably over time, suggesting both an increase in quality of students as well as an increase in the number of options that Chinese students have for studying in other countries. For example, 22 percent of Chinese students who recently received a PhD in physics and entered a U.S. PhD program between 1995 and 1999 graduated from a top-fifteen program; 30 percent of those getting a degree in engineering went to a top-fifteen program; and 29 percent of those getting degrees in chemistry went to a top-fifteen program in chemistry. Chinese students still appear to have difficulty getting into top programs in biochemistry. Only 12 percent of Chinese students entering graduate school in biochemistry between 1995 and 1999 succeeded in graduating from a top-fifteen program.[32] Similar quality patterns can be seen for international students from India, South Korea, and Taiwan.[33]

Foreign students have a tendency to enroll in PhD programs attended by other students from the same country.[34] Georgia Tech, for example, is such a common destination school for Turkish students that they jokingly refer to it as "Georgia Turk." A study of Chinese, Indian, South Korean, and Turkish students found that, regardless of the quality of the program, students are drawn to programs having other students from the same nationality. There is, however, a tipping point. After some critical mass, the probability declines.[35] There is also some evidence that students are attracted to institutions having faculty of the same ethnicity. The same study found evidence that Chinese students and Korean students are more likely to attend institutions with heavier concentrations of Korean and Chinese faculty. This is consistent with recent work that shows that Chinese students who receive a PhD from a U.S. university disproportionately have a dissertation chair who is Chinese.[36]

A clever piece of detective work established that foreign students are also more likely to work for faculty of the same ethnicity than to work for native-born faculty. The study paired labs in eighty-two departments of engineering, chemistry, physics, and biology directed by a foreign faculty member with labs in the same department directed by a "native" principal investigator (PI).[37] The mean paired difference in staffing patterns tells the story: the difference for Chinese students in a laboratory directed by a Chinese PI versus a laboratory directed by a native U.S. faculty member is 37.8 percent, for Korean students 29.0 percent, for Indian students 27.1 percent, and for Turkish students (small sample) 36.3 percent. The findings are consistent with the fact that it is the PI who makes staffing decisions, given that most research assistantships are paid for out of grants the principal investigator has obtained. Not surprisingly, some of these laboratories conduct the day-to-day business of the laboratory in the language of the principal investigator. There is a quality twist as well: affinity effects are more common in bottom-ranked departments than in top-ranked departments.

The majority of foreign students who come to the United States to earn a PhD stay. For example, in 2007 fully two-thirds of those who earned their degrees two years earlier were in the United States; 62 percent who received their degree five years earlier were in the United States, and 60 percent who received their PhD 10 years earlier were in the United States.[38] Stay rates have increased over time. The two-year stay rate, calculated in 1989, was 40 percent; the five-year stay rate was 43 percent. The ten-year rate was only 44 percent when it was first calculated in 1997.[39] Some foreigners leave and then return. In 2007, for example, 9 percent of the graduates who were in the United States five years after getting their degree appear not to have been in the United States for one or more of the intervening years.[40]

Whether a new PhD stays depends in part on U.S. policies and the overall economic environment. The cohort that graduated in the years 2001–2003, during the recession and when visa restrictions were particularly arduous, had lower stay rates than the cohort who preceded them or the cohort who followed them.

Country of origin is an excellent predictor of whether a newly minted PhD will stay. Over 90 percent of Chinese PhD recipients on temporary visas are in the United States five years later; 81 percent of Indian PhD recipients are here. But only 42 percent of PhD recipients from South Korea and Taiwan—both large source countries—are in the United States five years later. It is partly an issue of selection, partly one of economics. Chinese and

Indian students who come to the United States often come with an eye to staying. The high salaries in the United States relative to their home countries make staying especially appealing. If he were to return, a Chinese faculty member would earn at best 50 percent of what he would earn in the United States. Moreover, he may have access to better resources for research in the United States. Koreans and Taiwanese, on the other hand, often come for a PhD with the explicit idea of returning to a job in their native country.[41] Salaries are also higher in South Korea and Taiwan, on average, than they are in China or India. Other countries whose citizens have relatively low stay rates are Mexico (32 percent) and Chile (22 percent). The stay rate is only 7 percent for students from Thailand as well as for students from Saudi Arabia. These low stay rates reflect in part the terms of national fellowships that supported their study abroad.

Stay rates are also related to field of study. Computer scientists and electrical engineers are the most likely to stay. Indeed, almost three out of four trained in the two fields are working in the United States five years after graduation. Over two-thirds of foreign students who received a degree in the life sciences are here five years later. By way of contrast, only 46 percent of the individuals who receive their degree in agricultural sciences are in the United States five years after graduation. These patterns no doubt relate to the strength of demand in the United States as well as commitments that students have made to funding agencies to return home. They may also be related to country of origin. For example, Indians have a high probability of staying in the United States. They also are extremely likely to receive degrees in computer science and in electrical engineering.

There is also some evidence that graduates of top programs are less likely to stay than graduates from lower-tier programs. This undoubtedly reflects the wider set of opportunities the former have outside the United States. But it also relates to the fact that Chinese students have historically received their degrees from lower-tier programs, as have many Indian students.[42] Both nationalities have exceptionally high stay rates.

Postdoctoral Fellows

For more than twenty years, the number of temporary-resident postdoctoral fellows working in academic graduate departments in the United States has surpassed the number of citizen and permanent-resident fellows working at U.S. universities (Figure 8.2).[43] The gap widened considerably during the late 1990s and early in this century, and then, after a post-9/11 dip, stabilized. It has narrowed slightly in the last two years. In 2008, the last year for which data are available, 58.5 percent of postdocs at U.S. universities were temporary residents.

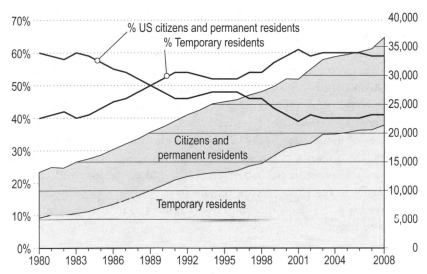

Figure 8.2. Number of science and engineering postdocs working in academe, 1980–2008, by citizenship status. *Source:* National Science Foundation (2011d).

While many foreign-born postdocs earn their PhD in the United States prior to applying for a postdoctoral position, a considerable number come to the United States with PhD in hand to take a postdoctoral position. A National Science Foundation (NSF) researcher extremely familiar with the data estimates that almost five out of ten postdocs working in academe in the United States earned their doctorate outside the United States and that four out of five postdocs with temporary visas earned their doctorate outside the United States.[44]

The vast majority of postdoctoral appointments are in the life sciences, and the largest increase in the absolute number of postdoctoral positions held by temporary residents in recent years has been in the life sciences.[45] By 2008, approximately 56 percent of postdocs working in the life sciences were temporary residents. But the percentage of foreign postdocs is even higher in other fields. In engineering, for example, nearly two out of three postdocs are foreign; the proportion is almost the same in the physical sciences.

Little is known about the country of origin of postdocs, because the NSF data on postdocs are collected from departments, not from the individuals. But a (nonrandom) survey of postdocs in 2004 found that the largest number of foreign postdocs came from China, followed next by India.[46]

At least three factors explain the large presence of foreign-born postdocs in the United States. First, and especially during the doubling of the NIH budget, funds have been readily available to support postdocs. Most of this funding—in contrast to funding available through traineeships—does not have visa restrictions attached to it. The opportunity to work in the United States with support at the level of $35,000 to $40,000 can be an appealing prospect for students who received their PhDs outside the United States. Second, the foreign born who receive their PhDs in the United States are more likely, other things equal, to take a postdoc position than are the native born. This undoubtedly reflects visa restrictions, which make it far easier to extend one's stay in the United States for purposes of training than for purposes of taking a job. It also reflects the ready supply of postdoctoral positions available in the United States. Third, the foreign born who receive PhDs in the United States remain in postdoc positions longer than the native born.[47] This, too, undoubtedly relates to relative opportunities and visa restrictions.[48]

Crowd Out?

Given the large and growing presence of foreign-born scientists and engineers in the United States, a natural question, particularly among U.S. citizens, is whether foreigners take slots away from natives in graduate school, depress salaries, and displace citizens from positions in academe. Answering such questions is tricky, largely because the counterfactual is difficult to establish. The debate can also be highly charged, especially in difficult economic times. In 1995, for example, the American Mathematical Society noted, "Immigrants won 40 percent of the 720 mathematics jobs available last year (1995) . . . and helped boost the unemployment rate into double digits among newly minted math Ph.D.s."[49]

Crowd out could occur for one group and not for another. For example, the foreign born could crowd citizen students out of doctoral programs but not out of faculty positions, or vice versa. Thus, it is important to look at the data for graduate students separately from that for university appointments.

The most straightforward way to study the question of crowd-out in graduate school is to see if the number of citizen-PhD degrees that are awarded declines at the same time that the number of foreign-born PhDs increases. When displacement is defined in such a manner, there is no evidence that citizens have been displaced by foreigners in S&E PhD programs.[50] The results are similar for men and women. (There is, however, evidence of crowd-out in nonscience fields.)

Why? First, there is evidence that graduate schools give preferential treatment to native applicants over foreign applicants.[51] Furthermore, many PhD programs—especially lower ranked programs—have a flexible definition of capacity, expanding when demand increases. And it is to just such programs that the foreign born have overwhelmingly gone.[52] Moreover, the dramatic increase in the foreign born studying in the United States started at a time when the number of PhD programs was expanding. During part of this period, there was also an increase in federal funds for research. In addition, the large influx of Chinese students in the early 1980s came when, because of the Cold War, there was considerable federal support for research in the physical sciences. As a physicist at a research university recounted, Chinese students met the need for research assistants, and their program expanded accordingly.[53]

This, of course, does not answer the question of whether the increased presence of the foreign born in S&E affects the career decisions of citizens. Here, the evidence points to a relationship, but the path is indirect: the increased presence of the foreign born in the U.S. S&E field depresses wages in S&E occupations. Earnings in S&E have fallen relative to other fields (Chapter 3 and 7). Money matters: future cohorts of U.S. students respond, and are less likely to enter these fields.

According to one estimate, a 10 percent increase in the supply of foreign doctoral scientists and engineers decreases earnings of scientists and engineers by about 3 to 4 percent.[54] It is postdocs who feel the biggest bite: the same research concludes that "about half of this adverse wage effect can be attributed to the increased prevalence of low-pay postdoctoral appointments in fields where immigration has softened labor market conditions."[55]

The scenario that has unfolded has a special dynamic: an increase in the number of foreign-born PhDs lowers wages, especially wages in postdoc positions. Indeed, in many fields a postdoc receives no more than the starting salary of those with a bachelor's degree. The resulting salaries are not that appealing to citizens, but they can be extremely attractive to many of the foreign born, especially since a postdoc position in the United States increases the odds of staying in the United States. The dynamic has been fueled further by increased funding for research in the biomedical fields. The ready supply of the foreign born allows faculty to staff their laboratories with cheap postdocs.

What about positions in academe? What is the evidence with regard to displacement there? One way to address this question is to engage in a thought experiment, comparing the actual growth in employment of a specific group (citizen or noncitizen) in a specific sector of the economy with growth predicted using the counterfactual of what would have happened to employment of U.S.-citizen/noncitizen S&E doctorates in different sectors *if their employment had grown at the overall growth rate for all doctorates combined, regardless of citizenship status.* It is a mouthful, but there is really no easier way of saying it![56]

Based on this type of analysis, one can conclude that some displacement has occurred in academe, but it is fairly minimal and concentrated primarily in postdoc positions. For example, across all fields, the displacement of citizens from academe can primarily be attributed to their displacement from postdoctoral appointments, not from faculty positions. Indeed, there is minimal evidence of displacement from faculty positions (−1.7 percent) for all fields taken together. But displacement is approximately three times this magnitude for citizens in postdoc positions. The analysis also suggests that citizens in the life sciences have actually fared better than noncitizens with regard to faculty appointments (+5.3 percent). This is not the case, however, in engineering nor in the physical sciences. For these fields, the displacement of citizens from academe is largely accounted for by their displacement from faculty positions and not from postdoctoral positions. But the effect is relatively modest: −6.1 percent for engineering, and −7.5 percent for the physical sciences.

The story is rather similar when one differentiates between permanent and temporary faculty appointments. For all fields taken together as well as for each subfield, the displacement from academe observed for citizens can be attributed primarily to displacement from temporary positions. There is scant evidence of displacement from permanent academic appointments (−0.6 percent). In the life sciences, citizens have fared relatively better than noncitizens in terms of holding permanent faculty appointments (+1.6 percent). There is, however, evidence that citizens have been displaced by noncitizens in permanent faculty positions in the fields of engineering and the physical sciences, but the substitution of noncitizens for citizens is less than 5 percent in each instance.[57]

What we do not know from this type of analysis is whether displaced citizens were, on balance, pushed out of academe by the heavy inflow of foreign talent or pulled out by better job opportunities elsewhere in the economy. The finding that displacement, to the extent it occurs, is primarily focused in postdoc and other temporary appointments is highly suggestive of pull. Specifically, citizens may have left these less desirable positions because they were attracted to better opportunities in the for-profit sector. Moreover, consistent with the pull interpretation is the fact that the only evidence of displacement from permanent positions is in the fields of engineering and the physical sciences—fields that experienced considerable growth outside academe during the period of analysis.

Publishing

The sample of papers published in *Science* discussed in Chapter 4 provides a lens for examining the degree to which the foreign born contribute to academic research, as measured by publication. But it is an imperfect lens because citizenship status can only be inferred from the name of the author.[58] This means that the methodology overstates the number of noncitizens from countries such as China and India which already have a substantial first- and second-generation presence in the United States. At the same time, the method understates the number of noncitizens from certain other countries by counting those with European and English names as citizens, despite the fact that a number of scientists from Europe and the United Kingdom train and work in the United States. Elsewhere I have shown that, because these biases come close to cancelling each other out, one can get a fairly reasonable overall count of the citizenship status of authors by keying on ethnicity of name, classifying authors with English and European last names as citizens and all others as foreign.[59]

Based on this methodology, 63.6 percent of the U.S. authors of the *Science* articles are citizens, and 34.4 percent are noncitizens.[60] Approximately one in six of the authors at U.S. institutions are Chinese. In light of the earlier discussion, it is not surprising that citizenship patterns vary by position. Fifty-nine percent of the postdoc authors are noncitizens, 40 percent of the staff scientists are noncitizens, and 39 percent of the graduate student authors are noncitizens. Only 21.8 percent of the faculty authors are noncitizens. When one limits the analysis to the first-author position—the author who generally does the heavy lifting—the percentage foreign is particularly striking: 44.3 percent.

The data suggest that the foreign born play a substantial role in academic research. This is not surprising, given their strong presence. But is there reason to argue that the foreign born disproportionately contribute to U.S. S&E, and if so, what is the evidence?

With regard to the first question, there are several reasons to argue that the foreign born may be more productive than their native counterparts. Some of these reasons apply specifically to graduate students and postdocs, but others do not. First, given the sacrifices immigration requires, immigrant scientists are likely to be highly motivated. Second, and depending upon the immigration laws in effect at the time of entry, a permit to work in the United States can require an employer declaration that the scientist is especially talented. Third, foreign-born scientists and engineers who come to the United States to pursue a PhD are typically among the most able of their cohort.

Often they have passed through several screens: they have been educated at the best institutions in their country (such as the Indian Institutes of Technology, Tsinghua University in China, and the University of Cambridge in the United Kingdom), withstood intense competition for a limited number of slots, and competed with applicants from other countries, including those from the United States, before being selected for training in the United States. The case for exceptional quality is particularly strong for students from developed countries who choose to study in the United States over numerous alternative options for excellent graduate training.

The quality issue is also relative. The foreign born may have an edge if the quality of U.S. graduate students is declining. And there is evidence that this is the case: a study sponsored by the Sloan Foundation found that the number of top U.S.-citizen graduate record examination (GRE) test-takers going on to graduate school in an S&E field had declined during the periods 1987–1988 and 1997–1998.[61]

So go the arguments. What does the evidence show? Unfortunately, the evidence is sparse. One recent study examines chemists who received their PhDs in the United States during the period 1999–2008. Compared with others in their cohort, chemists with Chinese names were the first author for a significantly larger number of papers than non-Chinese. There is one exception: U.S. citizens supported on NSF fellowships are generally more productive. The research also finds that Chinese students are even more productive when they are trained by a Chinese faculty member. There are at least two plausible explanations for this: Chinese advisors attract and/or select particularly talented students, or the cost of communicating between student and advisor may be lower when both are Chinese.[62]

The results are consistent with the idea that the foreign born contribute disproportionately to research. But, except for the NSF fellows, the data do not permit one to distinguish between Chinese and U.S.-born authors. Rather, the research only establishes that the Chinese are more productive than their classmates, regardless of where the classmates (or where the Chinese) were born.

A study that did determine the country of birth of all authors found strong support for the hypothesis that the foreign born contribute disproportionately to exceptional contributions in research.[63] But the study is now quite dated, using data from the 1980s and early 1990s. Nevertheless, given the paucity of work in the area, I report it here.

The study used three bibliometric indicators of exceptional research in S&E to test the hypothesis that the foreign born contribute disproportionately to scientific research conducted in the United States: authors of citation classics, authors of "hot papers," and authors who were among the top

250 most cited during the period of 1981 to 1990.[64] The study compared the percentage of foreign-born authors with the percentage of foreign-born scientists in the United States to determine if the proportion of foreign-born scientists making exceptional contributions was significantly different from the underlying benchmark population. The study found strong support for the hypothesis that the foreign born contributed disproportionately. In the physical sciences, the foreign born contributed disproportionately compared with the underlying benchmark for all three bibliometric measures. In the life sciences, the foreign born contributed disproportionately for two of the three measures.[65]

Policy Issues

Considerable concern has been expressed in recent years over the large number of claims in the form of debt that China holds against the United States. The instability that would be created if China were to liquidate part of this debt is widely discussed. A similar concern has been expressed that the foreign born might go home, leaving the United States short of a key input into the production of knowledge, one of the few things in which the United States has a comparative advantage these days.

The concern is real. As noted in Chapter 6, China is investing a considerable amount on research institutes and universities. And China managed to escape most of the financial woes that accompanied the 2008 meltdown. In terms of size, its economy is now second only to that of the United States; in terms of growth, it ranks at or near the top.

There is no doubt that China is aggressively seeking talented individuals to bring home. But to date the number returning has been relatively small. For example, only three of the 297 Chinese chemists holding a faculty position in the United States any time between 1993 and 2007 in a department that granted a PhD degree were found to have returned to China by 2009. (The comparable figure for India—out of 219—was one.)[66] As of 2007, the forty-five select universities, known in China as the "985," had a total of sixty-seven biology faculty trained in the United States.[67] This does not include U.S.-trained scientists working at research institutes or those who take visiting positions.[68] Nonetheless, one must conclude, at this point in time, that the number of the U.S. trained who have returned to China is low.

Another issue is whether foreign-born scientists and engineers will continue to come to the United States. There are three major career points at which the foreign born enter the U.S. system: as graduate students, as postdoctoral fellows, and as established scientists. By far, the largest entry point is graduate school. Will the foreign born continue to study in the United States? In the past, the foreign born have had limited alternative opportunities that provide financial support for graduate studies and employment at a relatively favorable salary after completion of graduate school. This has been particularly the case for students coming from less developed countries.

But the alternatives open to the foreign born are changing. Programs outside the United States are becoming increasingly competitive. The number of S&E PhD degrees awarded in Europe has exceeded the number awarded in the United States since the late 1980s. The number awarded in Asian countries surpassed the number awarded in the United States by the late 1990s. China is of special interest, given the large number of Chinese students who study in the United States. The number of PhD degrees that China is awarding went from virtually zero in 1985 to over 12,000 in 2003, the last year for which there is good data.[69]

To date, competition for Chinese students has not dampened the number of Chinese students studying in the United States, and the supply of Chinese students is likely to persist, at least in the near future, because of the tremendous growth in the number of bachelor's degrees awarded in China and the sheer magnitudes involved. It is estimated, for example, that in 2002 (the latest year for which data are available) there were 884,000 bachelor's degrees awarded in China in S&E compared with 475,000 in the United States. Moreover, the size of the potential supply is staggering: the number of 18 to 23 year olds in China is projected to be 118,562,000 in 2015—four times the number of 18 to 23 year olds in the United States.[70] Still, one must be concerned that, in the longer run, U.S. programs are at risk of becoming less attractive to foreign-born students, especially if financial support for university research does not increase significantly in the future.

Postdoctoral appointments are another path by which the foreign born enter the U.S. science and engineering enterprise: almost five out of ten postdocs come to the United States with a PhD in hand. Here, U.S. support has slackened, as federal funds for university research have remained relatively flat, in real terms, in recent years. In the last two years, there was a slight decline in the number of postdocs working on temporary visas in the United States. And while the American Recovery and Reinvestment Act (ARRA) provided considerable funds for postdoctoral positions, preliminary research suggests that few additional postdocs came to the United States as a result of ARRA—perhaps because of the requirement that funds be spent quickly and because of delays associated with getting a visa.

Universities in the United States have also benefited by hiring established foreign scientists into faculty positions. Their entry has often been facilitated by exogenous shocks. In the 1930s, the United States benefited from the dismissal of Jews from German universities. More recently, the United States has benefited from eased emigration policies that resulted from the collapse of the Soviet Union. Forecasting exogenous events is outside the scope of this book. Suffice it to say that it is not only shocks that brought these individuals to the United States. Resources played a role in their choosing the United States. And whether they continue to come depends in part upon whether the United States continues to fund scientific research at a competitive level.

There are several policy choices that the United States faces when it comes to the supply of scientists and engineers. First, there are ways that the United States could make coming to and staying in the United States more attractive to foreign scientists. One is to ease restrictions on certain fellowships and traineeships that are reserved exclusively for citizens and permanent residents. Another is to make it easier to stay in the United States after training is completed. The Obama administration is on record promoting such a strategy. In his 2011 State of the Union address, President Obama spoke of those who come from abroad to study, saying, "It makes no sense" that "as soon as they obtain advanced degrees, we send them back home to compete against us." He continued, "let's stop expelling talented, responsible young people who can staff our research labs, start new businesses, and further enrich this nation."[71]

The United States must also remember in setting policy that, when it comes to the supply of the foreign born, it is not (as President Obama's remarks suggest) a zero-sum game: not all is lost if they leave. Many continue to work on research with U.S. colleagues. A recent study found a strong and significant relationship between the fraction of U.S.-trained PhDs working in a top-twelve research country and the relative contribution by foreign authors in these countries to articles having at least one faculty author from a top U.S. university.[72] Some come back and forth to the United States. Moreover, in a broader sense, those who leave continue to contribute to innovation in the United States, as knowledge, once published, flows across international boundaries and, as embodied in patents, gives rise to new products and processes that affect productivity worldwide.

The United States could also implement policies to make careers in S&E more attractive to citizens by increasing financial support for graduate study and making a concerted effort to shorten the amount of time it takes to train. The discussion of Chapter 7 suggests that supply would be responsive to such actions. But it would require considerable resources and a will to change. The United States exhibited such will in the 1950s with the passage of the National Defense Education Act (NDEA), and students responded.[73]

A recent report recommends that the United States initiate a somewhat similar program whereby students pursuing a graduate degree in an area of "national need" would receive an annual stipend of $30,000 plus up to $50,000 more a year to cover tuition and other costs for up to five years. It remains to be seen whether the recommendation, which would provide support for 25,000 students and cost an estimated $2 billion in its first year and $10 billion when it reaches steady state in 2016 with 125,000 students, could possibly win congressional support.[74] Furthermore, it is unclear whether the market could absorb such a large increase in supply if the foreign born continue to come and stay. Perhaps not surprisingly, the report was sponsored by two groups, the Council of Graduate Schools and the Educational Testing Service, both of which have a stake in growing the number of graduate students. To return to a statement made in Chapter 7 regarding the U.S. need for scientists and engineers, "where one stands depends upon where one sits."[75]

Conclusion

The United States imports much of its academic talent. Some foreigners come for training and remain in the United States; others come already trained. Those who come for training play an important role in staffing labs while they are graduate students or postdocs. The rate at which the United States imports talent has increased over time, although it took a dip in the late 1990s, which was partly related to the East Asian economic crisis, and another dip early this century when visa restrictions were tightened after 9/11.

The foreign-born scientific workforce is highly productive. Indeed, there is some evidence that it contributes disproportionately to research. It is also younger on average than the population of U.S.-citizen scientists. Thus, in the future, the foreign born are poised to assume increased leadership roles in U.S. science.

CHAPTER NINE

The Relationship of Science to Economic Growth

IT IS ESTIMATED that per capita income, as measured by gross domestic product (GDP), grew by approximately 8 percent during the fifteenth century, 2 percent during the sixteenth century, about 15 percent during the seventeenth century, and 20 percent during the eighteenth century.[1] Not until the Industrial Revolution, which commenced toward the end of the eighteenth century, did a period of significant economic growth occur. In a short span of time, the steam engine was introduced, textile mills were mechanized and traveling by rail became a possibility. Despite its accomplishments, the industrial revolution did little to substantially alter daily life for most people, except in terms of what they wore and the ability to travel. Growth might have leveled off there,[2] but it did not.

Beginning in the mid-nineteenth century, the world—especially the Western world—enjoyed a period of sustained economic growth that persisted for much of the twentieth century. World economies, measured in per capita terms, grew by 250 percent in the nineteenth century. This was dwarfed by the 850 percent growth in per capita GDP in the twentieth century. Much of this growth—at least in the West—occurred during the first 70 years of the twentieth century.[3] After 1970, and until 1995, annual growth rates in the West declined, hovering around 2 percent per annum. Not an abysmal rate, but at 2 percent the standard of living only doubles every 36 years.[4] Then, in the mid-1990s, the United States, Canada, and several European countries experienced a burst of growth. From 1995 to 2000, per capita income grew at an annual rate of approximately 3 percent.[5] Many attributed the new growth to advances in information technology and its widespread adoption in a number of sectors in the economy.[6]

Clearly a number of factors contributed to the tremendous economic growth that began in the late eighteenth century. The Catholic Church, for example, had lost its monopoly in the West; the Protestant ethic had emerged. A changing political climate brought securer property rights, the freedom to engage in business, and the ability to sell goods at unregulated prices.[7] These, and other factors, undoubtedly played a large role. But many economists argue that the most important factor contributing to growth was that people learned to use science to advance technology. To quote Simon Kuznets, the father of national income accounts and the 1971 Nobel Prize winner in economics "for his empirically founded interpretation of economic growth," the West entered the "the scientific epoch."[8] People not only learned to use science to advance technology, they learned to use technology to advance science. In the terms of the economic historian Joel Mokyr, propositional knowledge (science) informed prescriptive knowledge (technology), and prescriptive knowledge informed propositional knowledge. The result: people learned how to invent in a systematic way.

Prior to the industrial revolution, a good deal of prescriptive knowledge was available. Examples abound: how to preserve meat, how to build a cannon, how to make glass, what to take (digitalis) to treat edema. But this knowledge was built on trial and error, not on an epistemological base. Some scientific knowledge existed as well. For example, Galileo confirmed the existence of the Copernican system, that the earth revolved around the sun. Newton described the laws of gravity and, together with Leibniz, developed the calculus. Considerable advances in science were made in the seventeenth and eighteenth centuries. But before the nineteenth century, prescriptive knowledge rarely built on scientific knowledge, although prescriptive knowledge did at times lead to advances in propositional knowledge. Galileo, after all, had a telescope. The breakthrough of the industrial revolution was that science and technology began to reinforce each other. "The mutual co-evolution of practical and theoretical knowledge set off an unprecedented wave of technological advance."[9]

An unprecedented period of economic growth—"more radical and spectacular in its technical and conceptual advances than perhaps any era in human history"—ensued.[10] In a relatively short period of time, advances in metallurgy, chemistry, electricity, and transportation occurred that changed the world. And much of it was accomplished by scientists and technicians building off each other's work. Successes in German chemistry (synthetic dyes, for example) built off research at German universities. Advances in the production of steel drew on a scientific base. Electricity could not have been tamed had it not been for the intertwining efforts of scientists and engineers.

The Importance of Growth

Economic growth is important to society. The case for growth, according to the economist Paul Romer, trumps all others: "For a nation the choices that determine whether income doubles with every generation, or instead with every other generation, dwarf all other policy concerns."[11] Growth offers a solution to problems such as debt, population explosion, and the means of supporting an aging population. The U.S. federal budget deficit disappeared in the late 1990s, not only because of an increase in taxes but because of a significant increase in economic growth.

Differential growth rates put countries on different trajectories. In 1960, the Japanese standard of living, as measured by per capita income, was approximately one-third of that in the United States. The standard of living in India was approximately one-fifteenth of that in the United States. During the period 1960 to 1985, per capita income in Japan grew at an annual rate of 5.8 percent; that in India grew at 1.5 percent. (Per capita income, by contrast, grew in the United States at 2.1 percent.)[12] As a result, the standard of living doubled in Japan almost every twelve years and increased by a factor of four during the twenty-five-year period. India, by way of contrast, was on a growth path that would take almost forty-eight years to double and, in the process, widen the income gap between itself and more developed economies. The tables have turned in recent years: the Japanese economy has grown on average at but 0.7 percent a year while India has averaged 5.5 percent. And China, which was hardly in the game in 1960, has been growing at more than 9.0 percent a year, a rate that led China to surpass Japan as the second largest economy in the world in 2010.[13]

The Role of the Public Sector

Much of the research that contributes to economic growth is performed in the public sector. This is not by accident; rather, it is by design. The reason: economic growth is fueled by upstream research—research that is years away from leading to new products and processes. Moreover, basic research has the potential of having multiple uses, contributing to a large number of areas. Theoretical work in physics is a case in point. It has contributed to multiple inventions including integrated circuits, lasers, nuclear power, and magnetic resonance imaging. Because of the multiuse nature of most basic research as well as the long time lags between discovery and application, it is unlikely that any one company or industry would support a sufficient amount of basic research to advance innovation. The economic incentives are not there. The findings would spill over, and other firms (including competitors) could use the knowledge to their advantage without paying for it. Knowledge, by its very nature, is not depleted with use. Spillovers are great for growth, but they are not a viable economic model to induce market-based institutions to invest considerable amounts in upstream research—hence the need to support research in the public sector.[14]

There are other reasons, in addition to a long time horizon and strong spillovers, for research to be performed in the public sector. First, basic research is risky. Results simply may not be forthcoming—at least in the foreseeable future. Physicists have been searching for years for the elusive Holy Grail—a quantum theory of gravity.[15] The lumpy nature of some research is another reason for public support. A tenth of an accelerator will not get you a tenth of a result. It is all or nothing. But the cost is so great (approximately $8 billion for the latest accelerator) that no one company, or in this case no one country, can rationalize supporting the effort.

As noted in earlier chapters, some research that occurs at universities and research institutes is of a dual nature in the sense that it focuses on a basic understanding of the laws of nature but in areas that can lead to practical applications. Research on acquired immunodeficiency syndrome (AIDS) and cancer are two specific examples. Research that has a dual goal is often referred to as falling in Pasteur's Quadrant, a name that appropriately recognizes Louis Pasteur's research on bacteriology, which set the standard in this regard.[16] His work not only helped the wine and beer industry solve the problem of spoilage. It also led to a fundamental understanding of the role that bacteria play in disease and provided a strong impetus for the investment in public water and sewer systems in the late nineteenth century—an investment that did more than anything else in human history to increase life expectancy.[17]

In many countries (and states), public institutions have also been given the job of providing the know-how for solving practical problems. Much of this occurs in engineering schools, which have a long history in both the United States and Europe. Although some of this research is of a basic nature, much is application oriented and solves problems of interest to local citizens. The Georgia Institute of Technology initially had a strong textile program, and the Colorado School of Mines, as the name implies, had a focus on mining, as did the École des Mines in France. Purdue University's hands-on engineering program arguably contributed to the university's athletic teams being called the Boilermakers.[18] The connection between research focus and local needs has become more attenuated in the United States since World War II.[19]

Research universities also contribute to economic growth by training students to work in industry. This is not an inconsequential contribution. Knowledge may be generated in the public sector, but—as the discussion above suggests—it rarely has instant economic value. A considerable amount of research and development (R&D) is involved in creating new products and processes. Universities supply industry with the workforce to do this.

Public Research and New Products and Processes

Examples of how research in public institutions has led to new products and processes abound: hybrid corn, which did much to increase the food supply, was first produced by a faculty member at (what is now) Michigan State University.[20] The World Wide Web, which has transformed the way we share and use knowledge, was invented by a scientist working at CERN. Lasers, which have had a profound impact on the fields of communication, entertainment, and surgery as well as defense, owe an intellectual debt to work done by Gordon Gould, a graduate student at Columbia University in the late 1950s (although much of the concept work was developed by physicists at Bell Labs working in conjunction with a faculty member from Columbia University who was consulting at Bell Labs at the time).[21] Bar codes can trace their origin to Rutgers University, where a curious graduate student overheard the president of a local food chain ask a dean to research a system that could automatically read product information during checkout.[22] The phenomenon of superconductivity, first discovered in 1911 at the University of Leiden, could potentially lead to the transmission of electricity at zero resistance and hence at no loss of efficiency.[23]

Nowhere is the contribution of public research more clear-cut than in the areas of pharmaceuticals. Three-quarters of the most important therapeutic drugs introduced between 1965 and 1992 had their origins in public sector research.[24] A more recent study finds that 31 percent of the 118 scientifically novel drugs approved by the FDA between 1997 and 2007 were developed first in a university. The estimate is a lower bound of the importance of university research to the development of new drugs because it measures contribution in terms of patents, not in terms of the origin of the fundamental research.[25] Yet, as we will see, firms acquire knowledge through a variety of paths, including reading articles written by university researchers.

Almost all of the drugs coming out of biotechnology companies originated at universities.[26] Some of these, such as synthetic insulin, have had a substantial impact on public health.[27] Virtually all important vaccines introduced in the past 25 years have come from research conducted in the public sector.[28]

Drugs make a difference. At least one-third of the reduction in mortality associated with cardiovascular disease is due to the development of non-acute cardiac medications that treat such conditions as hypertension and elevated levels of cholesterol.[29] The number of years of increased life expectancy is a non-trivial 1.7. Earlier in the century, penicillin and sulfa drugs contributed substantially to increased life expectancy.

More generally, the average new drug increases the life expectancy of people born the year it was approved by approximately six days. This may sound minimal—but spread it across 4 million new births a year, and it adds up: 63.7 thousand years of life for the birth cohort. When the effects are spread across other cohorts, the estimate is 1.2 million life years. And the cost? Somewhere between $416 and $832 per life year.[30]

Increased life expectancy comes not only from new medications, procedures, and devices but also from research that induces changes in the behavior of individuals. Much of this research is conducted in the public sector. Antismoking campaigns, as well as smoking bans in public places, have contributed considerably to improved health outcomes. One estimate is that behavioral factors have contributed to about a third of the life expectancy gains that have come about as a result of a decrease in mortality from cardiovascular disease.[31]

Universities and Economic Growth

It has become popular in recent years to stress the role of universities in contributing to economic growth. University presidents routinely conjure up the economic contributions of universities in their quest for funds; local communities lobby for "research" universities in the belief that a research university will lead to economic growth. College administrators publish accounts of the wonders that research universities have bestowed on the populace.[32] Universities commission studies that tout their contributions. The Massachusetts Institute of Technology, for example, provided the support for BankBoston's 1997 report *MIT: The Impact of Innovation.*

This view, of course, is not incorrect, but it is simplistic. As we noted earlier, much of the research of universities and public research institutions cannot instantly be transformed into new products and processes. It takes time. Lasers, when discovered in the late 1950s, were described as "a solution looking for a problem."[33] It took more than twenty years for them to become embedded in new products and processes. Hybrid corn was first produced in the late nineteenth century. It was not introduced commercially until the 1930s.[34] The science behind much of biotechnology is based on research findings dating from the 1950s. There are, of course, exceptions. The World Wide Web had an enormous impact almost from its inception; Google (with its Stanford origins) transformed the way in which we find information—and did so within five years of being founded.

It also requires considerable investment and know-how to translate basic research into new products and processes.[35] Universities and public research institutes are not organized or governed to excel in bringing new products and processes to market. Firms are. "In the realm of innovation a public research organization will never be more than a second rank institution."[36]

An exclusive focus on products and processes developed from university research also understates the role that university research plays in economic development. Basic research rarely produces direct economic benefits or tangible products. Instead it provides intermediate inputs that are "indispensable in the further research leading eventually to commercial innovations."[37] An exclusive focus on products and processes coming from universities also fails to recognize that subsequent research builds on failures as well as successes. Universities contribute to both streams of knowledge.

It is also important to note that just because the research was done at a public institution it does not follow that new products and processes would not have arisen had it not been for research conducted in the public sector—or research conducted at the particular public institution. Multiples, as noted in Chapter 2, occur in science. And their absence does not mean that a multiple was not in the making at the time the discovery was made. Looking for multiples, as noted there, is a classic case of censored data: those in the scientific hunt, so to speak, quit searching once others have made the discovery.

A focus on products and processes also ignores the important feedback that goes from industry to universities. University faculty get research ideas by interacting with individuals in industry. Barcodes are but one such example. University researchers also acquire new tools and instruments from industry. New academic disciplines and departments have grown out of the needs of industry for research and training. Electrical engineering and chemical engineering are just two cases in point.[38] Industry has also been known to press academe to create new programs and departments. Molecular biology is one such example.[39]

The remainder of this chapter examines how public research contributes to economic growth and describes the ways in which knowledge is transmitted from the public sector to the private sector and vice versa. In the process, several themes are further developed. First, while there is a strong link between research and economic outcomes, the lags are often quite long—sometimes in the neighborhood of twenty to thirty years. It is pure folly to think that the benefits will be reaped in the wink of an eye. Second, public research is not manna from heaven. Firms must invest considerable resources to bring new products and processes to market. Third, universities get a considerable amount in return: new ideas, funds for research, jobs for their students, and access to new equipment.[40] Fourth, the interaction between industry and universities is not new. Numerous examples can be found in the nineteenth and early twentieth centuries. But in recent years, the connection between industry and universities has intensified.

The Link between Public Research and Economic Growth

It is one thing to argue that public research contributes to economic growth. It is quite another to establish the extent to which scientific knowledge spills over from the public to the private sector and to measure the lags that are involved in the spillover process. Although the ratio of empirical evidence to theory is relatively low, there is a body of work that demonstrates a relationship. One line of inquiry examines the relationship between published knowledge and economic growth. Another surveys firms regarding the role that public knowledge plays in innovation. A third examines how the innovative activity of firms relates to the research activities of universities and links measures of innovation (such as patent counts) to university research. A fourth looks at whether firms with links to public research institutions perform better.[41]

Relationship between Published Knowledge and Growth

A clever piece of research by the economist James Adams uses the published knowledge line of inquiry to examine the relationship between research in science and engineering to multifactor productivity growth in manufacturing industries between 1953 and 1980.[42] The study is ambitious; it measures the stock of knowledge available in nine fields (such as chemistry) at a particular date by counting publications in the field over a substantial period of time, usually beginning before 1930. Publication counts are discounted to capture obsolescence—an article published thirty years earlier contributes less to the stock of useful knowledge than an article published ten years earlier. "Knowledge stocks" are calculated for each industry by weighting the knowledge stock measure in a discipline by the number of scientists employed in that field in each industry studied (publications in chemistry get more weight in industries employing more chemists; publications in physics get more weight in industries employing more physicists).

The goal is to see if there is a relationship between the stocks of knowledge and productivity growth in eighteen manufacturing industries over a period of twenty-eight years. Not surprisingly, there is: the stock of knowledge directly relevant to the industry accounts for 50 percent of growth in total factor productivity. But recent discoveries take many years to have an impact on productivity. The lags are on the order of twenty years. This is less true for research in the applied fields of engineering and computer science than in fields such as chemistry and physics.[43]

Long before public sector research has a measurable effect on economic outcomes, scientists and engineers working in industry have become aware of the research. The evidence: industrial researchers cite articles written by university faculty within two to four years of the research being published.[44] The lag is longest in computer science (4.12 years) and shortest in physics (2.06 years).

The amount of time it takes for university research to become embodied in a new invention for which a firm receives patent protection is considerably longer, on the magnitude of 8.3 years. The evidence comes from an examination of citations in patent applications to scientific papers published by University of California faculty.[45]

Such citations are not uncommon and provide another piece of evidence that a link exists between university research and innovation. In 2002, the last year for which the National Science Foundation collected data on patent citations, the average U.S. patent cited 1.44 science and engineering articles; when nonarticle material, such as reports, notes, and conference proceedings are included, the average patent cited 2.10 pieces of scientific literature. Perhaps more important, the trend over time has been one of increase, suggesting that the link between industry and academe has been increasing. By way of example, only ten years earlier the average patent cited only 0.44 articles and 0.72 pieces of scientific literature.[46]

Evidence from Surveys

Asking directors of R&D labs the extent to which they rely on research produced in the university sector provides another way to examine the degree to which industry builds on university research. Several studies have been conducted in recent years that do precisely this. Although the studies initially focused exclusively on the United States, in the mid-1990s European researchers developed a survey instrument, the Community Innovation Survey (CIS), to study, among other things, links between firms and public sector research.

The Carnegie Mellon University survey was administered to directors of R&D laboratories in the United States in 1994 with the goal of determining the extent to which public research is utilized by firms in their R&D activities.[47] For purposes of the survey, public research was defined to be research conducted at universities and in government labs. Respondents were asked to indicate whether information from a specific source either suggested new R&D projects or contributed to the completion of existing projects over the prior three years. A number of sources were included in the list in addition to public research, such as consultants, competitors, independent suppliers, customers, and own operations.

The survey found that public research is absolutely critical to R&D in a small number of industries. Pharmaceuticals head the list. In other manufacturing industries, public research is less critical but nevertheless plays an important role.[48] The general perception that public research provides the ideas for new products is not proved wrong, but the survey found that public research is even more likely to contribute to the *completion* of a project than to suggest a *new* project. Public research has more of an impact on large firms than on small firms, with one exception: start-ups (which are small) consistently report benefiting from public research.

Fields vary widely in terms of the importance industry ascribes to public research. Material science heads the list, followed by computer science, chemistry, and mechanical engineering. Biology is at the bottom in terms of importance across all manufacturing industries, although it plays an important role in the drug industry.

A smaller survey of firms in seven manufacturing industries took a longer view, asking firms to report the proportion of new products and processes that could not have been developed (without substantial delay) in the absence of academic research carried out within fifteen years of when the innovation was first introduced. The findings suggest that 11 percent of new products and 9 percent of new processes introduced in these industries could not have been developed in the absence of recent academic research.[49]

The relationship between firms and faculty is reciprocal. Interactions with firms enhance the productivity of faculty. A related study by the same economist asked firms to identify five academic researchers whose work contributed most importantly to new products and processes introduced in the 1980s. It followed up by surveying the university faculty identified by the firm as playing a key role.[50] The study found that academic researchers with ties to firms report that their academic research problems frequently or predominately are developed out of their industrial consulting, and that this consulting also influences the nature of the work they propose for government funded research. To quote an MIT faculty member, "It is useful to talk to industry people with real problems because they often reveal interesting research questions."[51]

The four Community Innovation Surveys administered in Europe generally find a smaller role for public research than do the U.S. surveys. But there is a reason for this: surveys in the United States have generally been directed at manufacturing firms with internal R&D facilities, but the CIS sample includes many firms that have absolutely no record of innovation and no internal R&D facilities. In these instances, there is virtually nothing for public research to contribute to![52]

Relationship of Innovative Activity to University Research

Another way to study the relationship between public research and innovative activity is to look at the degree to which innovative activity relates to the research expenditures of universities, which, as noted in Chapter 6, are considerable. This approach ignores the lags between university research and new products and processes, focusing instead on the extent to which spillovers exist between public research and private research and the degree to which they are geographically bounded; that is, to what extent does research performed at the University of Pennsylvania affect innovative activity in the greater Philadelphia area?

The rationale for expecting a relationship is based on the logic that one way that firms find out about new knowledge is through informal networks or by formal consulting or employment arrangements with faculty and students from local universities. Because some of this knowledge is of a tacit nature—especially in areas such as biotechnology, where techniques play a large role—face-to-face communication is quite important. It is not so much that knowledge is "in the air." It is more that the opportunities for acquiring the new knowledge are greater the closer one is to the source.

This line of inquiry was first initiated by Adam Jaffe in 1989, when he studied the relationship between patent counts and university research expenditures at the state level.[53] His findings suggest a strong relationship, particularly in the areas of drugs, medical technology, electronics, optics, and nuclear technology.

Jaffe's article sparked a new line of inquiry in economics, and a large number of studies followed in quick succession. Each had a slightly different angle, such as a different measure of innovation or a different definition of geographical proximity (standard metropolitan statistical areas versus state).[54] Almost without exception, the research has found a relationship between the measure of innovation and university research performed in close proximity.[55]

The geographic proximity story is given credence by case studies that show that certain universities—most importantly MIT and Stanford—have had a significant economic impact on their community. Much of the impact comes as a result of new firms that have spun off from the university—created either by students or by faculty. Stanford University estimates that in the past several decades over 4,668 companies have been founded by 4,232 members of the Stanford University community, including Yahoo, Google, Hewlett-Packard, Sun Microsystems, Cisco Systems, and Varian Medical Systems.[56] Most but not all of the firms are in the Stanford area. Stanford firms have had a particularly large impact in Silicon Valley, accounting for 54 percent of gross revenue generated by the 150 largest firms in Silicon Valley in 2008. While the "Silicon Valley 150" collectively lost $7.1 billion in 2008, the Stanford firms reported $19 billion in net income.[57]

The BankBoston study, which was completed in 1997, concluded that MIT graduates and faculty had founded approximately 4,000 companies; the companies employed 1,100,000 people in 1994. Massachusetts was not the top state benefitting from MIT-spawned job creation—rather, California was. But Massachusetts came in second, laying claim to approximately 125,000 jobs in MIT-related companies. Most of the firms are relatively new, having been founded in the past fifty years—many considerably more recently than that. But there are a few oldies, including Arthur D. Little, Inc. (1886), Stone and Webster (1889), Campbell Soup (1900), and Gillette (1901).[58] Of course, studies by universities that feature their successes must be taken with a grain of salt, but there is sufficient detail in these to make a reasonable case that the two universities have contributed substantially to new businesses, especially those in close geographic proximity to the university.

Firm Performance and Links to Public Research

There is also a line of research that shows that firms with links to researchers at public research institutions perform better than those without such links. For example, biotechnology firms that coauthor with a "star" university researcher in biotechnology perform better than firms that do not, whether performance is measured by products in development, products on the market, or employment.[59] Pharmaceutical firms that coauthor with university researchers have a higher research performance, measured in terms of "important patents."[60] Indeed, doing research with university faculty increases a firm's research productivity by as much as 30 percent. Even firm value is related to "connectedness." The market-to-book value of firms that cite published research in patent applications is greater than that of firms that do not.[61]

Mechanisms by Which Knowledge Is Transmitted from the Public to the Private Sector and Used by the Firm

The Paths of Transmission

Public research contributes to corporate R&D and subsequently to economic growth. This is beyond dispute. But how do firms learn about research that has been performed in the public sector?

It turns out that the priority system (see Chapter 2), which requires faculty to share their research in order to make the research theirs, is a powerful transmission mechanism: survey data show that the primary mechanism by which knowledge is transmitted from the public to the private sector is through the printed word. Firms learn about new research by reading articles and reports written by faculty. The second most important mechanism for transmitting knowledge is informal exchange, followed by public meetings or conferences and consulting. Firms place considerably less importance on the hiring of new graduates, joint and cooperative ventures, and patents as a way of learning about new knowledge arising in the public sector.

To be a bit more specific, the Carnegie Mellon survey, discussed previously, asked firms to report the importance to a recently completed "major" R&D project of each of ten possible channels of information on research performed in the public sector. Publications and reports were the dominant channel: 41 percent of the respondents rated them as at least moderately important. Informal information exchange, public meetings or conferences, and consulting had aggregate scores in the 31 percent to 36 percent range. Recently hired graduate students, joint and cooperative ventures, and patents had aggregate scores in the 17 to 21 percent range. Licenses and personal exchanges are the least important means by which the firms accessed public knowledge—having scores of less than 10 percent.[62]

A considerable amount of importance has been attributed to the large and growing number of patents that universities have received in recent years. And the number is impressive, nearly tripling in a ten-year period between 1989 and 1999, going from 1,245 to 3,698 per year.[63] Since then, the university patent frenzy has slowed a bit; in recent years, universities have been awarded on average approximately 3,300 patents a year. The low importance firms ascribe to licenses and patents in the Carnegie Mellon survey may reflect the fact that the survey was fielded when universities were patenting at a considerably lower rate than they are today. It may also reflect the fact that most university patents end up earning minimal licensing revenue for universities, suggesting that the vast majority are of limited economic value to the firm; only a handful produce substantial royalties.

The most direct way that university knowledge is transmitted to the private sector is through the formation of new companies by faculty and students based on research done in the university. The Carnegie Mellon survey did not ask directly about this mechanism of knowledge transfer, perhaps because of the almost tautological nature of the link and the relatively small number of start-up firms. However, because the number of start-ups from universities has generally been growing, one would expect the importance of this mechanism to have increased in recent years. By way of example, in 2004 the average number of new companies started by faculty and students at universities and medical schools was 2.2 per institution; in 2007 the average had increased to 2.9.[64]

What about geography? Do firms get their knowledge from universities in close geographic proximity? Or is the location of the source of knowledge inconsequential? The findings that patent counts and other measures of innovative activity are positively related to the research expenditures of universities in close geographic proximity to the firm suggest that local knowledge plays an important role. Face-to-face interaction, which is facilitated by proximity, is particularly important for the transmission of tacit knowledge. Knowledge arising in close geographic proximity may also be more readily transmitted informally. As noted earlier, informal information exchange is one of the mechanisms firms use for learning about university research.

But when it comes to hiring consultants or directly seeking expertise, the importance of geography depends in part upon the kind of expertise a firm is seeking. If the firm seeks expertise in basic research, distance is less relevant. Instead, firms seek the best research available regardless of location. But if the expertise the firm seeks is of an applied, problem-solving nature, the firm is more likely to use local talent.[65]

To elaborate a bit more on the role of geography, research shows that industrial laboratories that have a relationship with one or more of the top private research universities are located on average about 900 miles from the "source" university; those labs whose relationships are exclusively with lower-tier universities are located about 400 miles from the source.[66] The fact that top universities exert influence over a greater distance than most other universities does not preclude their having a large local influence as well: "A top university like MIT has greater influence at every distance" compared with lower ranked universities.[67] But local universities play an important role: firms spend about 50 percent more learning about academic research that is within 200 miles of the laboratory than they do learning about academic research that is farther away.[68]

The Role of the Firm

It is sometimes popular to portray the process by which knowledge moves from the public to the private sector as a waterfall, with public knowledge spilling over and being turned into new products and processes without cost by industry. This is not the case. There is considerable work on the receiving end. Before the knowledge can be transformed into new products and processes, it must first be "absorbed." This is not straightforward. Rather, it requires active researchers who stay abreast of scientific developments and are capable of understanding the research findings of others.[69]

The importance to industry of employing active researchers capable of absorbing new knowledge is one reason that scientists and engineers in industry publish papers in scientific journals. Absorptive capacity is nurtured by industrial scientists who are actively engaged in research, some of which is published. Sixty-two percent of PhD research scientists working in R&D in industry in 2004 reported that they had published one or more articles in the past five years.[70] The comparable figure in academe for those engaged in research is 92 percent. The contribution of industry

R&D scientists to the scientific literature is, however, considerably smaller than that of academics: during a five-year period, PhD scientists working on industrial research reported publishing 3.5 papers; those in academe, by contrast, published 12.0. At the macro-level, the number of articles published by scientists and engineers working in industry is relatively small. Collectively, industry contributed about 6.8 percent of the articles (fractional counts) published in the United States in 2008.[71]

In certain industries, such as pharmaceuticals, absorptive capacity is not enough. For the firm to fully benefit from public research, researchers working in the firm must be actively involved with researchers working in academe. "Connectedness" is important. Successful firms not only read the literature; their scientists actively work with colleagues in academe on research projects. And they publish with them as well. Slightly more than 50 percent of articles that have at least one author from industry also have an author from academe.[72] Firms that do so perform better, especially in pharmaceuticals, as the evidence presented above implies.[73]

Training

The lag between university research and innovation may be indirect and long in terms of knowledge spillovers, but when it comes to the training of people to work in industry, the link is direct and almost immediate in terms of economic benefits. And the impact is substantial. Approximately 225,000 scientists and engineers with doctoral training work in industry in the United States, many in R&D labs.[74]

Just how likely is it for PhDs to work in industry? How much does working in industry vary by field? What do PhDs in industry do? Do they stay in close proximity to where they were trained? That is, do Purdue engineers remain in Indiana, Stanford computer scientists in California, and MIT biochemists in Massachusetts?

Close to 40 percent of all PhDs trained in science and engineering work in industry in the United States.[75] Not surprisingly, the pervasiveness of industrial employment varies considerably by field, depending in part upon how applied the field is. For example, in 2006, approximately 55 percent of PhD engineers were working in industry; the proportion of PhD chemists working in industry was approximately the same. The percentage of computer and information scientists working in industry was somewhat lower (46 percent), and in physics and astronomy it was still lower (37 percent). Life scientists and mathematicians were the least likely to be working in industry: in both cases, only one out of four were employed in industry in 2006.[76]

Three fields—mathematics, computer and information sciences, and biological sciences—have witnessed a dramatic increase in the percentage of PhDs working in industry in recent years. There is likely an element of push as well as an element of pull here. Push arises in the sense that in recent years the academic job market has been overcrowded, especially in the biological sciences. Thus, despite the preferences of many new PhDs to work in academe (see the discussion in Chapter 7), they have been forced to look elsewhere for jobs. Pull arises in the sense that many jobs in industry are not unappealing to individuals with a preference for doing research: the researcher need not seek funding in order to do research, and, although jobs in industry provide for less independence than those in academe, researchers in industry report being reasonably satisfied with the amount of independence they enjoy. Moreover, as noted in Chapter 5, they often have access to better, more up-to-date equipment than researchers in academe. Nor does it hurt that jobs in industry pay significantly more than jobs in academe.[77]

A sense of the changing industrial employment patterns for PhDs can be obtained from Figure 9.1, which shows for selected time periods the percentage of PhDs who graduated five to six years earlier working in industry. (The 2006 number, for example, reports the percentage of those who received their PhD in 2000 or 2001 working in industry in 2006.) I choose five to six years after the degree to allow the new PhDs sufficient time to have settled into a more or less permanent position.

The figure shows that employment patterns for recently trained PhDs in industry vary considerably over time. Moreover, the overall trend is not always upward. This is especially the case in chemistry, where the probability that a recently trained PhD is working in industry was slightly lower in 2006 than it was in 1973. In the intervening years, the market for chemists experienced some ups and downs. There have been fluctuations in the market in industry for recently trained engineers as well: the market in the 1980s was particularly unwelcoming; by contrast, the market was considerably stronger in the 1990s. But overall, the percentage of recently trained engineering PhDs working in industry has grown by almost a third over the thirty-three-year interval.

The percentage of recently trained PhDs working in industry in the biological sciences has also increased substantially, although not in the last few years. The increase is partly due to the growth in pharmaceutical R&D expenditures during much of the period, as well as to the growth of employment

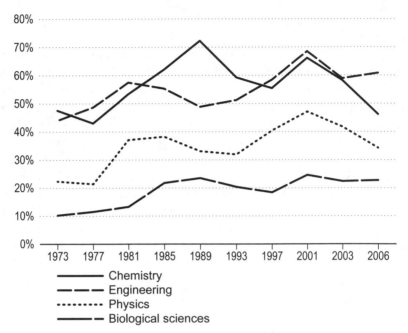

Figure 9.1. Percentage of PhDs working in industry by field, fifth and sixth year cohort, 1973–2006. *Source:* National Science Foundation (2011b). The use of NSF data does not imply NSF endorsement of the research methods or conclusions contained in this book.

opportunities in biotechnology firms. The overall trend of employment in industry for physicists has also been one of increase; but there have also been some bleak periods for young physicists. The job situation in industry was especially difficult soon after the dot-com bubble burst. When the 2008 data become available (in 2011), it is likely that the downturn in the market for physicists and chemists in industry will persist and that the market for recently trained PhDs in the biological sciences and in engineering will have deteriorated as well.

PhDs working in industry contribute to economic growth in a variety of ways. The most obvious means is through their work in R&D. But many innovative activities reside in functions not typically regarded as drivers of innovation and growth. Some of these functions have only developed in recent years. One such example is the assignment of scientific personnel to evaluate and seek R&D opportunities through mergers and acquisitions—a practice that has become particularly common in the pharmaceutical industry in recent years. Another example is the involvement of technically trained personnel in marketing and distribution. Inventories are controlled by sophisticated algorithms; computer scientists and mathematicians develop elaborate platforms for Internet marketing. A third example is the evolution of what is sometimes referred to as "service science," which relies on scientists and engineers to improve performance in the service sector. Examples include innovations in the way that passengers check in for flights, that truck routes are programmed to save drivers time and gas, and that networked sensors and analytic software are used to diagnose engine problems.[78]

New PhDs who go to work in industry have the potential of contributing in all these ways, but their placement is also an important means by which knowledge is transmitted from academe to industry. To quote the physicist J. Robert Oppenheimer, "The best way to send information is to wrap it up in a person."[79] This is particularly the case for tacit knowledge, which can only be transmitted by face-to-face interaction. Neither the technique of gene splicing nor the creation of transgenic mice could be learned by reading the literature; it required hands on participation. According to Bruce Alberts, former president of the National Academy of Sciences, former chair of the Department of Biochemistry and Biophysics at the University of California–San Francisco, and current editor-in-chief of *Science*, "the real agents of technology transfer from university laboratories" were the students from UCSF who took jobs in the local biotech industry.[80]

This may seem at odds with the Carnegie Mellon survey results, which found recently hired graduates to show up in the second cluster of mechanisms by which knowledge is transmitted from the public sector to the private sector—behind the cluster that includes printed articles and reports, informal information exchange, public meetings and conferences, and the use of consultants. But in reality it is less at odds than it appears for at least two reasons. First, recently hired PhDs contribute indirectly through networking to several pathways of knowledge transfer listed in the first cluster, such as informal information exchange, public meetings or conferences, and consulting. Second, the survey did not ask the method by which firms acquire tacit knowledge.

Firm Placements of New PhDs

Each year new PhDs are asked to fill out a survey administered by the National Science Foundation at or near the time they graduate.[81] The response rate is impressive—in the 92 percent plus range—perhaps because some students think it is a requirement for graduation. No matter why, the data provide an excellent snapshot of career plans of the new PhDs. Relevant to the current discussion is a set of survey questions regarding whether the respondent has definite plans upon graduating, and, if so, what those plans entail. For those going to work in industry, respondents are also asked to provide the name and location of the firm.

The firm placement data have been coded and analyzed for the four-year period of 1997 to 2000.[82] Even though the period is limited and the research is now a bit dated, a considerable amount can be learned about the placements of new PhDs in industry from this coding exercise. But before describing the main findings, it is important to recognize that the data have two limitations. First, they only describe outcomes for those with *definite plans at the time of graduation;* about a third more planned to work in industry but did not have definite plans at the time they filled out the survey. Second, the data also undercount placements of PhDs who eventually work in industry but initially took a postdoctoral position upon graduating. This is particularly the case in the biomedical sciences, where the percentage of new PhDs who take a postdoctoral training position upon graduation exceeds 50 percent; yet approximately one in three of these postdocs eventually end up working in industry.[83]

With these shortcomings in mind, we can learn four very useful things from the data for the 21,667 new PhDs in science and engineering who had definite plans to take a job in industry. First, a handful of U.S. universities train the lion's share of the new PhDs going to industry. The list is not random but rather includes some of the world's leading research universities. At the top is Stanford University, followed by the University of Illinois–Urbana/Champaign, the University of California–Berkeley, the University of Texas–Austin, Purdue University, the Massachusetts Institute of Technology, the University of Minnesota–Twin Cities, the University of Michigan, the Georgia Institute of Technology, and the University of Wisconsin. Combined, these ten train 40 percent of PhDs with definite plans to work in industry after graduation. It is worth noting that half of the top-ten training institutions are located in the Midwest.[84] Eight of the ten are public institutions.

Second, a surprisingly large percentage do not plan to work at an R&D-intensive firm, as measured by whether the firm is on the list of the top 200 R&D firms in the United States. Indeed, only 38 percent of the new talent has plans to work with an R&D-intensive firm.[85] The finding is consistent with the idea that much innovative activity in today's world is not restricted to the development of new products and processes in manufacturing but rather to innovations in other sectors of the economy. The finding also suggests that R&D expenditure data understate the amount of innovative activity taking place in the economy. To be a bit more specific, the top 200 R&D firms conduct 70 percent of all R&D in the United States, yet they hire less than 40 percent of new PhDs.[86]

Third, destinations are fairly concentrated; almost 60 percent of the newly minted PhDs going to work in industry plan to work in one of twenty U.S. cities. Heading the list is San Jose, California, which employed almost twice as many new scientists and engineers as Boston, the city in second place, or New York, the city in third place. California is a particularly popular destination: five of the top 20 cities are in California.[87] The Midwest, by contrast, is not a particularly popular destination: only three Midwestern cities are on the top-twenty list: Chicago, Minneapolis, and Detroit (recall that the data were collected before the recent woes of Detroit's auto manufacturers).

Fourth, certain states, many located in the Midwest, hemorrhage PhDs headed to work in industry: Iowa retains only 13.6 percent of those it trains who go to work in industry, Indiana retains only 11.8 percent, and Wisconsin only 17.7 percent. By way of contrast, the average state retains 37.1 percent, while California retains almost seven out of ten.

Policy Issues

The importance of publicly funded research to economic growth raises a number of policy issues, some of which have been discussed in previous chapters. For example, there is the question, raised in Chapter 6, of whether the country is investing enough in public R&D and whether the resources that are being invested are being deployed efficiently.

With regard to the first question, a case can be made, as was done in Chapter 6, that the United States (as well as other countries) underinvests in public R&D. With regard to the question of mix, there is the concern that the heavy emphasis on research related to health may jeopardize the future by failing to invest sufficiently in other areas that contribute to economic growth. The imbalance may even affect outcomes in the medical sciences. Magnetic resonance imaging and the laser, after all—two of the most important breakthroughs that have led to better health outcomes today— had their origins in physics.

There is also the question of whether universities and faculty act in ways that impede the diffusion of knowledge from the public to the private sector. Some of these issues were discussed in Chapter 3. By way of example, universities have executed exclusive licenses with firms that can deter the diffusion of knowledge. DuPont's licensing of Harvard's patent on the

OncoMouse and the aggressive terms that DuPont required of users are a case in point. Scientists have been known to withhold material from colleagues whom they perceive as competitors. Industrial sponsorship of public research can encourage secrecy and delay publication. There is the additional concern that universities have become overly aggressive in dealing with industry—structuring agreements that discourage industry from licensing the innovations or sponsoring university research. They fail to realize, according to Tyler Thompson, of the Dow Chemical Company, "that they are not the only game in town."[88]

There is also concern that public institutions—and their leaders—have become overly reliant on the growth story to promote their institutions. In doing so, they risk future financing and support. The public increasingly wants results now, not twenty to thirty years down the road. But the growth story they are promised takes time.

Finally, there is the concern that something has gone wrong with the research system, especially when it comes to pharmaceuticals. It is no wonder that the issue has been raised: only twenty-one new molecular entities were approved by the U.S. Food and Drug Administration in 2010—compared to fifty-three in 1996.[89] A recent study concludes that the "evidence suggests NIH R&D funding has little, if any impact on the number of drugs in Phase III clinical trials." The same study found a positive and significant relationship between funding levels and the number of drugs in phase I clinical trials.[90]

Francis Collins, the current director of the National Institutes of Health (NIH), is sufficiently worried about the slowed pace of new drugs coming from the pharmaceutical sector that he has pressed to create a $1 billion drug development unit within the NIH. If all goes as planned, the unit should open in October 2011. Collins, who led the NIH's participation in the Human Genome Project, sees the problem as lying with industry. He has publicly stated that he is tired of waiting for the "pharmaceutical industry to follow through" with discoveries that "look as though they have therapeutic implications."[91]

But others point out that at least part of the blame belongs with academe, noting that for "big" things to happen, scientists must quit working with their own group and begin working in interdisciplinary teams—that's where the gold lies. But the incentive system has not encouraged this. Rather, the grant culture, until recently, has encouraged scientists to specialize and create a niche. Their funding depends upon it, and their reputation also depends on it. One can get lost—or go unnoticed—in a large group. Nobel prizes are not, after all, awarded to groups. Neither are Kyoto Prizes or the Lemelson-MIT Prize. They are handed out one by one (or at most three at a time).

The slowed pace of drug discovery may also relate to the fact that biomedical researchers increasingly lack training in human biology and diseases. As a result, research results that look promising in the early stages often fail because the research focuses on but a small piece of the puzzle. An approach that looks promising on one level proves to be untenable within the larger system. Once again it is a case of misplaced incentives. NIH, which holds the monopoly on training grants in the United States, has not supported training in human biology and diseases.[92]

Conclusion

Much of the research that contributes to economic growth is performed in the public sector. This is not, as argued earlier, by accident; rather, it is by design. The multiuse nature of most basic research and the long time lags between discovery and application discourage any one company or industry from engaging in sufficient basic research to advance innovation. Instead, much of basic research and a considerable amount of applied research occur in universities and research institutes. The knowledge resulting from this research spills over to the private sector, contributing to the development of new products and processes as well as to helping industry complete projects currently in development. It is not, however, a one-way street. Knowledge, techniques, and instruments developed in industry contribute to research conducted in the public sector.[93]

Research universities also contribute to economic growth by training the scientists and engineers who work in industry. This is not an inconsequential contribution. Approximately 40 percent of all scientists and engineers trained in the United States work in industry today. In this respect, the university model of research has an edge on the institute model because the latter focuses exclusively on research while the former does both. There is considerable evidence that the strong connection that the training mission provides between academe and industry contributes to the development of new research ideas in both sectors.

Were we to end the story here, however, we would miss a great deal of what furthers economic growth. First, industry invests a substantial amount in R&D with considerable results.[94] Moreover, knowledge not only spills over from the public sector to the private sector. It also spills over within the private sector. It does so in a variety of ways: through informal gatherings (such as those that occurred at the Wagon Wheel—a popular watering hole in Silicon Valley in the 1960s, where semiconductor engineers exchanged technical ideas and information),[95] through employees changing jobs and taking knowledge developed in the firm with them,[96] and through the reverse engineering of new products and processes developed elsewhere. Some of the knowledge is transmitted through patents. Jack Kilby, the co-inventor of the integrated circuit, for example, tried to read every patent issued by the U.S. government. "You read everything—that's part of the job. You accumulate all this trivia, and you hope that someday maybe a millionth of it will be useful."[97]

The evidence—some of which is examined through the lens of geography—is fairly convincing: innovative activity of firms relates to R&D expenditures of other firms in close geographic and technological proximity, suggesting that firms appropriate the R&D of other firms.[98] The patents that firms cite in patent applications are in closer geographic proximity to the citing patent than they are to a sample of "control" patents that have the same temporal and technological distribution but are not linked through citation.[99] The market-to-book value of firms is related to the number of times its patents are cited by other firms—reflecting the value other firms place on knowledge developed in the firm.[100]

The growth story does not, however, end here. The new growth economics argues that knowledge spillovers are not only a source of growth; rather, the spillovers are endogenous and lead to increasing returns to scale.[101] The story goes something like this. In an effort to seek profits, firms engage in R&D. Certain portions of this R&D spill over to other firms, thereby creating increasing returns to scale and to long-term economic growth.[102]

This chapter has been devoted largely to a discussion of how research conducted in the public sector spills over to firms and affects economic outcomes. Does this spillover process mean that research in the academic sector is a component of the new growth economics? The answer depends upon the extent to which scientific research in the public sector is endogenous—that is, the degree to which it is affected by the actions of firms. If it is not, spillovers from the public sector to firms are important determinants of growth, but not as a component of the new growth economics.

Three aspects of public research developed in this book lead me to argue that an endogenous element of public research exists. First, companies, in an effort to maximize profits, support academic research. In 2008, this amounted to approximately $3 billion.[103] Second, the problems that academic scientists address often come from ideas developed through consulting relationships with industry. Third, government supports much of the public sector research (see Chapter 6), and the level of government support clearly relates to the overall state of the economy. The 2009 stimulus package was the first time that public funding for science had been countercyclical.

CHAPTER TEN

Can We Do Better?

I HAVE MADE THE CASE in the preceding chapters that economics plays a role in shaping science as practiced at universities and research institutes. Incentives and cost matter in science. But economics is also about the allocation of scarce resources across competing wants and needs, or to use the jargon of the profession, economics is also about whether resources are allocated efficiently. In this final chapter, I revisit the issue of efficiency. I begin by describing the research landscape that has emerged in the public sector in recent years. I then discuss issues of efficiency and, where the evidence is sufficiently convincing, a possible course of action that could make the public research system—particularly that in the United States—more efficient. Where evidence is insufficient, I, in the tradition of other researchers, encourage further research.

The Current Landscape

In many ways universities in the United States behave as though they are high-end shopping malls. They are in the business of building state-of-the art facilities and a reputation that attracts good students, good faculty, and resources. They turn around and lease the facilities to faculty in the form of indirect costs on grants and the buyout of salary. In many instances, faculty "pay" for the opportunity of working at the university, receiving no guarantee of income if they fail to bring in a grant. To help faculty establish their labs—their space in the mall—universities provide faculty start-up packages. After three years, faculty are on their own to get the funding to stay in business.

Faculty use the space and equipment to create research programs, staffing them with graduate students and postdocs who contribute to the research enterprise through their labor and fresh ideas. The incentives are to get bigger and bigger, employing more graduate students and postdocs, which in turn result in more publications, more funding, and more degrees awarded.

The shopping mall model carries some risk. Universities have put up some of the buildings "on spec," taking out loans on the assumption that the lease will continue to be paid through grants brought in by faculty tenants. Universities are severely threatened when funding for grants plateaus, or does not grow sufficiently to keep pace with the expansion. They face even more serious prospects when budgets decline in real terms. To quote an editorial by Bruce Alberts, the editor of *Science,* "the current trajectory is unsustainable, threatening to produce a glut of laboratory facilities reminiscent of the real estate bust of 2008 and, worse, a host of exhausted scientists with no means of support."[1]

In other respects, universities have found ways to minimize risk. They have hired faculty in non-tenure-track positions and have increased the proportion of adjuncts they hire. Medical schools have gone a step farther, employing people, whether tenured or nontenured, with minimal guarantee of salary. Faculty principal investigators staff their labs with graduate students and postdocs—temporary workers for whom the university has no long-term obligation.

The system that has evolved discourages faculty from pursuing research with uncertain outcomes. Lack of success can mean that one's next grant will not be funded. Proposals that do not look like a sure bet may be hard to get funded in the first place. To quote the Nobel laureate Roger Kornberg, "If the work that you propose to do isn't virtually certain of success, then it won't be funded."[2] Risk avoidance is particularly acute for faculty on soft money. As Stephen Quake, a Stanford professor of bioengineering, says, "The rubric for today's faculty has gone from publish or perish to 'funding or famine.'"[3]

What is inefficient about avoiding risk? First, it is pretty clear that if everyone is risk averse when it comes to research there is little chance that transformative research will occur and that the economy will reap significant returns from investments in research and development (R&D). Incremental research yields results, but in order to realize substantial gains from research not everyone can be doing incremental research. Second, recall from Chapter 9 that one rationale for government support of research is the notion that research is risky. As laid out by Kenneth Arrow, society has a tendency to underinvest in risky research without government support.[4] So it makes little economic sense for the public research sector to use the rubric of risk to garner resources and then create an incentive system that discourages risk.

The current university research system in the United States also discourages research that could disprove theories. To quote an official with a disease foundation, who asked not to be identified, "The way science careers are structured, big labs get established based on a theory or a target or a mechanism, and the last thing they want to do is disprove it and give up what they're working on. That's why we have so many targets [in studying this disease]. We'd like people to work on moving them from a 'maybe' to a 'no,' but it's bad for careers to rule things out: that kind of study tends not to get published, so doing that doesn't advance people's careers."[5] That kind of study can also be more difficult to fund. Researchers are rewarded for continuing a line of research. Renewals at the National Institutes of Health (NIH) are much more likely to be funded than are new grants.

The system can also discourage collaboration, especially across institutions and across disciplines. Incentives may be insufficient for encouraging interdisciplinary and interorganizational research. Questions arise as to whether faculty will get their fair share of credit—both monetary (the lease has to be paid) and reputational—in collaborative research projects. There can only be one first author and one last author, after all. It can be hard to stand out from the group when it comes to promotion and tenure time. Prizes are not awarded to groups; they are handed out one by one (or at most three at a time). There is also the problem that inter-organizational research can prove difficult to coordinate.[6]

The university research system also has a tendency to produce more scientists and engineers than can possibly find jobs as independent researchers. In most fields, the percentage of recently trained PhDs holding faculty positions is half or less than what it was thirty-three years ago; the percentage holding postdoc positions and non-tenure-track positions (including staff scientists) has more than doubled. In the biological sciences it has more than tripled. Industry has often been slow to absorb the "excess." A growing percentage of new PhDs find themselves unemployed, out of the labor force, or working part time.

Inefficiency arises from the fact that substantial resources have been invested in training these scientists and engineers. The trained have foregone other careers—and the salary that they would have earned—along the way. The public has invested resources in tuition and stipends. If these "investments" then are forced to enter careers that require less training, resources have not been efficiently deployed. Surely there are less expensive ways to train high school science teachers than to turn PhDs who cannot find a research position into teachers. Yet this is exactly what a recent report suggested.[7] Many of these PhDs may not even have characteristics that make them good teachers. Surely there are better ways to create venture capitalists with a knowledge of science than for PhDs to become venture capitalists—or better ways to create journalists who write about science than for PhDs to become journalists. Yet such careers are often put forward as appropriate alternatives for new PhDs. There is also the question of incidence, the term used by economists to refer to who bears the cost. The current system may be "incredibly successful" from the perspective of faculty, as a recent report described it, but at whose cost?[8] It is the PhD students and postdocs who are bearing the cost of the system—and the U.S. taxpayer—not the principal investigators.

How can universities continue to "overproduce" (especially in the biomedical sciences) year after year? Are potential students blind or ignorant to the negative signals being sent? Several factors allow the system to persist. First, there has been a ready supply of funds for graduate school support. The level of support makes studying for a PhD a particularly attractive prospect for the foreign born.

Second, money plays a role in who chooses a career in science, but other factors play a role as well. A taste for science is important. There are a considerable number of people with a sufficient taste for science—an interest in finding things out—who aspire to a research career. As I said in Chapter 7, dangle stipends that cover tuition and the prospect of a research career in front of students who find puzzle solving rewarding—and who have been stars in their undergraduate pond—and it is not surprising that some of them keep coming, discounting the all-too-muted signals that all is not well in the research community. Overconfidence also likely enters in: they perceive themselves to be considerably above average.[9] Others may not make it, but they will.

Third, when it comes to promoting PhD programs, faculty are good salesmen. Their lifeblood depends on recruiting new talent to staff their labs. The most effective recruits are those who aspire to a research career. There is a moral hazard here: faculty lack the incentive to provide straightforward information regarding job outcomes—and they don't. As David Levitt, a professor of physiology, University of Minnesota, puts it: "There's no honesty at all in recruiting PhDs. . . . There's not a hint that there's a shortage of jobs."[10] PhD programs do not make placement information readily available; in the rare cases when they do provide information, it is about postdoctoral placements—a temporary position in what may turn out to be a series of temporary positions throughout one's career—rather than about permanent placements.

There are other inefficiencies in the system. Ups and downs in research funding—especially funding from the federal government— can play havoc with careers. Getting a job—and the resources with which to do research—often depends upon the luck of the draw in terms of when one comes of age scientifically. Such variability in funding means that certain cohorts undergo a minimum of seven years of training only to find upon graduation that the funding spigot has been slowed to a dribble and the prospects of getting a research job are substantially lower than they had anticipated when they chose to get a PhD. The scars of coming of age during a period of tight resources linger throughout the career. Initial placements make a difference for years to come. Moreover, variability in funding not only causes problems for those who have a degree—it also sends a negative signal to those thinking of getting a PhD. It can also reap havoc on the research programs of established scientists, who must cut back on their research agendas and terminate persons working in their labs.

More generally, stop-and-go funding for research wastes resources. Anyone who has ever driven a car knows that a sure way to waste gas is to alternate between speeding up and slowing down. Moderate acceleration to a constant speed saves gas: one can go further on the same tank. Funding is the gas that keeps the research enterprise going. The enterprise could go further if the funds were more prudently and gradually deployed. Instead, and at least for the last 50 years, there have been periods of rapid acceleration followed by periods in which the enterprise is left virtually to run on fumes. This does not promote the health of the research enterprise.

Possible Solutions

Before discussing possible solutions, two caveats are in order. First, when it comes to assessing recommendations, one should be leery of those coming from groups who have a vested interest in keeping the system the way it is. Thus, for example, participants in the most recent evaluation of the NIH National Research Service Awards (NRSA) program had a vested interest when it declared the system to be "incredibly successful."[11] Committee members were faculty and deans—not students and postdocs who could not find jobs.[12]

Second, one must also recognize that universities and faculty do not respond to recommendations that lack teeth, such as two made by the Tilghman Committee in 1998, concerning (1) restraint in the growth of the number of graduate students in the life sciences and (2) dissemination of accurate information on career prospects of young life scientists. It is not in the interest of the institution (or the faculty member) to be the only one that cuts back or provides up-to-date placement information.

But institutions and faculty do respond to incentives and costs. That's good news: change the rules of what is fundable and what is not fundable, what can carry indirect costs and what cannot, and one will get a response. But one must do it carefully. The bad news regarding incentives is that if one does not get the incentives right, one can get unintended responses that considerably diminish the effectiveness of the system.

Here, I make seven suggestions for change that I believe could lead to a more efficient allocation of resources, especially with regard to the performance of research. Some are directed specifically at ways to alter the university research environment. Others are directed more broadly at ways to more efficiently use resources for research.

First, require universities to report placement data as part of all research grant applications. Do not merely require that they report the information—use the outcome data in scoring proposals.

Second, place limits on the amount of faculty time that can be charged off grants, thereby dulling the incentive for universities to hire faculty on soft money. This may seem radical, but others, including Bruce Alberts, have raised the possibility. Indeed, in the above mentioned *Science* editorial, Alberts suggested that NIH consider requiring "at least half of the salary of each principal investigator be paid by his or her institution, phasing in this requirement gradually over the next decade."[13] This would discourage universities from putting up buildings on spec and filling them with faculty on soft-money positions. Universities would no longer be able to export as much of their risk to their employees. It might encourage researchers to adopt more uncertain lines of research: their livelihood would be sufficiently divorced from their research outcomes. It would also diminish the demand for graduate students and postdocs to staff labs. But the change would have to be made gradually. There are simply too many people funded on soft money for the system to change overnight.[14]

Third, lessen the coupling between research and training. While effective training requires a research environment, effective research can be done outside a training environment. Yet in the United States, the majority of research in the public sector occurs at universities and medical schools. Labs are staffed by graduate students and postdocs. Thus research and training go hand in hand. And, while the model has much to recommend it, there are no incentives to engage in birth control when it is the dominant research model. The needs of the researcher come before the job prospects of the trainee.

One way to lessen the coupling between research and training is to encourage the establishment of more research institutes that are decoupled from universities or only loosely coupled. Institutes could employ postdocs, but they would not be in the business of training PhDs. Abstinence, after all, is the most effective form of birth control! This is common practice in certain areas of physics where, because of the scale of the equipment, national labs play a prominent role. Postdocs go to national labs to work after receiving their degree, but Argonne, Brookhaven, and Fermilab are not PhD mills.[15]

Research institutes have additional characteristics that make them attractive. They can create administrative structures that encourage interdisciplinary research and collaboration, minimizing the costs of coordination. They may be able to make more efficient use of equipment. And, if properly funded and endowed, they can discourage the hiring of scientists on soft money. They also have the possibility of creating environments in which staff scientists can find permanent employment with satisfying career outcomes. But buyer beware: institutes can also promote "senioritis," where research agendas are selected and directed by an aging, and perhaps less flexible, staff who keep young researchers under their thumb.

Fourth, try to determine once and for all the most effective way to support graduate students and rebalance funds toward means that are more effective. Many believe that fellowships and training grants produce better outcomes for students than do graduate research assistantships. They decouple students from advisors and lead to competition among institutions for students.[16] But the lack of proper control groups with which to compare outcomes makes the advocacy of one form of support over another more faith based than evidence based.

The argument goes something like this: if more money were put into fellowships, such as the NSF doctoral fellows program, universities would have to compete with each other in order to attract fellows.[17] The quality of the training experience and the outcomes of the department with regard to placements arguably should affect their success in doing so. The expectation is that such a move would enhance the research experience of graduate students. To quote Thomas Cech, former president of the Howard Hughes Medical Institute and a Nobel laureate in chemistry, "The real power of an individual fellowship is that it empowers a young scientist to act in a more independent manner, on something creative and for which they have a passion."[18]

The argument for putting more funds into training grants and fewer into graduate research assistantships is closely aligned with that for fellowships. Such a move gives departments the incentive to provide a high-quality training experience, because the quality of the training experience is considered in the application for renewal of the training award. At least one metric of quality must be placement outcomes.

Fifth, monitor existing science policies and develop new policies with the understanding that policies can affect the practice of science and, by extension, research outcomes. Policies that level the playing field by making resources available to new groups of researchers can lead to an increase in output and potentially an increase in the diversity of approaches. The establishment of Biological Resource Centers, for example, led to an increase in the number of individuals working with specific materials. The lifting of onerous restrictions on the use of certain patented mice expanded the mouse research community. The adoption of the Internet increased productivity among those at lower-tier institutions—and women.[19]

Sixth, if collaborative research really produces better research (see the discussion to follow), change the reward system. Encourage the creation of prizes to be awarded to groups of scientists. Status, as the Nobel Peace Prize so aptly demonstrates, need not be conferred on one person at a time.[20]

Finally, convince advocacy groups and the Congress to drop the doubling rubric. Instead of asking for a doubling of research funds, set goals: for example, spend 0.5 percent of the GDP on federally supported university research. (Politicians often give lip service to GDP-benchmarked goals, but this is about as far as it goes.) Such a policy is friendly to careers. It also eliminates inefficiencies caused by stop-and-go funding.

Three Other Efficiency Questions

Three more general efficiency questions are far easier to raise than to answer. They are:

1. Is 0.3 to 0.4 percent of gross domestic product (GDP) the right amount to spend on university R&D? Should it be more? Less?
2. Is the current allocation of federal funding for R&D, which gives two-thirds of the funds to the life sciences and one-third to everything else, the most efficient?
3. Are grants structured in an efficient way in terms of size, duration, criteria for evaluation, and number of people? A related question, is it more efficient to fund "big" projects, such as the Human Genome Project and the Protein Structure Initiative, or are a large number of small projects more efficient?

These are difficult questions. Little research has been done that is sufficiently thorough to warrant definitive answers. Some questions, due to
problems of measurement, may never be answerable. It is difficult, for
example, to measure the spillovers—and spillovers are an important part
of the story.

Amount

With regard to amount, studies have shown fairly impressive long-run returns to investment in public R&D despite the fact that distant outcomes
carry little weight in estimating rates of return. Yet many of the outcomes
are years away. But the studies are far from perfect. They often suffer, as
pointed out in Chapter 6, from comparing the benefits from winning outcomes with the costs of winning outcomes, ignoring in the calculation the
costs of all the dry holes that were sunk along the way. They also are prone
to exclude from the calculation the increased costs associated with people
living longer, focusing instead on the benefits associated with a longer life.
At times, the benefits are estimated by groups who have a vested interest in
showing substantial returns, such as the 2000 report by Funding First, *Exceptional Returns: The Economic Value of America's Investment in Medical Research.*[21] The organization, which has since been disbanded, lobbied
for increased resources for medical research in the United States. The lobbying void was soon filled by United for Medical Research, a coalition of
patient and health advocacy groups, universities and industry that issued
the 2011 report *An Economic Engine: NIH Research, Employment, and
the Future of the Medical Innovation Sector.*

The question regarding amount also concerns the future. But the evidence that can be assembled relates to the past. Thus just because the world
has reaped tremendous benefits from research conducted in physics over
the past hundred plus years (some physicists are wont to boast that 40
percent of the economy is due to advances in quantum mechanics) or because medical research—in a very short span of time—provided an effective way to prevent pregnant women from transmitting HIV to their children, it does not necessarily follow that research will continue to deliver at
the same rate that it has in the past. It could produce richer outcomes; but
it also could produce a string of duds.

The answer to the efficiency question regarding the right amount for the
United States to spend on research in the public sector is thus difficult to
answer. But one is on safer ground if the question is rephrased to ask
whether the amount being spent should be increased. We may never know
the right amount—but given the fairly healthy returns to previous invest-

ments in public research, the right amount is likely to be greater than 0.3 to 0.4 percent of the GDP. And surely the economy could afford it. We spend almost two times that amount drinking beer each year and more than twelve times that amount on defense.[22]

Allocation

What about mix? Is it efficient to spend two-thirds of the university R&D budget on research in the life sciences, a third on everything else? If the federal government were to reallocate the resources that it is spending, putting more on the physical sciences and engineering and less on the life sciences, the vast majority of which is for biomedical research, would the GDP grow at a faster rate? The economics test is to estimate the marginal benefit coming from another dollar spent on the biomedical sciences and compare it with the marginal benefit coming from another dollar spent on the physical sciences. If the former is lower, the portfolio would benefit from rebalancing. The fourteen year increase in life expectancy in the past seventy years makes a good case that research in the biomedical sciences has a high marginal product. But the slowed rate at which new drugs are being brought to market makes one wonder whether the marginal productivity of resources spent in the biomedical sciences is diminishing. The research discussed in Chapter 9 makes a good case that spillovers from the physical sciences have made significant contributions to the economy. Some of these contributions are even in the area of health—such as the laser and magnetic resonance imaging technology. But none of the analysis is sufficiently precise to calculate whether the portfolio is seriously out of balance.

Three observations, however, make one question whether the current balance is efficient. First, the heavy investment that the United States has made in the biomedical sciences has created a lobbying behemoth composed of universities and nonprofit health advocacy groups that constantly remind Congress of the importance of funding health-related research. There is no comparably well-established lobbying group on the part of other disciplines. Thus, the public hears much more about the benefits from research in the biomedical sciences than it does about the benefits arising from research in other disciplines.

Second, portfolio theory leads one to think that the current allocation might be out of balance. A basic tenet of investing is to rebalance one's portfolio if a change in market valuations results in a change in the composition of the portfolio that the investor is holding. Thus, when bond prices rise, an investor can, without intent, find that he is overinvested in bonds and underinvested in other assets, such as equities. The disciplined investor will generally sell bonds and buy more equities, bringing balance back to the portfolio. It is not a new principle, just a variant of the old adage of not putting all your eggs in one basket. The same logic could be extended to the national research budget, which became more tilted to the biomedical sciences as a result of the doubling of the NIH budget. When it comes to R&D, the argument for diversity is not new. Years ago, Kenneth Arrow wrote a seminal article on military R&D in which he argued that a goal of the government should be to invest in multiple lines of research.[23] More recently, Daron Acemoglu, a professor of economics at MIT and the 2006 winner of the John Bates Clark Medal in Economics, has argued that the government needs to promote diversity of the research that is undertaken.[24]

Finally, the mix of support for research—especially support from the federal government—affects the life of universities in a number of ways. For example, the NIH doubling was accompanied by a large increase in the construction of research facilities on campuses for research in the biomedical arena. This has consequences for facilities in other disciplines which got pushed to the back of the queue. It also has consequences for hiring. Moreover, these are long-run consequences, because much of the funding for these buildings was raised from the sale of bonds, and universities are not reaping the indirect cost they had expected. Other disciplines will end up footing part of the bill. One needs to take these types of unintended (but predictable) consequences into consideration when thinking about mix.

Grants

Are grants structured in an efficient way in terms of size, duration, criteria for evaluation, and number of people? This is something everyone has an opinion about, but again, the evidence is a bit thin. One study, for example, shows that researchers supported by the Howard Hughes Medical Institute (HHMI), which purports to support "people" rather than "projects," produce high-impact papers at a much higher rate than the control group. The study also found evidence that the direction of the HHMI investigators' research can change, compared with that of the control group.[25] The finding is intuitively pleasing; there are lots of people, including those at the Wellcome Trust, who believe that the HHMI model produces better science. Not only does it choose people over projects; it also forgives failure and provides for a longer period of secure funding. It also requires less administrative time on the part of the investigator (although the HHMI does not discourage its investigators from seeking other, additional sources of funds). Are the results due to these characteristics of the HHMI funding process? Or are the results due to the fact that, in spite of the study's effort to compare apples to apples, the HHMI researchers come from better stock—and thus the effect may be due to selection rather than from the way they were funded?

What about size? Is it better for lots of principal investigators to have $250,000 in grant money or for a third as many to have $750,000 in funds? Ignoring discipline, and there are major discipline differences in cost, an analysis done by the National Institute of General Medical Sciences suggests that the marginal product of allocating another dollar to an investigator is close to zero. Recall from Chapter 6 the finding that the amount a faculty member received in grants was only loosely correlated with more output.[26] At a more aggregate level, recall that Frederick Sacks found for the period during the NIH doubling no "upward jump" in publications in the biomedical fields from U.S. labs relative to publications from labs outside the United States.[27]

What about investing in megaprojects? Is it better to spend $3 billion on the Human Genome Project (HGP) or to support 6,000 researchers, each to the tune of $500,000? We just don't know. To the best of my knowledge, no one has attempted to do the calculations. Proponents of the HGP argue that there have already been substantial benefits and the best is yet to come. They also point to the advances in technology that the HGP has encouraged. Critics argue that the HGP was overhyped and will never live up to expectations. In one sense, both may be right. Large projects such as the HGP, the experiments that are ongoing at the Large Hadron Collider at CERN, and the Protein Structure Initiative do not necessarily provide answers. Rather, they provide inputs for more research down the road. Thus, they are especially difficult to evaluate.

What about collaboration? Is the heavy focus on collaboration— especially collaboration across countries, which the European Union requires in its Framework initiatives—an efficient way to allocate resources? Would one get better outcomes if the resources had not been structured in such a way? Again, it is hard to know. And, of course, the European Union has the goal not only of increasing research output but also of integrating the European research community. There is clear evidence, summarized in Chapter 4, that papers that are coauthored lead to better science. But there is little evidence regarding the marginal product of an additional investigator from an additional country. The research that has been done suggests that coordination can be problematic across multiple research sites.[28]

All of these are powerful questions. Remember that resources can only be said to be efficiently allocated if one cannot increase the size of the proverbial pie by reallocating them. It follows that, if resources are not efficiently allocated, one can get more through reallocation. There are those, of course, who will be hurt by the reallocation, but the system will benefit. In an era of tight resources, efficiency concerns are especially important.

Thus, it is particularly important at this juncture to begin to address some of these efficiency questions. Partly by design, and partly by luck, the "Science of Science and Innovation Policy" initiative is underway at the National Science Foundation (NSF).[29] Many of the questions I have raised are questions that researchers affiliated with the initiative are trying to answer. The initiative is also investing in data tools and databases that will facilitate answering some of these questions. Thus, answers to some of these hard questions may be forthcoming, but not tomorrow or next year. And some—as noted above—are likely to remain unanswerable.

Encouraging Trends

There are some encouraging trends. A number of research institutes have opened in recent years that are only loosely affiliated with a university. The Janelia Farm Research Center, opened by the HHMI in Ashburn, Virginia, in 2006, is an example of such an institute. It has the goal of employing about 250 resident investigators in positions of group leader or fellow. The farm also employs a number of postdocs. The newly formed Lieber Institute for Brain Development in Baltimore is another.[30] The Institute for Systems Biology that Leroy Hood helped found in Seattle is yet another.

There is also evidence that Washington has become more attuned to some of these questions. The Science of Science and Innovation Policy initiative was set in motion by John Marburger when he served as science advisor to President George W. Bush and as director of the Office of Science and Technology Policy. Congress and the administration are on record supporting more research funding for the NSF, the National Institutes of Standards and Technology (NIST), and the Department of Energy (DOE); research budgets of the three agencies have grown relative to those of NIH in very recent years.

Last, but certainly not least, Francis Collins, the director of the NIH at the time this book was completed, has gone on record that there is a need for the NIH to develop better models to guide decisions about the optimum size and nature of the U.S. workforce for biomedical research.[31] Nine months later, Collins followed up by appointing Shirley Tilghman, the current president of Princeton University and a staunch advocate of the importance of balancing student outcomes with faculty needs, to chair a committee on workforce issues.[32] Collins has also gone on record, stating, "A related issue that needs attention, though it will be controversial, is whether institutional incentives in the current system that encourage faculty to obtain up to 100 percent of their salary from grants are the best way to encourage productivity."[33]

Appendix

This appendix describes five databases referred to in this book available through the National Center for Science and Engineering Statistics of the National Science Foundation.

National Survey of College Graduates (NSCG)

This is a longitudinal survey that is updated each decade. The last update was in 2003. The 2003 survey respondents were individuals living in the U.S. during the reference week of October 1, 2003, holding a bachelor's or higher degree in any field, and under age 76. The survey included a sample of individuals drawn from the 2000 Decennial Census long form who indicated they had a BA degree or higher. The 2003 survey also includes cohorts from earlier NSCG surveys. The survey collects information on a wide variety of variables, including field of degree, type of degree, highest degree, salary, employment status, sector of employment, age, gender, race, citizenship status, country of birth. See National Science Foundation, 2011a and http://www.nsf.gov/statistics/showsrvy.cfm ?srvy_CatID=3&srvy_Seri=7.

Survey of Doctorate Recipients (SDR)

The Survey of Doctorate Recipients is conducted every two years and follows individuals until age 76. The survey is restricted to those who received a research doctorate degree in the United States in a field of science, engineering or health and are living in the United States the week of the survey. The survey began in 1973. The sampling frame is drawn from the SED. The survey collects information on a variety of key variables such as sector of employment, primary and secondary work activity, salary, date of birth, gender, marital status, and geographic place of employment. National Science Foundation 2011b and http://www.nsf.gov/statistics/srvydoctor atework/.

Survey of Earned Doctorates (SED)

The Survey of Earned Doctorates is administered to all individuals in the United States at or near the time of receipt of a research doctoral degree. The survey has been administered since 1957. It has a response rate of over 90 percent. It collects information on key variables such as institution conferring the degree, field of degree, employment plans, birth year, race, gender, country of birth, citizenship, marital status, education of parents, and source of support while in graduate school. See National Science Foundation 2011c and http://www.nsf.gov/statistics/srvydoctorates/.

Survey of Graduate Students and Postdoctorates
in Science and Engineering (GSS)

An annual survey of all academic institutions in the United States awarding research-based graduate degrees, conducted by the National Science Foundation. It provides information on enrollment data for graduate programs as well as information on the number of postdoctorates. See National Science Foundation, 2011d and http://www.nsf.gov/statistics/srvygradpostdoc/.

Survey of Research and Development Expenditures
at Universities and Colleges

The annual survey collects information on research and development expenditures by source of funds and by academic field. The survey began in 1972. National Science Foundation 2011e and http://www.nsf.gov/statistics/srvyrdexpenditures/.

Summary data from these surveys can be obtained through the NSF WebCASPER data system. In the case of the SDR and the SED, individuals working at qualified organizations in the United States can apply for the organization to have a site license for use of the data. See National Science Foundation, 2010c and https://webcaspar.nsf.gov/ for a description of the WebCASPER system.

Notes

1. What Does Economics Have To Do with Science?

1. National Science Board 2010, appendix, table 5-12.
2. The exact figure is 58.7 percent and includes basic research performed at universities and at university-affiliated Federally Funded Research and Development Centers (FFRDCs). If university-affiliated FFRDCs are excluded, approximately 56.1 percent of basic research is performed at universities and medical schools. National Science Board 2010, appendix, table 4-4.
3. The collider first came on line September 10, 2008, but within two weeks was shut down for repairs necessitated when a mechanical failure triggered a helium leak. See Meyers 2008.
4. See Chapter 6. Henry Sauermann and Michael Roach, in a survey conducted in 2010, found the median lab size across disciplines in science and engineering to be eight (personal correspondence, Henry Sauermann).
5. See discussion in Chapter 6.
6. See discussion in Chapter 5.
7. See Britt 2009.
8. Organization for Economic Cooperation and Development 2010, Main Science and Technology Indicators, 1.
9. See discussion in Chapter 5. The E-ELT has a 42-meter-diameter aperture. The OWL was to have had an aperture 100 meters in diameter.
10. Clery 2009c, 2009d. It is difficult to know the exact costs of ITER because the seven countries involved committed to contributing specific components, not a specific amount of funds; however, it is a sure bet that ITER will cost considerably more than the 5 billion euros originally estimated to build and the 5 billion estimated to operate it for 20 years.

11. Clery 2010b.
12. http://lhc-machine-outreach.web.cern.ch/lhc-machine-outreach/faq/lhc-energy -consumption.htm. An exception was made during the winter of 2009/2010 to make up for delays experienced due to the shutdown of 2008. See Large Hadron Collider, 2011, *Wikipedia*, http://en.wikipedia.org/wiki/Large_Had ron_Collider#Cost.
13. The criteria for the Archon X Prize for Genomics also stipulate that the sequencing have "an accuracy of no more than one error in every 100,000 bases sequenced, with sequences accurately covering at least 98 percent of the genome, and at a recurring cost of no more than $10,000 per genome." Archon Genomics X Prize website. Prize Overview: A $10 Million Prize for the First Team to Successfully Sequence 100 Human Genomes in 10 Days. Available at http://genomics.xprize.org/archon-x-prize-for-genomics/prize-overview.
14. Heinig et al. 2007.
15. University of North Carolina at Chapel Hill 2010.
16. Franzoni, Scelatto, and Stephan 2011. The research uses submission data supplied by the journal *Science* and relates it to different types of incentive schemes adopted by countries in recent years. The research controls for a number of variables, including the annual stock of research resources, lagged one year, the national composition of the editorial board, and the extent of international collaboration.
17. John Simpson 2007; Eisenstein and Risnick 2001.
18. Salary is for those at the 90th decile. SDR 2006 data. See Chapter 3, National Science Foundation 2011b and the Appendix.
19. European University Institute 2010.
20. "Investment Banking: Salaries," 2010, Careers-in-Finance website, available at http://www.careers-in-finance.com/ibsal.htm. Bonuses are included in the calculation.
21. The earnings reported are median and are expressed in 2006 dollars. The mean salary of those who had been out 10 or more years and who had started in banking was $815,914. See Bertrand, Goldin, and Katz 2009, table 2.
22. For example, see the *h*-index tracker on Zhong Lin Wang's webpage at the Georgia Institute of Technology, http://www.nanoscience.gatech.edu/zlwang/.
23. "Richter Scale," 2010, *Wikipedia,* http://en.wikipedia.org/wiki/Richter_mag nitude_scale.
24. Norwegian Academy of Science and Letters 2010.
25. A Rand report suggests that "universities recover between 70 and 90 percent of the facilities and administrative expenses associated with federal projects" (Goldman et al. 2000, xii).
26. Economists were not the first to note the public nature of knowledge. Almost 200 years ago, Thomas Jefferson wrote, "If nature has made any one thing less susceptible than all others of exclusive property, it is the action of the thinking power called an idea, which an individual may exclusively possess as long as he keeps it to himself; but the moment it is divulged, it forces itself into the possession of every one, and the receiver cannot dispossess himself of it. Its peculiar character, too, is that no one possesses the less, because every other

possesses the whole of it. He who receives an idea from me, receives instruction himself without lessening mine; as he who lights his taper at mine, receives light without darkening mine" (Jefferson 1967, vol. 1, 433, sec. 4045).

27. In 1848, Mill used the lighthouse as an example of a public good: "no one would build lighthouses from motives of personal interest, unless indemnified and rewarded from a compulsory levy by the state" (1921, 975). Coase (1974) reviewed the British lighthouse system and showed that during certain periods lighthouses were constructed by the private sector.

28. Arrow 1987, 687.

29. See the discussion in Chapter 5.

30. The government can also encourage private firms to engage in research by providing research and development tax credits or, in exchange for disclosure, awarding monopoly rights in the form of a patent or copyright to the inventor.

31. In 1940, the life expectancy of a U.S. male at birth was 60.8 years; today it is 75.1. In the same seventy-year interval, the life expectancy for women has risen from 65.2 to 80.2 years. Data for 2006 come from U.S. Census Bureau 2011, table 104, Selected Life Table values, available at: http://www.census .gov/compendia/statab/cats/births_deaths_marriages_divorces/life_expec tancy.html. The data for 1950 come from Information Please Database 2007.

32. Murphy and Topel 2006. The authors use a "willingness to pay" methodology to compute the value. Approximately half of the value comes from reduced mortality from heart disease.

33. For a discussion of computers, see Rosenberg and Nelson 1994.

34. The LHC re-creates the conditions of the universe just after the Big Bang in order to understand why the matter of the universe is dominated by an unknown type called dark matter. If the constituents of dark matter are new particles, the ATLAS detector at the LHC should be able to discover them and elucidate the mystery of dark matter. See Lefevre 2008.

35. Public sector research contributed in other ways to the development of the global positioning system (GPS). For example, Friedwardt Winterberg, a theoretical physicist at the University of Nevada, Reno, in 1956 proposed a test of general relativity using accurate atomic clocks placed in orbit in artificial satellites. Brad Parkinson, a professor of aeronautics and astronautics at Stanford, led the military team that developed GPS.

36. "Atomic Clock: History," 2010, *Wikipedia,* http://en.wikipedia.org/wiki/ Atomic_clock#History.

37. "Heterosis," 2010, *Wikipedia,* http://en.wikipedia.org/wiki/Heterosis.

38. See the discussion in Chapter 9.

39. Ellard 2002. In the late 1930s, Isidor Rabi had come across nuclear magnetic resonance but considered it to be an artifact of his experiment.

40. "The Nobel Prize in Physics 1952: Felix Bloch, E. M. Purcell," http://nobelprize .org/nobel_prizes/physics/laureates/1952/.

41. See Chapter 9.

42. Superconductivity is a phenomenon of exactly zero electrical resistance.

43. "High Temperature Conductivity," 2010, *Wikipedia,* http://en.wikipedia.org/ wiki/High-temperature_superconductivity. See also Cho 2008.

44. See Kong et al. 2008.
45. Two independent studies reported in 2008 that gene therapy had partially restored the sight of four young adults who were born with severe blindness (Kaiser 2008e).
46. Clery 2010b.
47. Bhattacharjee 2008a.
48. Service 2008.
49. Couzin-Frankel 2009.
50. To quote Rosenberg and Nelson (1994, 323), "Industry is more effective in dealing with problems that are located close to the market place."
51. See the discussion in Chapter 6.
52. See the discussion in Chapter 9.
53. In a natural experiment the treatment is random and not by design. To state it differently, the treatment is administered "by nature" and not by the experimenter. Natural experiments can be helpful when a well-defined subpopulation experiences a change in treatment. By way of example, one can compare use of certain research mice by researchers before certain patent restrictions were removed with their use after the restrictions were removed to see how patents affect the use of mice. See Natural Experiments, 2011. *Wikipedia* http://en.wikipedia.org/wiki/Natural_experiment.
54. Hunter, Oswald and Charlton 2009.
55. Data are for research 1 institutions. See Winkler et al. 2009.
56. Stokes 1997. The distinction between basic and applied research used here, as well as Stokes' definition of Pasteur's Quadrant, depends on the goals of the researcher, rather than the outcomes from the research. However, the distinction between basic and applied research is often made based on outcomes, not motives, and Pasteur's Quadrant is also loosely used to describe the nature of the research outcomes, not the motives, of the researcher. In this book the terms are used in both senses, depending upon the context and data source.

2. Puzzles and Priority

1. Richard Feynman, in the context of explaining why "I don't have anything to do with the Nobel Prize . . ." (which he won in 1965), wrote, "I don't see that it makes any point that someone in the Swedish Academy decides that this work is noble enough to receive a prize—I've already got the prize . . ." (1999, 12).
2. Kuhn 1962, 36. Kuhn goes on to say that what challenges a scientist "is the conviction that, if only he is skillful enough, he will succeed in solving a puzzle that no one before has solved or solved so well" (ibid, 38).
3. Hagstrom 1965, 65.
4. Hull 1988, 306.
5. Hull 1988, 305.
6. Letter from Joshua Lederberg to Sharon Levin, September 21, 1992.
7. Roberts 1993.
8. Reid 1985. Kilby was working at Texas Instruments at the time he invented the integrated circuit. Robert Noyce, the other inventor of the integrated

circuit, was working at Fairchild Semiconductor in California when he invented an integrated circuit several months later. The two are often referred to as the "coinventors" of the integrated circuit.

9. McKnight 2009.
10. Feynman 1985.
11. From a psychologist's point of view, an interest in puzzle solving is what motivates scientists. The "aha" moment is the reward for solving the puzzle. In the discussion, however, I speak of puzzles as a reward—given the common practice among scientists to speak of them as such.
12. See "Power of Serendipity," 2007.
13. Ainsworth 2008.
14. Sauermann, Cohen, and Stephan 2010. The NSF survey asks scientists to report on a 5-point scale the importance of and satisfaction derived from nine job attributes: opportunities for advancement, degree of independence, contribution to society, salary, intellectual challenge, benefits, job security, job location, and level of responsibility. Sauermann, Cohen, and Stephan examine the first five.
15. Harré 1979, 3.
16. Attributed to Napoleon by Menard 1971, 195.
17. In a series of articles and essays begun in the late 1950s, Merton (1957, 1961, 1968, 1969) argued convincingly that the goal of scientists is to establish *priority of discovery* by being first to communicate an advance in knowledge, and that the rewards to priority are *the recognition awarded by the scientific community for being first.* See Dasgupta and David 1994 for a discussion of the role of priority.
18. Merton 1969, 8. A tension that exists between experimentalists and theorists in physics is the "awkward matter of credit." That is, "Who should get the glory when a discovery is made: the theorist who proposed the idea, or the experimentalist who found the evidence for it?" (Kolbert 2007, 75).
19. See Lehrer 1993. The song, which suggests that Lobachevsky endorsed plagiarism, was not, according to Lehrer, "intended as a slur on [Lobachevsky's] character," and the name was chosen "solely for prosodic reasons" (quoted in "Nikolai Lobachevsky," *Wikipedia,* http://en.wikipedia.org/wiki/Nikolai_Lobachevsky). See further discussion of Lobachevsky in the section on multiples to follow.
20. Merton argues that "far from being odd or curious or remarkable, the pattern of independent multiple discoveries in science is in principle the dominant pattern rather than a subsidiary one" (1961, 356).
21. Rivest, who wrote up the paper, listed the authors alphabetically. Adleman, a number theorist, reportedly objected, stating that he had not done enough work to warrant inclusion as an author. But Rivest objected, and Adelman reportedly reconsidered, on the condition that he be listed last, out of alphabetical order, to reflect what he saw as his minimal contribution (Robinson 2003). The RSA algorithm was first presented to the public by Martin Gardner, in an article in *Scientific American* in August 1977. The authors published their paper later that year (Rivest, Shamir, and Adleman 1978).

22. The five groups were led by Ruddle at Yale, Brinster and Palmiter at the universities of Pennsylvania and Washington, Costantini at Oxford, Mintz at Fox Chase, and T. E. Wagner at Clemson University. The five papers, with the group leader often holding the last author position—a common practice in the biomedical sciences—are as follows: Gordon et al. 1980; Brinster et al. 1981; Costantini and Lacy 1981; Wagner, E. F., et al. 1981; Wagner, T. E., et al. 1981. See Murray 2010.

23. The field of computer sciences is an exception. In this field, the preferred way of establishing priority is through presentations at conferences and subsequent publication in conference proceedings.

24. Stephan and Levin 1992.

25. Applied Physics Express (APEX) advertisement. *Science*, 2008. The web link for this journal http://apex.jsap.jp/about.html says, "Papers for APEX will be published online within 2 weeks, in the fastest case, from receipt to online publication."

26. Agre 2003.

27. See Fox 1994.

28. Damadian is clearly not the only individual to have felt that he had been wronged in not being included in the prize, but very few scientists go public with the complaint. Damadian's claim had teeth in the sense that he had already been awarded the 2001 Lemelson-MIT Award for Lifetime achievement. The award described Damadian as "The man who invented the MR scanner." See Tenenbaum 2003. More than one hundred years ago, another scientist "whined" after being omitted from the prize—with some effect. The physicist Philipp Lenard "vainly and disingenuously" claimed credit for the discovery of X-rays (1901 Nobel Prize in physics) and the discovery of the electron (1906 Nobel Prize in physics). In spite of this, he won the 1905 Nobel Prize in physics for experiments on the photoelectric effect (ibid).

29. Each of the three Nobel Prizes in science can be given to at most three individuals.

30. *Science* (2008) 322:1765.

31. Edelman and Larkin 2009.

32. Honorary or guest authorship is distinct from "ghost authorship," the practice of not naming as an author an individual who has made a substantial contribution to a piece of research.

33. Such lists are not without error. The presence of common names, especially among the Asian community, means that attribution can be incorrect; thus, such rankings must be used with caution and carefully monitored.

34. Formally, the *h*-index is defined "as the number of papers with citation number higher or equal to h." See Hirsch 2005.

35. Hirsch (2005) demonstrated that the *h*-index has high predictive value for such honors as the Nobel Prize and membership in the National Academy of Sciences.

36. The *h*-index can readily be computed using several sites. The Thomson Reuters Web of Knowlege generates an *h*-index as part of its citation reports. The *scHolar Index* (Roussel 2011) and *Publish or Perish* (Harzing 2010) software

compute *h*-indexes based on the Google Scholar database. For other programs based on Google Scholar, see Whitton 2010. The resulting *h*-indexes are generally larger than those computed using the Web of Knowledge, because Google Scholar covers a wider set of journals than that covered by the Thomson Reuters Web of Knowledge. A number of variations of the *h*-index have been proposed. For example, the *g*-index can discriminate between two authors having the same *h*-index—when one of them has a blockbuster publication and the other does not. See Egghe 2006; for a review, see Alonso et al. 2009.

37. In "Slice of Life," *Science* (2008) 320, April 18.

38. Recognition is also awarded by attaching a scientist's name to a building, professorship, or lecture series, although this form of recognition usually comes after the death of the scientist, while eponymy can occur during the scientist's life. Not all discoveries are named for the scientist who was first to make the discovery. Benford's law, for example, was first discovered by Simon Newcomb in 1881. It was rediscovered by Frank Benford in 1938. See "Benford's Law," 2010, *Wikipedia,* http://en.wikipedia.org/wiki/Benford's_law. For a discussion, see Stigler 1980.

39. Such prizes are distinct from inducement prizes, discussed in Chapter 6, which offer a reward for the first individual or team to accomplish a specific goal. An early example of such a prize is that created by the British government in 1714 to be awarded to the first person to solve the longitude problem.

40. The Jeantet and Koch Prizes are examples of prizes dedicated to supporting the winners' labs, although, in the case of the Jeantet Prize, 100,000 of the 700,000 CHF are given to the researcher personally. In some instances, awards are for a position—such as the $10 million Polaris Award, which was created to recruit "world leaders in health science research to Alberta" (announcement in *Science* [2008] 322, October 24).

41. Zuckerman 1992. Rate of growth computed from National Science Foundation 1977 and National Science Foundation 1996.

42. It garnered considerable attention in 2007 when one of the four recipients of the medal, Grigory Perelman, honored for his proof of the Poincaré conjecture, refused the prize. The Fields Medal is awarded to up to four mathematicians under the age of 40.

43. The Foundation ran a full-page advertisement in 2009, "Congratulations to Elizabeth H. Blackburn," in *Science* congratulating Elizabeth H. Blackburn, a 2008 winner of the award, for her Nobel Prize in physiology or medicine in 2009, with Carol W. Greider and Jack W. Szostak.

44. In rare instances scientists are elected to all three academies. For example, in 2008, Frances Arnold became the first woman and eighth living scientist to be elected to all three of the U.S. national academies (*Science* [2008] 320:857, May 16).

45. Her reason: her husband and long-term collaborator Neal Copeland was not on the list. In her letter to the Academy she wrote, "It is impossible to separate my contributions from Neal's as we did everything together on an equal basis . . . Someday if both of us have a chance to accept this honor together,

it would be the highlight of our scientific careers" (Bhattacharjee 2008b). Although Richard Feynman did not initially decline membership, he later resigned from the National Academy of Sciences (Feynman 1999).

46. Research findings only become a public good when they are codified in a manner that others can understand. The distinction, therefore, is often drawn between knowledge, which is the product of research, and information, which is the codification of knowledge (Dasgupta and David, 1994, 493).

47. Stephan 2004.

48. Merton 1988, 620.

49. Merton 1988, 620. Partha Dasgupta and Paul David—in a classic case of multiples—express the private-public paradox exceedingly well, although a year later than Merton's lecture. "Priority creates a privately-owned asset—a form of intellectual property—from the very act of relinquishing exclusive possession of the new knowledge" (1987, 531).

50. Dasgupta and David 1987, 530 and Dasgupta and David 1994.

51. Merton 1957.

52. Ziman 1968; Dasgupta and David 1987.

53. Small-world networks are characterized by a high degree of clustering and a low degree of separation between members of the same network. In the case of publishing, clustering measures the probability that two of a scientist's collaborators have themselves collaborated. The concept of separation, which was made famous in John Gaure's play *Six Degrees of Separation,* is a measure of the number of "hops" one would have to take to move from one node in a network to another. See Uzzi, Amaral, and Reed-Tsochas 2007; Newman 2004.

54. Kohn 1986.

55. The papers in which he reported his results were retracted by the journal *Science* in 2006.

56. Charges were originally brought in 2006 (Couzin 2006, 1222). See also Office of Research Integrity, U.S. Department of Health and Human Services, http://ori.hhs.gov/misconduct/cases/Goodwin_Elizabeth.shtml. Goodwin resigned soon after the university began its investigation in 2006.

57. See Coyne 2010.

58. Miller 2010, 1583.

59. Agin 2007.

60. Lacetera and Zirulia 2009.

61. David and Pozzi 2010.

62. Eisenberg 1987.

63. Increased funding from industry for academic research has also led to a delay in the publication of research results or a withholding of results. See the discussion in Chapter 6.

64. DuPont placed two other onerous conditions on use of the mouse: it would not allow scientists to follow the traditional practices of sharing mice or breeding extensively from the mice, and it required that scientists fulfill annual disclosure requirements, reporting annually on their published (and unpublished) findings (Murray 2010).

65. A few months earlier, a similar MOU had been signed regarding Cre-lox mice.
66. The authors find that citations to OncoMouse articles increased by 21 percent after the MOU. Citations to Cre-lox mice articles increased even more (34 percent). The differential effect is likely explained by the fact that the Cre-lox MOU came first, and thus the OncoMouse MOU was in all likelihood anticipated—plus the fact that Jackson Laboratory (JAX), the nonprofit lab that breeds and distributes most mice used for research, had already made an informal commitment to make the OncoMouse available to researchers (Murray et al. 2010).
67. Murray and Stern 2007.
68. Walsh, Cohen, and Cho 2007.
69. Von Hippel 1994.
70. Wagner, E. F., et al. 1981.
71. Murray 2010, 21.
72. Francesco Lissoni (personal correspondence) points out that baseball and other team sports do not provide as good an analogy as individual sports, such as golf or tennis, for two reasons. First, the reward system in science addresses the individual, not the team. Second, in individual sports such as golf and tennis, all professionals are ranked according to their past and recent performance, very much like scientists are ranked (implicitly) according to bibliometric indicators. In team sports, this does not occur.
73. The review process at the NIH begins at the study section. Each section meets three times a year; more than 175 sections exist, and a proposal is generally assigned to only one section. Scientists rarely change study sections and often refer to the section that reviews their work as "my section."
74. Strictly speaking, the panel does not make the award but makes recommendations to the NSF program officer. The NIH study sections assign a score to a proposal; proposals are referred to "council," there being one council for each NIH institute. The "payline" (the cutoff score for funding) is determined institute by institute. The question could be raised as to whether there have become too many niche contests in science. I address this in Chapter 6.
75. Edward Lazear and Sherwin Rosen, the fathers of the tournament model, show that under certain conditions tournament models produce an efficient allocation of resources. If science is a tournament model, this would suggest that inefficiencies are not an issue. But the scientific tournament is not like other tournaments: tenure makes a difference. Rock stars, opera singers, and soccer players do not have tenure; professors do. This means that creative scientists, despite their demonstrated creativity, may find it difficult to secure a lab of their own, especially when the number of tenure-track positions does not grow and the number of people seeking such positions does (Lazear and Rosen 1981). We will return to this in Chapter 7.
76. "The Nobel Prize in Chemistry 2008: Osamu Shimomura, Martin Chalfie, Roger Y. Tsien." 2011. *Nobelprize.org.* http://nobelprize.org/nobel_prizes/chemistry/laureates/2008/.
77. Lotka's law states that if k is the number of scientists who publish one paper, then the number publishing n papers is k/n^2. In many disciplines this works

out to some 5 or 6 percent of the scientists who *publish at all* producing about half of all papers in their discipline (Lotka 1926). Although Lotka's Law has held up well over time and across disciplines, Paul David shows that other statistical distributions also provide good fits to observed publication counts (David 1994).

78. de Solla Price 1986; David 1994.
79. Weiss and Lillard 1982 find that not only the mean but also the variance of publication counts increased during the first 10–12 years of the career of a group of Israeli scientists. Research shows that the distribution of output is also characterized by having a fat tail (Veugelers 2011).
80. Merton 1968, 58. The title comes from the Bible, the book of Matthew, Chapter 13, verse 12: "For whosoever hath, to him shall be given, and he shall have more abundance: but whosoever hath not, from him shall be taken away even that he hath." From an economist's perspective, the Matthew effect expresses the endogenous nature of reputation in science.
81. Allison and Stewart 1974. Cole and Cole 1973.
82. Allison and Long 1990.
83. Allison, Long, and Krauze 1982.
84. Stephan and Levin 1992, 30.
85. David 1994.
86. Frank and Cook 1992, 31.

3. Money

1. Wolpert and Richards 1988, 146.
2. Rosovsky 1991, 242.
3. In 2008–2009, full professors on average earned $192,600 at Harvard, $142,100 at the University of Michigan–Ann Arbor, and $92,500 at Central Michigan (American Association of University Professors 2009).
4. This is not to say that the gender gap in salaries can be entirely explained by mobility. Nor is it to say that discrimination does not play some role in the lives of women faculty. There is a vast body of work on gender differences in pay, promotion, and productivity among scientists. For pay, see Toutkoushian and Conley 2005. For productivity, see Xie and Shauman 2003. For promotion, see Ginther and Kahn 2009.
5. The figures cover full-time members of the instructional staff excluding those in medical schools. The salaries are adjusted to a standard nine-month work year (American Association of University Professors 2010).
6. Byrne 2008. It is a bit too soon to know how the financial meltdown of 2008–2009 will affect the gap, although the gap between privates and publics narrowed ever so slightly (going from 31.6 to 31.0 percent) between 2008–2009 and 2009–2010.
7. University of North Carolina at Chapel Hill 2010.
8. American Association of University Professors 2010.
9. The survey has been conducted by the Office of Institutional Research and Information Management at Oklahoma State University (OSU) since 1974.

10. The 1974–1975 salaries come from Bound, Turner, and Walsh 2009.
11. Universities need not match the salary offered by industry, however, given that a process of selection occurs whereby those who care more about salary work in industry, and those who care less about salary and more about independence take positions in academe. See Sauermann and Stephan 2010.
12. For example, other things being equal, top economics departments paid lower starting salaries to new assistant professors in economics in the late 1970s than did lower ranked departments (Ehrenberg, Pieper, and Willis 1998).
13. Graves, Lee, and Sexton 1987.
14. The statistic, named for Corrado Gini, who devised the measure early in the twentieth century, provides another example of eponymy (see Chapter 2). For more information, see "Gini Coefficient," 2010, *Wikipedia,* http://en.wiki pedia.org/wiki/Gini_coefficient.
15. "Income Inequality in the United States," *Wikipedia,* http://en.wikipedia.org/ wiki/Income_inequality_in_the_United_States.
16. Diamond 1986.
17. Levin and Stephan 1997. The study uses panel data and thus can control for individual fixed effects.
18. By contrast, there have been a number of studies looking at the relationship of publication to salary in economics and management. See, for example, Hamermesh, Johnson, and Weisbrod 1982; Gomez-Mejia 1992; Geisler and Oaxaca 2005.
19. Toutkoushian and Conley 2005. The estimate quoted is for all sciences. It is unpublished and was provided by Toutkoushian.
20. Another indication of the relationship that exists between productivity and salary comes by examining salary differences among fields *within* institutions. One might initially think that differences between fields would be the same for all institutions. For example, if chemistry professors earn 17 percent more than English professors at one institution, they would earn 17 percent more at another institution. But such is not the case. Research shows that salary differentials between fields vary across universities. The differences can be explained in part by how highly rated the department is in terms of quality of graduate education, where the rating variable is publication based. For example, the premium enjoyed by chemistry professors relative to English professors is greater the higher the rating of the chemistry department. See Ehrenberg, McGraw, and Mrdjenovic 2006. Note that the equations also control for the ranking of the English department.
21. See National Institutes of Health 2009a.
22. The data are for the 119 medical schools that offer tenure to basic science faculty; the data were collected in 2005 (Bunton and Mallon 2007).
23. Mallon and Korn 2004. The numbers in the text are from Bunton and Mallon 2007.
24. Lissoni et al. 2010.
25. Franzoni, Scellato, and Stephan 2011.
26. A number of factors are included in the rankings, but publications constitute the core. The 2008 RAE graded publications into one of four categories.

Departments were then given an overall "quality profile" based on the grades. (See Research Assessment Exercise 2008). Funds for research are distributed to departments based on the quality profile. Australia and New Zealand drew on the RAE to put in place major policy reforms for funding academic institutions whereby better performing institutions receive more funding than lower performing ones and thus have more resources for competing in the job market for scientists. Prior to the reforms, the national budgets were largely distributed on the basis of the number of students and the number of research personnel. Norway, Belgium (Flanders), Denmark, and Italy started similar policies during the past decade for allocating a share of the budget. Other countries focus on incentives directed at individuals rather than at institutions (Franzoni, Scellato, and Stephan 2010). The RAE will be replaced by the Research Excellence Framework (REF), to be completed in 2014. The REF is exploring the allocation of research outputs based on publication address rather than location of employment at the time the data were collected. If adopted, publications will only count toward the assessment if the faculty member was actually employed at the university at the time the article was published. See Imperial College London, Faculty of Medicine 2008.

27. Hicks 2009. Recent reforms in Germany ostensibly were designed to provide for performance-based salary increases for highly productive faculty, although they arguably may not succeed in accomplishing this goal. A major component of the change from the "C" to the "W" system is the way in which base salaries are negotiated for senior faculty. Under the (old) C system, faculty with a competing job offer could negotiate a higher salary at their home institution. The resulting raise was permanent and included in the base used for the computation of pensions. Under the "W" system, the base salary has been lowered with the idea that performance-based supplements would be possible. But the supplements are in principle for a limited period of time. Only if they have been granted for five or more years do they become permanent, although the latter is subject to negotiation (Stephan 2008).

28. Franzoni, Scellato, and Stephan 2011.

29. Mowery et al. 2004, 59.

30. Jones worked at the Connecticut Agricultural Experiment Station. Thimann and Galinat, 1991.

31. For early U.S. university patent data, see figure 3.2, "University Patents, 1925–80," in Mowery et al. 2004. For more recent years, see various issues of Science and Engineering Indicators. USPTO statistics are from U.S. Patent and Trademark Office 2010.

32. Data come from the Survey of Doctorate Recipients. See National Science Foundation 2011b and Appendix.

33. As a result of the act, patentable inventions arising from federal funding are considered university property rather than the property of the U.S. government. Virtually all universities have adopted a similar standard of ownership for patents arising from corporate-sponsored research. In some cases, universities grant ownership to sponsors who cover all costs of research (Jensen and Thursby 2001).

34. The term, developed by Donald Stokes (1997), contrasts such research to research that exclusively seeks basic understanding (Bohr's Quadrant) and research that is exclusively use-oriented (Edison's Quadrant). See discussion Chapter 1.

35. Mowery and coauthors, who have written one of the definitive works on the subject, conclude that "Bayh-Dole accelerated the growth of university patenting and resulted in the entry into patenting and licensing by many universities during the 1980s. But the 'transformation' wrought by the 1980 Act followed trends that were well established by the late 1970s" (2004, 36).

36. Ibid., 90.

37. Bok 1982, 149. Bok's comments relate to ways in which universities could share in revenues stemming from ideas developed in their labs.

38. Quoted in Mowery et al. 2004, 45.

39. Ibid., 70.

40. National Science Board 2000; see Chapter 6 for 1989–1990 income. Information concerning licensing income has been collected periodically since 1991 by the Association of University Technology Managers (AUTM). The survey initially included 98 universities. Over the years it has been augmented and now includes 194 universities and research institutes, some of which are in Canada.

41. Data were provided by Henry Sauermann, Georgia Institute of Technology, Atlanta, and are for 205 universities.

42. There are two exceptions in which universities pay faculty a larger percentage, rather than a smaller percentage, as the amount of royalties increase.

43. There is variation in the rate paid by these institutions. Ten out of seventy-eight schools for which the rate is not fixed give 100 percent of the first $10,000 to the inventor; twenty-two universities give more than 50 percent of the first $10,000. But there are also some "cheap" schools that give less than 35 percent of the first $10,000. Data provided by Henry Sauermann, Georgia Institute of Technology, Atlanta.

44. Jensen and Thursby's (2001) survey of the licensing practices of sixty-six universities finds that the top five inventions licensed by each university accounted for 78 percent of gross license revenue. Scherer reports similar findings for Harvard inventions, and Harhoff et al. have reported similar results for German patents (Scherer 1998; Harhoff, Scherer, and Vopel 2005).

45. Bera 2009.

46. The 5–4 *Diamond v. Chakrabarty* decision allowed patents on "anything under the sun that is made by man" (Feldman, Colaianni, and Liu 2007). The first patent was granted in late 1980, the second in August 1984, and the third in April 1988.

47. Bera 2009. The inventors' estimated share is based on Stanford's current policy of sharing one-third of all royalty income with the university inventor.

48. Butkus 2007a.

49. Vilcek was born in Bratislava. He and his wife Marica left Bratislava in 1964 after being allowed, "probably by mistake," to visit Austria. He joined the faculty of the NYU Medical School in 1965. "NYU gave me a faculty position

when I came to this country. I was 31 and had no prior experience anywhere outside communist Czechoslovakia. It was a courageous thing for NYU to do. They took a risk and I think it worked out" (Kelly 2005).

50. Florida State University, Office of Research, 2010.

51. National Science Board 2010, appendix, table 5-41.

52. Data are from the 1996 Association of Technology Managers (AUTM) survey.

53. University of Chicago, Office of Technology and Intellectual Property [2007].

54. AUTM 2007 data. The figure also excludes the $700 million received by Northwestern late in the year.

55. Ninety-one percent of the licensing income reported by U.S. institutions responding to the fiscal year 2004 Association of University Technology Managers survey came from institutions having one or more licenses that yielded $1 million or more a year in revenue.

56. This is the weighted average for the "fixed" rate and the marginal rate above $1 million.

57. There is generally a close correlation between the number of licenses and the number of patents, but a patent can be associated with more than one license, and universities license nonpatented intellectual property such as software and "marked" items (for example, Gatorade, as mentioned in the text).

58. Ducor 2000.

59. The "blockbuster" inventors represent approximately 0.4 percent of the 92,000 faculty in S&E doing research (See Appendix Tables 5-15 and 5-17, National Science Board 2010). Yet approximately 13 percent of faculty reported on the 2003 Survey of Doctorate Recipients that they have been listed on a patent application in the past five years. National Science Foundation 2011b and Appendix. Survey of Doctorate Recipients.

60. Lach and Schankerman 2008. The research controls for university characteristics such as size, academic quality, and research funding. It also uses the number of patent counts for an earlier period that falls outside the window of analysis to control for endogeneity.

61. Sauermann, Cohen, and Stephan 2010.

62. Hendrick 2009.

63. The nonprofit organization Principalinvestigators.org also sees things differently: The subject heading of a July 28, 2010, e-mail was "IP & Patent Laws—Sitting on a Gold Mine." It went on to say "You could be sitting on a potential gold mine! It's right under your nose, in the form of intellectual property created by you & your lab. Don't let your invention representing millions in potential revenue sit idle simply because you aren't aware [of] IP & patent protection laws and other key aspects of moving innovation from your lab to the market."

64. Jensen and Thursby 2001. It should also be noted that many universities allocate a part of the licensing fees to help support the faculty member's lab or department.

65. Strictly speaking, one should speak of the expected utility of the sum, not the expected value.

66. Trainer 2004.

67. Lissoni et al. 2008.
68. Czarnitzki, Hussinger, and Schneider 2009.
69. Markman, Gianiodis, and Phan 2008; Thursby, Fuller, and Thursby 2009.
70. See Waltz 2006.
71. Couzin 2008.
72. Buckman 2008.
73. See Institute for Systems Biology 2010.
74. A fact that is listed on Hsu's curriculum vitae (2010).
75. Wilson 2000.
76. Ibid. See also "Inktomi Corporation," 2010, *Wikipedia,* http://en.wikipedia
    .org/wiki/Inktomi_Corporation.
77. Brewer founded the Federal Search Foundation, a 501-3(c) organization fo-
    cused on improving consumer access to government information in 2000 and
    helped create USA.gov, the official portal of the federal government, which
    was launched in September 2000. See his online biographical sketch, "Prof.
    Eric A. Brewer, Professor of Computer Science, UC Berkeley," at http://www
    .cs.berkeley.edu/~brewer/bio.html.
78. Edwards, Murray, and Yu 2006.
79. This is clearly an upper bound of the value of the portfolio for several rea-
    sons. First, in the initial days of trading, the ability of insiders to trade is re-
    stricted. Second, the market for these stocks is thin; the knowledge by the
    market of an insider making a large sale could have a significant negative
    effect. Third, in many instances the scientists must exercise an option before
    a sale can be made. In some instances, the option price is miniscule ($0.001);
    in other instances, it can be more than $10.00. It should also be noted that in
    some instances stock is not held by the scientist but instead is in trust either
    for relatives or for a nonprofit institution (Stephan and Everhart 1998).
80. The company was developing resveratrol, a substance found in grapes and in
    red wine. See "Money Matters" 2008. Glaxo suspended mid-phase 2 trial of
    SRT501 (a formulation of resveratrol) in May 2010 in patients with multiple
    myeloma after a number of patients developed a complication generally as-
    sociated with the disease, which is a type of blood cancer. See Hirschler, 2010.
81. Kaiser 2008a, 35.
82. See Hsu 2010.
83. See Levy 2000.
84. Wilson 2000.
85. Ding, Murray, and Stuart 2009.
86. Stephan and Everhart (1998) studied 52 firms that made an initial public of-
    fering in the early 1990s. They found that 67 percent of the forty-six compa-
    nies that had SABs for which the form of compensation could be determined
    had offered stock options to the members.
87. One academic director, with a strong equity position in a biotechnology firm
    that made an initial public offering, received $68,500 in consulting fees in
    one year; another received around $5,000 (ibid.).
88. Litan, Mitchell, and Reedy 2008.
89. Goldfarb and Henrekson 2003.

90. Although it is somewhat country specific, there are considerably fewer faculty start-ups in Europe. Some attribute this to an incentive system that penalizes faculty for attempting to commercialize science coming out of their research. For example, it has become quite common in the United States to grant faculty a leave of absence to start a firm, but it is considerably harder to get a leave of absence in Europe, and faculty risk losing their academic appointment. See Goldfarb and Henrekson 2003; Gittelman 2006.
91. Zucker, Darby, and Armstrong 1999.
92. See Frankson 2010.
93. Butkus 2007b.
94. Mowery et al. 2004.
95. Ibid.
96. Saxenian 1995.
97. Cohen, Nelson, and Walsh 2002.
98. Mansfield 1995. A study of 210 life science companies in 1994 found that 90 percent indicated that they used academic consultants (Blumenthal et al. 1996).
99. Mansfield 1995.
100. Agrawal and Henderson 2002, 58.
101. Markman, Gianiodis, and Phan 2008; Thursby, Fuller, and Thursby 2009.
102. Thursby, Fuller, and Thursby 2009 identify approximately 6500 patent-inventor pairs at 87 PhD-granting departments at Research I universities. They find considerable variation on patent assignments by discipline: patent-inventor pairs in engineering are far more likely to be assigned to industry (30.5 percent) than in the biological sciences (14.2 percent). The practice in the physical sciences, 28.7 percent, is much closer to that in engineering. It is also interesting to note that there is considerable variation across universities: almost 50 percent of the Stanford patent-inventor pairs are assigned to industry compared with 17 percent at Michigan and Princeton, and 33 percent at Northwestern. The university with the lowest percentage assigned to the university was University of Arizona (25 percent); that with the highest was Columbia University (88 percent).
103. Mansfield 1995.
104. Jensen and Thursby (2001) found in a survey of technology transfer offices that over 75 percent of inventions licensed were no more than a proof of concept: for 48 percent, no prototype was available; for another 29 percent, only a laboratory-scale prototype was available at the time of licensing.
105. See Figure 6-1, Chapter 6.
106. Heller and Eisenberg 1998.
107. Argyres and Liebeskind 1998; Slaughter and Rhoades 2004.
108. For the relationship between the number of patents and the number of articles, see Carayol 2007; Wuchty, Jones, and Uzzi 2007; Stephan et al. 2007. For the relationship between the number of articles and the number of patents, see Franzoni 2009; Azoulay, Ding, and Stuart 2009; Fabrizio and Di Minin 2008; Breschi, Lissoni, and Montobbio 2007.
109. Another reason for complementarity relates to the fact that instruments and materials developed in the course of doing research are sometimes patented.
110. Thursby and Thursby 2010a. There is also no evidence that faculty who place a higher weight on monetary incentives, as measured by an interest in salary,

are more likely to engage in applied research (Sauermann, Cohen, and Stephan 2010).

111. A large number of neurological disorders such as Alzeimer's, Huntington's, and Parkinson's diseases are thought to be associated with problems "in the folding process, the protein misalignments that arise and the strange protein structures that subsequently arise" (Thursby and Thursby 2010b).

112. Ibid.

113. Thursby and Thursby 2006;Thompson 2003. Universities may also have over-invested in technology-transfer efforts. The goal of developing a strong technology transfer program is much like the goal of building a strong football team. The program is expensive, and only a few universities reap sufficient rewards to even cover the cost of the TTO.

114. Krimsky et al. 1996.

115. Kaiser and Kintisch 2008.

116. Kaiser and Guterman 2008.

117. Ross et al. 2008.

118. There are, however, instances of "salary inversion," where young faculty earn more than more senior faculty who have either not been exceptionally productive or who are not highly mobile.

119. Mansfield 1995.

4. The Production of Research: People and Patterns of Collaboration

1. Giacomini 2011.

2. IceCube Project presentation made by Francis Halzen, conference at Hitotsubashi University, March 25, 2010. Also see "IceCube Neutrino Observatory," 2010, *Wikipedia,* http://en.wikipedia.org/wiki/IceCube_Neutrino_Observatory. The project involved transporting more than 1 million pounds of cargo on over fifty flights of a C-130 plane.

3. "David Quéré," 2010, *Wikipédia,* http://fr.wikipedia.org/wiki/David_Quéré.

4. See Interfaces & Co 2011.

5. See Berardelli 2010, and "Roberto Carlos, the Impossible Goal," http://www.youtube.com/watch?v=ZnXA0PoEE6Y. The final score was 1-1. Carlos scored at minute 22 and a French striker at minute 60.

6. Wang 2011.

7. Serendipity also plays a role in the production of knowledge. Although serendipity is sometimes referred to as the "happy accident," this is a misnomer. True, Pasteur "discovered" bacteria while trying to solve problems that were confronting the French wine industry. But his discovery, although unexpected, was hardly "an accident." Distinguishing between the unexpected and the "accidental" is especially difficult when research involves exploration of the unknown. The analogy to discovery makes the point: Columbus did not find what he was looking for—but the discovery of the new world was hardly an accident. (I thank Nathan Rosenberg for the analogy.)

8. Smartness was second, mentioned by 25 percent (Hermanowicz 2006).

9. *Science* (2008) 310:393.

10. *Science* (2008) 320:431.
11. Shapiro's patent was for a process related to synthetic diamonds (Dimsdale 2009).
12. Coyle 2009.
13. Simonton 2004.
14. Sauermann, Cohen, and Stephan 2010. The data come from the 2003 Survey of Doctorate Recipients. See National Science Foundation 2011b and the Appendix.
15. Long hours can, of course, also reflect a lack of administrative skill. A successful scientist who has worked in the nonprofit sector, academe, and industry once commented to me that academic science requires great administrative skills and that academic scientists who work exceptionally long hours lack such skills.
16. Freeman et al. 2001b.
17. Rockwell 2009; Kean 2006. Paul Rabinow and Martin Kenney's earlier work, which estimated that 30 to 40 percent of a faculty member's time is spent on the grant application process, is consistent with the survey's results. Rabinow 1997, 43–44; Kenney 1986, 18.
18. *Science* (2008) 320:431.
19. Harmon (1961) reports that PhD physicists have an average IQ in the neighborhood of 140. Cox, using biographical techniques to estimate the intelligence of eminent scientists, reports IQ guesstimates of 205 for Leibnitz, 185 for Galileo, and 175 for Kepler. Roe (1953, 155) summarizes Cox's findings.
20. Summers's remarks, which included the statement that the underrepresentation of women in science could be due to the "different availability of aptitude at the high end," received a considerable amount of attention in the media. The comment may have contributed to his stepping down as president of Harvard the following year. For a verbatim copy of the remarks, see Summers 2005.
21. Ceci and Williams 2009.
22. Induced pluripotent cells are adult stem cells that have the ability to grow into a variety of tissues in the same way that embryonic stem cells can. They could ultimately lead to the capacity to cure certain diseases using a patient's own cells.
23. Wolpert and Richards 1988, 107.
24. Another reason is the belief that research experience is one of the best ways to encourage undergraduate students to aspire to careers in science and engineering (see Chapter 7). Research productivity can also represent a "distinction" for scientists in settings other than research universities, distinguishing those who do not do research (the large group) from those who do (a small group) in these settings. See Fox 2010.
25. Stephan and Levin 1992.
26. There is literature suggesting that individuals coming from the margin— "outsiders," if you will—make greater contributions to science than those firmly entrenched in the system (Gieryn and Hirsch 1983). The incentives to stay current in one's field may also decrease over the career as one gets closer

to retirement and the present value of benefits from learning decrease. Other reasons for a relationship between age and productivity are explored by Stephan and Levin (1992). In studying Nobel laureates, they concluded that, although it does not take extraordinary youth to do prizewinning work, the odds decrease markedly by midcareer. There are substantial differences by field: 54.5 percent of the physicists did their prize winning work before the age of 35. The comparable figure for chemists was 43.6 percent and for those winning the prize in medicine/physiology it was 43.2 percent. Stephan and Levin 1992 and 1993.

27. Hull 1988, 514.
28. Stephan 2008.
29. Details regarding research and staffing are available for seventeen of the twenty-six via laboratory webpages. Three other faculty have webpages for their laboratories that are not fully developed. For the other six, one can find reference to the name of their laboratory when searching the Internet.
30. MIT Museum 2011.
31. Pines Lab 2009, specifically http://waugh.cchem.berkeley.edu/people.html.
32. White Research Group 2011.
33. The laboratory also has five graduate students, four undergraduate students, two research scientists, two staff scientists, and six technical associates. There are also a lab manager, two administrative assistants, one lab administrator, and one project manager. See Lindquist 2011.
34. Stephan, Black, and Chang 2007. Laboratories in other disciplines can be somewhat smaller. The Science and Engineering PhD and Postdoc Survey (SEPPS) conducted by Michael Roach and Henry Sauermann, fall 2010, found the average lab size across disciplines in S&E to be ten; the median to be eight (personal correspondence, Henry Sauermann).
35. Data come from the 2006 SDR. Relative to staff scientists, postdocs earn the most in the life sciences and the least in engineering. Calculations assume that non-tenure-track scientists who report research to be their primary activity and do not have a professorial title are staff (or research) scientists. See National Science Foundation 2011b and Data Appendix.
36. Penning 1998.
37. Mervis 1998.
38. The NIH guidelines in 2010 called for a minimum salary of $37,740 for post-docs with one or fewer years of experience, rising to $47,940 for postdocs in the fifth year. See Stanford University 2010a.
39. Tuition at Stanford University for graduate school in 2010–2011 was $12,900 per quarter for students taking 11 to 18 units (Stanford University 2010c). Note that institutions cannot always recoup all tuition costs from a funding agency. NIH, for example, pays 60 percent of tuition and fees up to $16,000 per year on training grants. See National Institutes of Health 2010. Costs of graduate stipends vary by field. See *Chronicle of Higher Education* (2009) for a 2008–2009 survey of graduate research assistant stipends in several fields. The University of Wisconsin–Madison's 2004 study of the costs for 50 percent RA appointments among "Big 10+" institutions in engineering

found that the median full cost (exclusive of indirect) was $29,000; the high was $48,000, and the low was $17,000. See Tuition Remission Task Force 2006.

40. Postdocs in the life sciences on average reported earning $41,255 a year in 2006 and working 2,643 hours a year. This results in an hourly wage rate, before fringe benefits, of about $15.60. The average wage rate (including tuition) for a research assistant who works thirty hours a week, fifty weeks a year, and attends a private university is about $31.00. That for a research assistant who attends a public institution is about $20.00.

41. Lindquist 2011.

42. In doing so, the PI assumes some risk, because if the postdoc does not receive a fellowship, the PI is implicitly obligated to support the postdoc for a period.

43. See Hill and Einaudi 2010. The count excludes postdocs in the social sciences and psychology as well as postdocs in health. It is restricted to those working in academic graduate departments.

44. Specifically, 59 percent in the life sciences, 21 percent in the physical sciences (including mathematics and computer science), and 15 percent in engineering.

45. The actual number is 94,584 (the health sciences are excluded). Data come from the *Survey of Graduate Students and Postdoctorates in Science and Engineering*. National Science Foundation 2011d. Also see Data Appendix.

46. Black and Stephan 2010. Articles were assigned to a U.S. university on the basis of the address of the last author. Internet searches were used to determine the status of all authors on papers having ten or fewer authors and of first and last authors for papers with more than ten authors. Articles are for a six-month period in 2007.

47. See Chapter 8 for a discussion of the role of the foreign born in U.S. science.

48. "When you're in the university and you're the PI, you are 'God in your realm,' she [Joan Rhodes] said (using a common formulation)" (Shapin 2008, 259).

49. Stephan and Levin 2002.

50. Davis 2005.

51. Marx 2007.

52. The only field not to have experienced an increase in the number of coauthors was marine engineering—which went from 1.25 authors to 1.22 authors (online supplementary material for Wuchty, Jones, and Uzzi 2006).

53. The "top" institutions are defined by the Institute for Scientific Information (ISI) in terms of publication counts, and they make up what are referred to as "Science Watch" institutions.

54. As is somewhat common practice when so many authors are involved, authors are listed in alphabetical order. See the Fermi LAT and Fermi GBM Collaborations (Abdo et al. 2009).

55. Growth occurred in 168 of the 172 S&E subfields studied (Jones, Wuchty, and Uzzi 2008).

56. National Science Board 2010, appendix tables 5-21 and 5-22, are computed from the 2006 SDR.

57. Carely 1998.

58. Cochrane reviews refer to review articles coming out of the Cochrane Collaboration Review Groups, which support authors in "preparing and main-

taining systematic reviews according to a common methodological framework" (Mowatt et al. 2002, 2769). In some instances, the ghost authors were editors.

59. Since 1985, the International Committee of Medical Journal Editors (2010) has published and updated the criteria.

60. Authorship on papers from the IceCube project is alphabetical, not by order of contribution. The group initially tried the latter for the first twenty authors but found it to be too difficult and time consuming to establish the order.

61. According to U.S. patent law, one should be listed as an inventor if one has contributed to the initial conception of the invention (Section 35 of U.S.C 102(f)).

62. Lissoni and Montobbio 2010.

63. Systems biology studies the relationship between the design of biological systems and the tasks they perform.

64. Levi-Montalcini 1988, 163.

65. Jones 2009.

66. Wuchty, Jones, and Uzzi 2007, 1037.

67. Jones, Wuchty, and Uzzi 2008. Work by Fox and Mohapatra (2007) finds that productivity, measured by counts of publication, is positively and significantly related to collaboration within one's department and collaboration outside one's university. Note that although teams can enhance productivity through the specialization and collective knowledge they bring to bear on a problem, they may underperform on certain tasks due to social network and coordination losses. See discussion in Jones, Wuchty, and Uzzi 2008.

68. The IT data are collected for the universe of 1,348 four-year colleges, universities, and medical schools that have not undergone substantial organizational change since 1980. See Winkler, Levin, and Stephan 2010.

69. Ding et al. 2010.

70. Agrawal and Goldfarb 2008.

71. Overbye 2007.

72. "PubChem," 2009, *Wikipedia,* http://en.wikipedia.org/wiki/PubChem#Data bases.

73. Kolbert 2007, 68.

74. National Institutes of Health 2009g.

75. National Institute of General Medical Sciences 2009b. NIGMS discontinued the Glue Grants in the fall of 2009.

76. National Institute of General Medical Sciences 2011.

77. Bole 2010.

78. European Commission 2007b, 2010. By way of contrast, the European Research Council (ERC), which was established in 2006, does not see fostering collaboration as its primary goal. Instead, it stresses economies of scale that could emerge in selecting research projects across countries.

79. By way of example, MIT and Stanford generate more than twice as many patent applications a year as does Harvard and report two or more times as many start-ups. They also receive considerably more funding from industry for research and significantly more licensing income (Lawler 2008). Harvard also committed funds to create new departments that foster collaborative

research. A case in point: the commitment of $50 million in 2007 to begin a department of developmental and regenerative biology (Mervis 2007a, 449).

80. Office of the Executive Vice President 2010. Earlier, in February 2009, President Drew Faust announced that the construction of the facility would proceed at a "slower pace" (Marshall 2009). Also see Groll and White 2010.

81. I thank Francesco Lissoni for suggesting this line of argument.

82. Some programs, such as the Medical College at the University of Pennsylvania, have relaxed this rule and now consider for promotion individuals who continue to work with their mentor. The practice of awarding bonuses to faculty receiving grants is also incentive-incompatible with the increase in multi-investigator research projects—given that the bonus is generally awarded to the PI rather than to the members of the group.

83. The scientist may, of course, still be listed as an author on an article but is increasingly unlikely to play a leading role in the research.

84. Ben Jones deserves the priority for the idea. See Jones 2010b.

5. The Production of Research: Equipment and Materials

1. Gierasch was a professor in biophysical chemistry at the University of Delaware. "As her research became increasingly biological she was attracted to a setting where her collegial interactions would offer top-notch biomedical research thrusts. Adding to this had been her continuing difficulty obtaining funds to purchase a high-field NMR instrument in a setting where her lab would be the only major user." She got an offer from Alfred Gilman, chair of the Department of Pharmacology at the University of Texas Southwestern Medical Center, who had become aware of her efforts to obtain a high-field NMR. He informed her that she would have access to the equipment she needed at UT Southwestern. "Plainly speaking," she says, "I was wooed by an NMR machine." In addition to the NMR, UT Southwestern offered a strong environment for her research (Biophysical Society 2003).

2. Vogel 2000. Per diems are for a cage holding five mice. The Institute offered the researcher a rate of $0.18 per cage.

3. *Science* (2008) 321:736, August 8.

4. Galison (2004, 46) points out that, although Switzerland's technological infrastructure came late, "when Switzerland inaugurated its rail, telegraph, and clock network, synchronized time there was a very public affair—and Bern was its center."

5. Quoted by Rosenberg 2007, 96.

6. de Solla Price 1986, 247.

7. Galison 2004. Quote is from Everdell 2003.

8. Cho and Clery 2009.

9. Lemelson-MIT Program 2003. Hood's interest in tools and cutting-edge research was instilled in him by his mentor William Dreyer, who reportedly told the then Caltech doctoral student "If you want to practice biology, do it on the leading edge and if you want to be on the leading edge, invent new tools for deciphering biological information" (Lemelson-MIT Program 2007).

10. National Science Foundation 2009d; fiscal year 2009, table 78, http://www
.nsf.gov/statistics/nsf10311/pdf/tab78.pdf. The NSF survey that collects the
information asks universities to report the portion of current fund expenditures that went for the purchase of research equipment.

11. Ibid.

12. McCray 2000. It is estimated that a night on each Gemini scope is worth
about $40,000; see "Gemini Observatory," 2011, *Wikipedia,* http://en.wiki
pedia.org/wiki/Gemini_Observatory.

13. Normile 2008. To subsidize the scientific work, the Japanese agency that
equipped the ship leases it to an oil-exploration operation.

14. The W. M. Keck Foundation funded the project, called the W. M. Keck Observatory. The observatory is managed by the University of California and
the California Institute of Technology (W. M. Keck Observatory 2009).

15. SLAC's focus has changed from studying high energy physics to understanding the properties of materials, such as protein structures, using the Linac
Coherent Light Source (LCLS), an X-ray laser that came online in April 2009.
See Cho 2006. According to the LCLS home page, the machine "produces
ultrafast pulses of X-rays millions of times brighter than even the most powerful synchrotron source" (http://lcls.slac.stanford.edu). Also see SLAC National
Accelerator Laboratory 2010.

16. See the discussion in Chapter 4.

17. The calculations include fringe benefits and are based on average twelve-
month salaries.

18. Ehrenberg, Rizzo, and Jakubson 2007.

19. It is not only a question of finding others to help share the cost of the equipment. Faculty also want to share equipment in order to preserve their own
space and to minimize responsibility for paying for personnel to operate the
equipment and for maintenance costs.

20. The cost includes some funds for operations and maintenance. In an NSF
competition, the hosting institution must also pay for utilities, which can run
into millions of dollars a year. Currently, the most powerful supercomputer at
a U.S. academic institution is the University of Tennessee's Kraken, which was
funded with a $65 million grant from the NSF. The supercomputer is housed
at the Oak Ridge National Laboratory. The location was chosen because it
had the necessary power supply, trained personnel, and appropriate space to
house the computer. NSF-funded supercomputers must allocate time to the
NSF user base.

Many supercomputers are funded either by the state or by a business–
university alliance, as in the case of Rensselaer Polytechnic Institute (RPI),
which has a partnership with IBM. NSF funds only a minority of supercomputers, although it has funded most of the most expensive supercomputers
on university campuses. See TOP500 (2010) for a list of supercomputers by
location and source of funding.

One can think of a "supercomputer" as a state-of-the-art high-performance
computer, the architecture of which can take many different forms. The definition of "high performance" also varies. One rule of thumb used by the

supercomputer community is the "top 500 list" (ibid.), which ranks high performance computers in terms of how they perform on a set of linear algebra benchmarks. This is a fairly narrow (and often unrepresentative) metric of performance, but the top ten machines on the list are considered the fastest in the world (correspondence with Fran Berman, September 8, 2009, and conversation with Fran Berman September 14, 2009). In the fall of 2010, China introduced the Tianhe-1A, which displaced the Jaguar XT5 system at Oak Ridge National Laboratory as the number one supercomputer on the "top 500 list." See Stone and Xin 2010 and Top500 2011.

Note that the term supercomputer is fairly fluid; most work that was done on a supercomputer in the 1990s can now be done on work stations costing less than $4,000. Because many problems carried out by supercomputers are suitable for parallelization—splitting the problem into smaller parts to be performed simultaneously—traditional supercomputers can be replaced by "clusters" of computers that can be programmed to act as one large computer.

Supercomputers today are most likely to be used for high-calculation-intensive tasks, such as those in quantum mechanical physics. They are also used for molecular modeling. The Anton (by D. E. Shaw Research) is an example of a supercomputer that is used for simulating molecular dynamics. The Anton currently costs approximately $13 million. See "Anton (Computer)" 2009, *Wikipedia,* http://en.wikipedia.org/wiki/Anton_(computer).

21. Advertisement in *Science* (2008) 319, March 28.
22. The Human Genome Project (HGP) was first envisioned in 1985. In 1986, the U.S. Department of Energy decided to start funding research into genome mapping and sequencing. In 1988, the National Research Council recommended the initiation of the HGP. James Watson was appointed the director of the NIH component of the effort in 1988. The actual sequencing effort began in earnest in 1990. Twenty centers in six countries (China, France, Germany, Great Britain, Japan, and the United States) contributed to the effort. Five large centers played a dominant role: the Sanger Institute in the United Kingdom, the Department of Energy's Joint Genome Institute in Walnut Creek, California, and NIH-funded centers at Baylor College of Medicine, Washington University School of Medicine, and the Whitehead Institute. See Collins, Morgan, and Patrinos 2003.
23. Because the Sanger method outperformed the Maxam and Gilbert method in terms of efficiency and also used fewer toxic chemicals and lower amounts of radioactivity, it quickly became the method of choice. See "DNA Sequencing" 2011.
24. Interview with Michael Hunkapiller (Dolan DNA Learning Center 2010).
25. Nyrén 2007.
26. Lemelson-MIT Program 2003.
27. Biotechnology Industry Organization 2011.
28. Collins, Morgan, and Patrinos 2003.
29. Jenk 2007.
30. Stephan 2010a.
31. Stephan 2010a. Not all of the increased efficiency was due to advances in sequencing technology. Improvements in library production, template prepa-

ration, and laboratory information management meant that "less human intervention was required" (Collins, Morgan, and Patrinos 2003, 289).

32. Cohen 2007.
33. Wade 2000. *Science* in February 2001 featured Mike Hunkapiller and his team at Applied Biosystems as one of the unsung heroes of the HGP for having "developed the lightning-speed PE Prism 3700 machine" ("The Human Genome. Unsung Heroes" 2001).
34. Collins, Morgan, and Patrinos 2003, 288. Applied Biosystems was called PE Biosystems for a time but in 2000 reverted to being known as Applied Biosystems.
35. Competition also played a role in accelerating the time it took to map the genome. The HGP was a public effort, funded by various governments and nonprofit organizations. But in 1998 Craig Venter and the company he helped to found, Celera, entered the race to sequence the human genome, relying on the Prism 3700 machine when it became available in 1999. When the announcement was made that a working draft of the genome had been compiled in June 2000, it was joint—issued by the HGP and Celera. When the genome was published in February 2001, it was published simultaneously by the two groups.
36. The company was a subsidiary of CuraGen, a company that Rothberg had founded earlier. Rothberg lost control of 454 in 2007 when CuraGen sold it to Roche for $140 million (Herper 2011).
37. *Science* (2009) 323:1400. Accuracy issues mean that faster does not necessarily mean cheaper. See Church 2005. Length does really matter: the longer the stretch of bases in each fragment, the easier it is to assemble a complete genome.
38. Cohen 2007.
39. "454 Life Sciences," 2011, *Wikipedia,* http://en.wikipedia.org/wiki/454_Life_Sciences.
40. Ibid.
41. Rothberg Institute for Childhood Diseases 2009.
42. Wade 2009.
43. Cohen 2007.
44. Stephen Quake, quoted by Wade 2009.
45. Illumina (2009), Genome Analyzer IIx.
46. Herper 2011.
47. Earlier in the year, scientists working at Complete Genomics, along with scientists at Harvard and Washington University, published a paper in *Science* describing their sequencing platform. See Drmanac et al. 2010.
48. Bowers 2009.
49. The machine costs $500 per run (Pollack 2011).
50. The *RS* sells for $695,000 (*The Scientist* Staff 2010).
51. J. Craig Venter Institute 2008.
52. McGraw-Herdeg 2009.
53. X Prize Foundation 2011.
54. Collins 2010a.
55. *New York Times* Editors 2010.

56. Paynter et al. 2010.
57. Berg, Tymoczko, and Stryer 2010.
58. See National Institute of General Medical Sciences 2007a, 1–2. The concern about the lack of biological relevance led NIGMS to redirect the PSI. Rather than solve any structure, the new initiative, known as "PSI: Biology," seeks to solve the structure of proteins nominated from the biological research community and considered of great biological interest (National Institute of General Medical Sciences 2009c).
59. Correspondence from Thermo Scientific.
60. See "X-Ray Crystallography," 2011, *Wikipedia*, http://en.wikipedia.org/wiki/X-ray_crystallography.
61. Work on protein structure has been rewarded by a number of Nobel Prizes. For example, Roger Kornberg won the Nobel Prize in 2006 in chemistry for solving the three-dimensional structure of RNA polymerase. Rod MacKinnon won the 2003 Nobel Prize in chemistry for publishing in 1998 "the first high-resolution structure of an ion channel, a member of the class of proteins that facilitates the transport of ions through cellular members and thus makes nerve impulses and other key biological processes possible." John Kendrew and Max Perutz shared the 1962 Nobel Prize in chemistry for being first to publish high-resolution protein structures. Aaron Klug won the Nobel Prize in chemistry in 1982 for his 1964 work showing that "the principles of structure determination by X-ray diffraction could be used to develop crystallographic electron microscopy, enabling scientists to solve quite complex structures, including those of intact viruses." See National Institute of General Medical Sciences 2009a.
62. RCSB Protein Data Bank 2009 (http://www.rcsb.org/pdb/).
63. An interesting account of the "longitude problem" is provided by Sobel 1996.
64. McCray 2000, 691.
65. More recently, NASA joined the partnership, when the second facility (Keck II) became operational.
66. McCray 2000.
67. Sloan Digital Sky Survey 2010 (http://www.sdss.org).
68. Cho and Clery 2009.
69. Bhattacharjee 2009.
70. TMT Project 2009.
71. Bhattacharjee 2009.
72. The GMT got a considerable boost when the University of Chicago pledged $50 million to the project in the summer of 2010 (Macintosh 2011).
73. Current technology limits the size of a single mirror to about 8 meters. It is anticipated that the E-ELT will fit together mirrors of approximately this dimension to attain the 42-meter diameter mirror. (The Gran Telescopio Canarias and the Southern African Large Telescope use hexagonal mirrors fitted together to make a mirror of more than 10 meters.) See European Southern Observatory 2010; "European Extremely Large Telescope," 2010, *Wikipedia*, http://en.wikipedia.org/wiki/European_Extremely_Large_Telescope.

74. Bhattacharjee 2009.
75. Center for High Angular Resolution Astronomy 2009.
76. Radio astronomy provides an example of serendipity. In the 1920s, Bell Labs asked Karl Jansky to determine the source of static on transatlantic radiotelephone service. Jansky was provided with a rotatable antenna for the work. In 1932, Jansky published a paper that reported that he had discovered three sources of noise: local thunderstorms, more distant thunderstorms, and a third source that he described as "a steady hiss static, the origin of which is not known." Jansky labeled it "star noise"; it opened the era of radio astronomy. See Rosenberg 2007.
77. "Arecibo Observatory," 2011, *Wikipedia,* http://en.wikipedia.org/wiki/Arecibo_Observatory.
78. Martin 2010.
79. Cho and Clery 2009, 334.
80. Clery 2009a; SKA 2011.
81. Koenig 2006.
82. SKA 2011. The SKA antennas would extend to New Zealand if Australia is selected; they would extend to the Indian Ocean Islands if South Africa is chosen.
83. The Herschel Space Observatory, operated by the European Space Agency, provides another example of an instrument having an extremely long gestation period. The idea for the mission first arose in a workshop in 1982. The observatory was launched twenty-seven years later, during May 2009 (Clery 2009b).
84. BLAST (Balloon-borne Large Aperture Submillimeter Telescope) is another example of a "space" telescope. In this instance, the telescope hangs from a high-altitude balloon. It is supported by a multiuniversity consortium headed by the University of Pennsylvania and the University of Toronto. BLAST's disastrous landing after its third flight highlights the fragility of such a project: the parachute failed to release, and the telescope was dragged along the surface of Antarctica for 24 hours.
85. Cho and Clery 2009, 334.
86. Because of human evolutionary history, yeast can also be used to study how certain genes work together to affect specific behaviors. In 2010, for example, by studying yeast Edward Marcotte and colleagues at the University of Texas–Austin identified five human genes that are essential for blood vessel growth. The research could prove useful in developing a drug to kill tumors by stopping the growth of blood vessels that feed the tumors (Zimmer 2010).
87. Charles Darwin reported in *The Voyage of the Beagle,* published in 1839, that he transversely cut planarian and observed regeneration. Planarian play a prominent role in Thomas Hunt Morgan's book *Regeneration,* published in 1901. Morgan eventually abandoned the study of regeneration, saying that "we will never understand the phenomena of development and regeneration" (Berrill 1983). See also Sánchez Laboratory 2010. Morgan won the Nobel Prize in medicine or physiology in 1933 for his discoveries concerning the role that chromosomes play in heredity.

88. Children's Memorial Research Center 2009; Minogue 2009. Zebrafish are also relatively inexpensive to keep. The daily charge at the University of Iowa for an entire tank of zebrafish is $0.37.
89. Critser 2007. Critser wittily points out that Clarence Little bore "no relation to Stuart." (2007, 68).
90. Ibid.
91. Murray et al. 2010.
92. Mouse innovations occurred initially in the 1980s when five teams, working independently, developed transgenic mice. See the discussion in Chapter 2.
93. Anft 2008.
94. Malakoff 2000.
95. The 80-million figure comes from Critser 2007. It includes rats, a distinct minority of all rodents used in research, in part because before 2009 knockout rats did not exist. The 20 to 30 million figure comes from Anft 2008.
96. Anft 2008.
97. Murray et al. 2010.
98. Clarence Little recognized this, describing the relation of mice and humans in terms of "the age-old enmity of [man] and the Muridae." See Critser 2007, 68, footnote.
99. Correspondence with James E. Yeadon, PhD, Technical Information Scientist, Jackson Laboratory, September 14, 2009.
100. Anft 2008.
101. Boston University Research Compliance, 2009, "Animal Care: Per Diem Rates," http://www.bu.edu/animalcare/services/per-diem-rates/.
102. Animal Research, Institutional Animal Care and Use Committee, 2009, "Per Diem Rates," Office of the Vice President for Research, University of Iowa, Iowa City. http://research.uiowa.edu/animal/?get=per_diem_rates.
103. Vogel 2000.
104. A survey of journal articles published in 2009 in which mammals were used in research showed that in five of ten fields studied, male animals were preferred over female animals; in two fields, the gender was not reported in the majority of articles; and in two other fields, a significant number of males as well as females were used. See Wald and Wu 2010.
105. Ibid.
106. Bolon et al. 2010.
107. Manufactured by APJ Trading Co., Inc., as advertised in *Science* (2006) 312, June 9.
108. VisualSonics has marketed an analog machine for approximately eight years. In 2008, it introduced a digital ultrasound, the Vevo 2100. The basic price for the Vevo 2100 is $195,000; the machine takes up to 1,000 frames per second. (Information gathered from interview with Larry McDowell of VisualSonics.)
109. Hagstrom 1965.
110. LaTour (1987) provides a detailed account of how academics use exchange to nurture their expertise.
111. Walsh, Cohen, and Cho 2007. The authors define academic researchers broadly to be those working in universities, nonprofits, and government labs.

112. There is the closely related anticommons issue of how multiple property rights claims, sometimes in the hundreds, dampen research by requiring researchers to bargain among multiple players to gain access to foundational, upstream discoveries (Heller and Eisenberg 1998). Walsh, Cohen, and Cho (2007) asked academic respondents reasons that may have dissuaded them from moving ahead with a project. Lack of funding (62 percent) or being too busy (60 percent) were the most commonly reported reasons. Scientific competition (29 percent) was also an important reason given for not pursuing a project. Technology control rights related to terms demanded for access to inputs (10 percent) and patents (3 percent) were significantly less likely to be mentioned.

113. Nelson-Rees 2001.

114. Vogel 2010.

115. Furman and Stern 2011.

116. Walsh, Cohen, and Cho 2007.

117. Murray 2010.

118. If scientists were to develop either a Cre-lox mouse or an OncoMouse under the DuPont license, they could share it with another scientist only if they complied with four terms: both parties signed the license and paid a fee, used a formal Material Transfer Agreement, committed to making annual disclosures to DuPont regarding their experimental progress, and granted DuPont reach-through rights on any follow-on commercial applications. See Murray 2010.

119. Murray 2010.

120. Furman, Murray, and Stern 2010.

121. Wenniger 2009.

122. National Science Foundation 2007d, table 4.

123. Heinig et al. 2007.

124. Adjusted by the Gross Domestic Product: Implicit Price Deflator, 2005 = 100. In nominal terms, the budget was $27.2 billion in 2003, and $30.3 billion in 2009.

125. Timmerman 2010.

126. One would not, of course, expect to see perfect competition in a market for an emerging technology that is not general purpose.

6. Funding for Research

1. The figures are projections for the fiscal year beginning in July of 2009. All figures include indirect costs. The Stanford figure excludes direct funds for the SLAC National Accelerator Laboratory. See Stanford University 2009c, p. 19, University of Virginia 2010, p. 12 and Northwestern University 2009, p. 1.

2. There is the added efficiency concern that if she did have something to sell, the efficient price would be zero because, given the nonrivalrous nature of knowledge, the marginal cost of another user is zero.

3. See Dasgupta and David 1994.

4. Costs for graduate students and postdoctoral fellows come from Pelekanos 2008.

5. Jayne Raper, a professor at New York University School of Medicine, reports that "each person in the laboratory spends $1500 per month on average [on supplies]." See Pelekanos 2008.

6. There is a quid pro quo in the patent system in the sense that the inventor is only given monopoly rights by publicly disclosing the invention in the patent document. Another way the government can stimulate research and development, albeit in the private sector, is to provide R&D tax credits to firms.

7. Williams 2010. The author finds that Celera's actions affected outcomes even after the intellectual property restrictions were removed when the genes were resequenced by the HPG.

8. Gans and Murray (2010) refer to these as the selection view and the disclosure view.

9. For a biography of Keith Pavitt see http://en.wikipedia.org/wiki/Keith_Pavitt. Pavitt's statement was recounted by Richard Nelson at the NBER conference celebrating the fiftieth anniversary of the publication of *The Rate and Direction of Inventive Activity,* held at the Aerlie Conference Center, Warrenton, Virginia, September 30–October 2, 2010.

10. It may be easier for universities to keep track of research funds that are external to the university than those that are internal. Thus, the contributions that universities themselves make to research may be undercounted.

11. Stephan and Levin 1992, 95.

12. National Science Foundation 2007b.

13. The exception is the recession of 2001. Because of the commitment to double the NIH budget, federal funds for university research continued to grow.

14. Mervis 2009a.

15. The "greedy" attitude of universities may also contribute to the decline. In an effort to increase licensing revenues, universities have become more aggressive in protecting intellectual property arising from industry-funded projects, and negotiations between firms and universities have become more difficult. "Even if we come in with the ideas and the money, we are expected to pay a licensing fee for the product of research that we already paid for," says Stanley Williams, a computer scientist at Hewlett-Packard Laboratories in Palo Alto, California. "Then we get into a negotiating dance that can take 2 years, by which time the idea is no longer viable" (Bhattacharjee 2006). See also Thompson 2003 and Thursby and Thursby 2006.

16. Pain 2008. The 25 percent comes from a survey done by Eric Campbell in 1995 and is discussed by Pain.

17. Blumenthal et al. 1986. Restrictions regarding publications are not an exclusive U.S. concern. Recent research finds a strong relationship between restrictions on publication and industrial sponsorship among German academic scientists and engineers. Specifically, 41 percent of the researchers with industry support reported a partial or complete ban on publishing compared to 7 percent of those with no funding from industry (Czarnitzki, Grimpe, and Toole 2011).

18. See Olson 1986.

19. Campbell 1997.

20. "Jonas Edward Salk, 1914–1995, American Virologist and Physician," *BookRags.com,* http://www.bookrags.com/research/jonas-edward-salk-scit-071234/.
21. BA Biology, "DNA Double Helix Discovery by Crick, Watson and Franklin," http://www.coledavid.com/dnamain.html.
22. A list of its grants, a number of which have been made to universities, can be found at the Bill and Melinda Gates Foundation site, http://www.gatesfoundation.org/grants/Pages/search.aspx.
23. The foundation was established by Lawrence Ellison, the founder of Oracle. In 2009, it paid out $41 million in grants. See "Foundation Data: Ellison Medical Foundation (Bethesda, Md.)," Chronicle of Philanthropy (website), http://philanthropy.com/premium/stats/foundation/detail.php?ID=356780.
24. Grimm 2006. Unlike most foundations, the Whitaker Foundation was not set up to last forever. Its founder, U. A. Whitaker, had a disdain for bureaucracy, and hoped the foundation would fold within 40 years of his death in 1975.
25. *Science* (2007) 318:1703, December 14.
26. Howard Hughes Medical Institute 2009c.
27. Howard Hughes Medical Institute 2009a.
28. Kaiser 2008c.
29. Howard Hughes Medical Institute 2009b. Note that the $700 million does not show up in the government's accounting because officially the faculty members who are investigators are employees of HHMI, as are their support staff. Investigators are permitted to spend 25 percent of their time in teaching, administration, or other activities that benefit the "host" institution.
30. Howard Hughes Medical Institute 2009e.
31. Howard Hughes Medical Institute 2009d.
32. Kaiser 2008d.
33. Couzin 2009.
34. The actual share that universities contribute to research declined in the early 2000s due to the tremendous increase in NIH funding. Since 2003, however, it has risen again.
35. An Internet-based search by Martha Lair Sale and R. Samuel Sale (presented at the 2009 Academic and Business Research Institute in Orlando, Florida) of the policies of thirty-one private doctoral/research universities in 2004 found the average indirect rate to be 54.4 percent. The situation is somewhat different for public universities, which typically have had lower indirect rates due partly to their reliance on state governments to build research facilities. However, in recent years, with declining state support for operating budgets of universities, public universities have paid more attention to indirect rates, and their average rate has actually increased slightly.
36. Goldman et al. 2000, 33.
37. Ibid., xii. Despite widespread complaints that indirect does not cover costs, universities continue to push faculty to bring in grants.
38. Lerner, Schoar, and Wang 2008.
39. Ehrenberg, Rizzo, and Jakubson 2007. The effects, however, are modest. Increased internal research expenditures led the student/faculty ratio to increase

during the period by about 0.5 at private institutions and by about 0.3 at publics. Tuition increased by less than 1 percent at privates in response to increased expenditures for research and by another 2 percent in response to the increased size of graduate programs. Public university students ended up paying about $50 more in tuition in response to the growth in graduate programs.

40. See Geuna 2001 and Geuna and Nesta 2006.

41. See McCook 2009.

42. See Enserink 2006.

43. National Science Board 2010, table 4-11.

44. China data come from European Commission (2007a, table 2-7) and National Science Board (2010, table 4-11). The U.S. and Japan data come from National Science Board (ibid.).

45. Grueber and Studt 2010.

46. National Science Board 2010, table 4-19.

47. Grueber and Studt 2010. U.S. data come from National Science Board 2010, chapter 4.

48. Xin and Normile 2006.

49. Wines 2011. Each cage holds a maximum of four or five animals. To put Xu's facilities in perspective, Johns Hopkins keeps 200,000 mice in ten research facilities.

50. Shi and Rao 2010.

51. Goldin and Katz 1998, 1999. Also see Rosenberg and Nelson 1994.

52. Leslie 1993, 12.

53. The program will be developed jointly by the University of Georgia (2010) and the Medical College of Georgia.

54. Center on Congress at Indiana University 2008.

55. Congressional Quarterly 2007, vol. 2: xx, 1606, 54.

56. After 44 years of serving in various capacities as an elected Republican, Senator Specter switched parties in April 2009. He ran and lost in the Democratic primary for Senate in 2010 and retired from the Senate in January 2011. Specter has battled a brain tumor twice (1993 and 1996) and was diagnosed with Hodgkin's lymphoma in 2005, which reoccurred in 2008. http://cancer.about .com/b/2008/06/01/arlen-specter.htm.

57. See National Science Board 2002, 2004, 2006, 2008, 2010.

58. Enserink 2008a.

59. Faculty in "hard-money" positions can use grant funds to buy out part or all of their teaching time and cover their summer salary. Faculty in "soft-money" positions are expected to get grants to cover part if not all of their salary.

60. The calculation is based on expenditure data for 2008 and comes from National Science Board 2010, appendix, table 5.7. Data are adjusted for the fact that not all NIH funds received by universities come through a grants mechanism.

61. National Institutes of Health 2009a.

62. See Austin 2010.

63. National Institutes of Health 2009f. Excluded from the discussion are institutes that received fewer than 500 proposals, as well as the NCCAM (National

Center for Complementary and Alternative Medicine). Success rates are for all grants. Those for R01s are generally slightly higher.

64. Ibid.

65. Scheraga, who was born in 1921, may be the oldest NIH investigator. In March 2009, he wound down another NIH grant for experimental work and in the process freed up laboratory space for a new faculty member. See Kaiser 2008b.

66. Approximately 10 percent of proposals are reviewed exclusively by mail. The use of mail-only review has declined considerably. National Science Foundation (2009c, fig. 21).

67. Ibid., 27.

68. The figure excludes proposals for centers, facilities, equipment, and instrumentation and are for the period 2001–2008 (ibid., fig. 6).

69. Ibid, 5.

70. See discussion by Freeman and Van Reenen 2009, 24.

71. Vogel 2006.

72. The process of how reviewers are chosen varies considerably by country and by agency. By way of example, applicants to NSF can specify individuals that they wish to be excluded from the possible reviewers, but they may not suggest reviewers. In the United Kingdom and in Flanders, however, individuals can propose reviewers before submitting the application and, at least informally, can contact the reviewers to ask if they will agree to review if asked.

73. Approximately 25 percent of university research funds are distributed through the RAE. See Katz and Hicks 2008; Clery 2009d; Franzoni, Scellato, and Stephan 2011.

74. De Figueiredo and Silverman 2007, 52; Mervis 2008b.

75. Robert Rosenzweig, former president of AAU. Mervis 2008b, 480.

76. Ibid.

77. De Figueiredo and Silverman 2007, 40. D-Pennsylvania means "Democrat from Pennsylvania."

78. Ibid., 43.

79. Idem.

80. Mervis 2009c.

81. Mervis 2010. The Davidson Academy is a division of the Davidson Institute.

82. Hegde and Mowery 2008.

83. Pennisi 2006. The Canadian company Archon Minerals donated the money for the prize. Craig Venter is on the board of the X Prize Foundation. 454 Life Sciences was an early entrant ("454 Life Sciences," *Wikipedia*, http://en .wikipedia.org/wiki/454_Life_Sciences.

84. Advertisement by FoundAnimals, *Science*, 7 November 2008.

85. McKinsey & Company 2009.

86. Kalil and Sturm 2010.

87. Lipowicz 2010.

88. Cameron 2010.

89. In February 2011, Dr. Seward Rutkove, a professor of neurology at Massachusetts General Hospital, won a $1 million prize from Prize4Life for "his

development of a novel tool to track the progression of the disease Amyotrophic Lateral Sclerosis (ALS)." http://www.prize4life.org/. The method developed by Dr. Rutkove can be used as a tool to screen drugs to see if they affect survival. Dr. Rutkove's work was underway before he heard of the prize and has been supported by public funding. But, according to Dr. Rutkove, the prize turned his attention to the specific task of lowering the costs of clinical trials (Venkataram, 2011).

90. X Prize Foundation 2009a. The Carnegie Mellon group is participating through a university spin-off created by faculty member Reid Whittaker, Astrobiotic Technology, Inc. (X Prize Foundation 2009b).

91. It can be challenging, for example, for young Japanese researchers to become independent, given the "monopoly full professors have on most laboratory space and a recruitment and promotion system that remains patronage-based" (Kneller 2010, 880).

92. Enserink 2008b. Elias Zerhouni, who chaired the panel making recommendations for reform of life science research in France, made this clear when he stated that although there are numerous exceptions "the large bulk [of journal articles] is published in lower-tier journals." The panel also expressed concerns regarding the amount of paperwork that French researchers must cope with, as well as problems related to the diffusion of responsibility and authority. The panel recommended setting up a unified agency to fund all of the life sciences.

93. Hicks (2009) reports that between 2002 and 2006 the number of British faculty earning more than £100,000 grew by 169 percent. The RAE will be replaced by the Research Excellence Framework (REF). The REF is currently exploring the possibility of allocating publications to institutions based on the publication address (that is, employment at time of publication) rather than employment at the time the assessment data are collected. See Imperial College London, Faculty of Medicine 2008.

94. Xin and Normile 2006.

95. Butler 2004.

96. Kean 2006, Rockwell 2009.

97. Scarpa 2010. There is also the considerable cost associated with travel. During 2010 alone, more than 19,000 scientists went to NSF headquarters in Arlington, Virginia, to take part in review panels. The NSF and other agencies cover the cost of travel and pay an honorarium, but it does not come close to covering the value of the reviewers' time. It is no wonder that in recent years NIH has begun to experiment with video conferencing for reviews (Bohannon 2011).

98. Using the average-hours data reported in Chapter 4 and salary data for senior faculty reported in Chapter 2, the hourly wage computes to be about $57.

99. National Science Foundation 2007c, vi.

100. In an effort to ease the reviewer burden, members of the NIH study sections can now serve out their twelve meetings over a six-year period rather than a four-year period. Those who go for the long haul are to be rewarded: after participating in eighteen study-section meetings, they will receive a grant extension of up to $250,000, or about nine months of funding. Scientists with three or more grants must serve as a reviewer, if asked.

101. Alberts 2009.
102. Lee 2007. Kornburg continued, "And of course, the kind of work that we would most like to see take place, which is groundbreaking and innovative, lies at the other extreme."
103. Kaiser 2008b.
104. American Academy of Arts and Sciences 2008, 27.
105. Quake 2009.
106. Young investigators have also experienced difficulties at NSF. For example, the funding rate for established investigators went from 36 percent in 2000 to 26 percent in 2006, a decrease of 28 percent. During the same period, the funding rate for new investigators, went from 22 percent to 15 percent, a decrease of 32 percent. American Academy of Arts and Sciences (2008, 14).
107. Garrison and McGuire 2008, slide 54. The figure is for PhDs receiving first-time R01 equivalent awards.
108. American Academy of Arts and Sciences 2008, 12. Ben Jones argues that one reason people are older than in the past when they get an academic position is that advances in science and the accumulation of knowledge mean that scientists must spend more time in training. See Jones 2010a.
109. Another concern regarding the peer review system is one of balance. When proposals are picked one by one, the research portfolio of an agency can be skewed toward specific topics. Daniel Goroff has drawn the analogy between this practice and the practice of choosing stocks one by one.
110. Sousa 2008.
111. Kaiser 2008f.
112. Garrison and Ngo 2010.
113. A GAO report released in September of 2009 found that 18.5 percent of all R01 grants (or 1,059) fell below the payline. Approximately half of these were awarded to new investigators. See Kaiser 2009a.
114. National Institutes of Health 2011.
115. National Institutes of Health 2008.
116. National Institutes of Health 2009c. For the number of applicants, see La Jolla Institute for Allergy and Immunology 2009.
117. National Science Board 2007.
118. National Institute of General Medical Sciences 2007b. Statistics are for the period that permitted two resubmissions. Current policy allows for only one resubmission.
119. Kaiser 2008d.
120. Sacks 2007.
121. Peota 2007.
122. Garrison and Ngo 2010. Data are for R01 equivalent awards.
123. Ibid.
124. Ibid. The figure excludes indirect.
125. Although there is not an earlier basis for comparison, Sally Rockey, the Deputy Director for Extramural Research, NIH, presented data from several sources suggesting that the overall range of salaries derived from soft-money is from 30 to 50 percent (Sally Rockey 2010). An Association of American Medical Colleges study (Goodwin et al. 2011) found that medical school

faculty with external research support received an average of 36 percent of total salary support from grants in fiscal year 2009. The proportion of salary derived from grants ranges from 14 percent to 67 percent at different medical schools. The average is 29 percent for MD and 49 percent for PhD faculty.

126. National Institutes of Health 2009b. The Consumer Price Index was calculated from U.S. Bureau of Labor Statistics 2011a.

127. Note that many federal agencies set limits on the amount of tuition that can be covered from a grant.

128. Garrison and Ngo 2010. The decline put investigators whose grants were up for renewal during the low years at a particular disadvantage. By 2009, there had been a slight increase, and the NIH had $2.4 billion for competing R01s.

129. Data come from the NIH's Research Portfolio Online Reporting Tools (Re-PORT) site (http://report.nih.gov). See the Frequently Requested Reports (http://report.nih.gov/frrs/index.aspx) by fiscal year (Research Grants).

130. Davis 2007.

131. The number of first-time investigators who received R01 (or equivalent funds) from the NIH went from 1,439 in 1998 to 1,559 in 2003 (National Institutes of Health 2009e). There was a considerable increase, however, in the number of R03 and R21 awards made to new investigators. Both are small in terms of funding. (The R03 is for $50,000 for two years; the R21 is for two years and cannot exceed $275,000 in direct costs.)

132. National Institute of General Medical Sciences 2007b.

133. See Marshall 2008.

134. Challenge grants were designed to jump-start research in fifteen specific areas designated by NIH.

135. Danielson 2009. See also Kaiser 2009b.

136. Basken 2009.

137. National Institutes of Health 2009d. NIH spent 16 percent of the ARRA funds on "GO" grants to "support high impact ideas that lend themselves to short-term funding, and may lay the foundation for new fields of investigation." They spent less than 2 percent on P30 grants to support new faculty hires.

138. Somewhat surprisingly, only 3,000 of the rejected Challenge applications reappeared in the spring of 2010 as an application for a different grant. One will have to wait to see if more appear in the next round of review.

139. The stimulus funds also posed an excessive burden on reviewers as well as NIH staff, who worked long hours during the late summer and early fall of 2009 to get the grants out.

140. To quote David Mowery and Nathan Rosenberg, "In fact, the difficulties in precisely identifying and measuring the benefits of basic research are hard to exaggerate" (1989, 11).

141. "Atomic Clock," 2010, http://en.wikipedia.org/wiki/Atomic_clock.

142. Alston et al. 2009.

143. Cutler and Kadiyala 2003. Half of the costs are attributed to the $3 billion that NIH spent on all factors related to cardiovascular disease between 1953 and 1993; the other half represents an estimate of what individuals have

spent on doctor visits over their life. Benefits are estimated to be $30,000 per person. The intent of the authors is to estimate the rate of return to investments in behavioral changes regarding cardiovascular health.

144. National Science Foundation 1968, ix.

145. Hall, Mairesse, and Mohnen 2010.

146. Mansfield 1991a. An approach that is more inclusive estimates social rates of return to publicly funded R&D using the production function approach laid out by Zvi Griliches (1979) and subsequently implemented and expanded upon by a number of economists. However, virtually without exception, such studies examine returns to federal R&D performed by firms–not to that performed by public institutions. The exception is a 1991 study by Nadiri and Mamuneas that estimates rates of return to government financed R&D regardless of the sector performing the R&D for 12 industries. The authors find social rates of return of 9.6 percent.

147. Mansfield 1991b, 26.

148. Berg 2010. The study did not control for quality of the publications and may, of course, have reached different conclusions if quality were controlled for.

149. Azoulay, Zivin, and Manso 2009.

150. Kaiser 2009c.

151. Ignatius 2007.

152. E-mail to Paula Stephan, February 24, 2009, with draft of comments for March 1 conference.

153. See Stephan and Levin 1992, 1993. Ben Jones (2010a) shows that the age at which scientists make exceptional contributions has increased over time.

154. Freeman and Van Reenen 2008. The authors also point out that research support not only produces knowledge but also contributes to the human capital of the people doing the research. This is another reason for supporting young researchers.

7. The Market for Scientists and Engineers

1. The fact that gas prices fell also helped by lowering demand.

2. Borjas and Doran use records from the American Mathematical Society to show that the unemployment rate among new doctorates in mathematics granted by U.S. institutions more than quadrupled between 1990 and 1995 while the employment rate of newly-minted PhDs at a PhD-granting institution declined by a third (See Borjas and Doran 2011, Figure 4).

3. Davis 1997, 2.

4. Ibid., 4.

5. Data are for 2003 and come from the National Survey of College Graduates. The academic count includes those working at four-year colleges and universities, medical schools, and research institutes. The count excludes the social and behavioral sciences. Only those in the labor force who are age 70 or younger are counted. See National Science Foundation 2011a and the Appendix.

6. Data come from the National Science Foundation's Survey of Earned Doctorates, which is administered to all PhDs at or near the time of graduation

and has approximately a 92 percent response rate. See National Science Foundation 2011c and the Appendix.

7. The decline among U.S. men receiving PhDs in science and engineering occurred disproportionately at less prestigious, smaller PhD-granting institutions. The increase among women occurred disproportionately at less prestigious institutions. See Freeman, Jin, and Shen 2007.

8. Ryoo and Rosen 2004, figure 4.

9. Gaglani 2009.

10. Ibid.

11. Application data come from a survey administered by the Council of Graduate Schools (2009, 14). Eighty-four percent of all doctoral institutions reported an increase in applications from U.S. citizens and permanent residents. For all doctoral institutions, the average increase was 10 percent. Enrollment data are for enrollment trends at doctoral institutions. They represent the average increase by institution, not the percentage increase for all institutions (ibid., 15).

12. For purposes of comparison over time, I restrict the analysis to men. The category "physical sciences" includes math and computer sciences.

13. See National Opinion Research Center (2008, table 10) for median number of years to doctorate award, by broad field of study, for selected years. Time to degree is measured as time since starting graduate school.

14. Average starting salary 2009 for bachelor's degrees was $49,000 (Campus Grotto 2009 from the National Association of Colleges and Employers Salary Survey). The $42,300 assumes that bachelor's starting salaries grew by 3 percent between 2004 and 2009.

15. See Lavelle 2008. The figure is for full-time programs that participated in the *Business Week* survey and is for 2006, calculated on the basis that 2006 was 9 percent less than those reported for 2008 in the article. It is an average that excludes bonuses.

16. Salary increases are computed using the Current Population Survey Outgoing Rotation Group (CPS ORG) data. See footnote 18.

17. Salary is for 2008 grown at 3 percent to 2011 and comes from 2008–2009 Faculty Salary Survey (University of Oklahoma), discussed in Chapter 3. Salary is for new hires in biology and biomedical sciences at research universities.

18. The present-value calculations assume a discount rate of 3 percent. The shape of the age-earnings profile for MBAs is drawn from CPS ORG data for years 2003–2005 and is based on the age-earnings profiles for all managers with a master's degree, aged 24 to 64 years, working full time, with reported weekly earnings on primary job in excess of $180. The top coded weekly earnings of $2,885 have been assigned an estimate of the mean earnings above the cap based on a Pareto distribution above the median. The shape of the experience-earnings profile for PhDs is based on the experience-earnings profiles of individuals working full time (in nonpostdoctoral positions) at medical schools and four-year universities and colleges in the field of the biological sciences in the 2006 Survey of Doctorate Recipients. All calculations assume that individuals retire at age 67. See National Science Foundation 2011b and the Appendix.

19. Groen and Rizzo 2007, 190.
20. The inference regarding MBAs in finance comes from Bertrand, Goldin, and Katz 2009, table 2. The inference regarding PhD pay comes from University of Oklahoma's Faculty Salary Survey discussed in Chapter 3.
21. Freeman et al. 2001a.
22. NIH provides $20,976 for a predoctoral training-grant stipend. Stanford graduate fellows receive $32,000 (Stanford University 2010b).
23. The "power" of the stipend depends upon the discount rate. Freeman, Chang, and Chiang (2009, note 2) estimate that a scientist who does ten years of study and postdoctoring will earn 29 percent of her lifetime earnings during her study and postdoctoral years. The calculation assumes a discount rate of 5 percent.
24. Ibid.
25. Chiswick, Larsen, and Pieper 2010.
26. Groen and Rizzo (2007) show that the PhD propensity for men, as measured by the number of PhDs awarded, divided by the number of individuals at risk to get the degree, increased from 6 percent in 1963 to 10 percent in 1971 and then declined to 3.2 percent. See also Bowen, Turner, and Witte (1992) for evidence concerning how draft deferment policies inflated the number of men getting PhDs during the early years of the war.
27. Jacobsen 2003; Halford 2011.
28. Phipps, Maxwell, and Rose 2009, figure 1.
29. Hoffer et al. 2011. Note that at the time of this writing data have not yet been released that permit a calculation of unemployment rates during the 2007–2009 recession.
30. Freeman 1989, 2.
31. *Nature Immunology* Editor 2006.
32. Romer 2000, 3.
33. For each field, the top-ten programs, as ranked by the National Research Council in 1995, were surveyed, as well as the five programs rated 21–25 (Stephan 2009b).
34. The webpage goes on to say "Other areas . . . include medical school, teaching, science publishing, investment banking, patent law and venture capital" but provides no specific placement information (Stephan 2009).
35. Mervis 2008a. The article follows the career outcomes of twenty-three members of the entering class of 1991 who earned PhDs. It finds that only one held a tenured faculty position in 2008.
36. National Survey of College Graduates data were used for the most recent period. Earlier data come from Stephan 2010, table 2. National Science Foundation 2011a and the Appendix.
37. Stephan et al. 2004. Data come from figure 2 and are for those who received a PhD in the United States and have been out five years or longer.
38. Baccalaureate institutions send a disproportionate share of their graduates on to get a PhD. More than half of the top-fifty U.S.-origin undergraduate institutions, measured in terms of the percent of students who go on to get a PhD, are baccalaureate colleges. Harvey Mudd heads the list among this group of institutions, followed by Reed, Swarthmore, Carleton, and Grinnell. Private

research institutions also play an important role. The California Institute of Technology heads the list in terms of the propensity of undergraduates to obtain a PhD in science and engineering; MIT, the University of Chicago, and Princeton are not far behind (Burrelli, Rapoport, and Lehming 2008).

39. The comic strip is the brainchild of Jorge Cham and was started when he was a graduate student at Stanford in response to a call from the student newspaper for a new comic strip (Coelho 2009).

40. In an effort to improve information flows, Geoff Davis (2010) has created the website (http://graduate-school.phds.org) that provides information on a number of dimensions of graduate school programs, such as the percentage of recent graduates with definite plans.

41. Richard Freeman estimates that 70 percent of the increase in the ratio of women to men getting PhDs is due to growth in the ratio of women receiving bachelor's degrees relative to men receiving bachelor's degrees. Likewise, 63 percent of the increase in the ratio of underrepresented minorities to non-minority PhDs is due to growth in the ratio of minority to non-minority bachelor's degree recipients. Source: Freeman's tabulations from data obtained from the Survey of Earned Doctorates (National Science Foundation 2011c and the Appendix) and the U.S. Department of Health, Education, and Welfare. See Stephan 2007b.

42. Stellar economic talent was drawn to the question of shortages after the launch of Sputnik. First, Blank and Stigler (1957) published a book on the demand and supply of scientific personnel; then Arrow and Capron (1959) wrote an article concerning dynamic shortages in scientific labor markets.

43. The working draft was titled "Future Scarcities of Scientists and Engineers: Problems and Solutions, Division of Policy Research and Analysis," National Science Foundation. The report was eventually published (National Science Foundation 1989).

44. NSF Director Neal Lane in testimony before the NAS Committee on Science, Engineering, and Public Policy, July 13, 1995 (Subcommittee on Basic Research 1995).

45. Quoted in Teitelbaum 2003.

46. Stephan 2008.

47. See Ryoo and Rosen 2004.

48. In response to forecast error, a National Research Council Committee was created to examine issues involved in forecasting demand and supply. The committee was chaired by Daniel McFadden, who shared the 2000 Nobel Memorial Prize in Economics the year the report was issued. The report should be mandatory reading for anyone tempted to enter the forecasting arena. The committee concluded that forecast error could occur from: (a) misspecification of models, including variables, lag structure, and error structure; (b) flawed data, or data aggregated at an inappropriate level; (c) unanticipated events. Even if model specification and lag structure are improved upon, unanticipated events continue to plague the reliability of forecasts. Both the fall of the Berlin Wall and the events of 9/11 had profound effects on scientific labor markets and would have been difficult to incorporate into any forecasting model. National Research Council 2000.

49. Teitelbaum 2003.
50. This is not to say that all universities or their administrators buy into the "shortage" idea. The 1998 National Academy of Sciences Committee on the Early Careers of Life Scientists concluded that "recent trends in employment opportunities suggest that the attractiveness to young people of careers in life-science research is declining." See National Research Council (1998) and discussion to follow. In 2002, when more recent data showed that career problems were persisting, Shirley Tilghman, the chair of the committee report, and currently the president of Princeton University, told *Science* magazine that she found the 2002 data "appalling." She went on to say that the data the committee reviewed had looked "bad," but compared with today "they actually look pretty good." See Teitelbaum 2003, 45.
51. See Freeman and Goroff 2009, appendix.
52. The report was written and released in a period of ten weeks, which may account for the numerous errors that it contains. By way of example, the report says that there were fewer physics majors in 2004 than in 1956, when in actual fact the U.S. awarded 72 percent more undergraduate physics degrees in 2004 than in 1956 ("Fact and Fiction" 2008). The first edition of the report (since then corrected) overstated by a considerable margin the number of engineers graduating each year in China and India (ibid.).
53. National Academy of Science 2007, 3.
54. The report called for stronger research and development tax credits to encourage private investment in innovation and the provision of tax incentives for U.S.-based innovation. It also recommended an increase of 10 percent each year over the next seven years in the federal investment in long-term basic research.
55. The count excludes postdocs working in industry, in government, academic departments without graduate programs, and at FFRDCs.
56. Data come from National Science Foundation 2011d. Also see the Appendix.
57. Tuition at Stanford University for enrollment in most graduate programs was approximately $13,000 a quarter in 2010. Stanford University 2010c.
58. These estimates are based on a comparison of counts from the NSF Survey of Doctorate Recipients and the NSF Survey of Graduate Students and Postdoctorates in 2001. (See National Science Foundation 2011b and 2011d and the Appendix.) For example, in 2001, when there were just under 29,500 postdocs working in the United States, 17,900 academic postdocs with temporary visas were reported through the Survey of Graduate Students and Postdoctorates, while only 3,500 postdocs with temporary visas were reported in the Survey of Earned Doctorates, which only collects data on doctorates earned in the United States. Mark Regets (2005) attributes the difference in these counts to postdocs with PhDs earned outside the United States.
59. Until the early 1980s, the United States produced the majority of PhDs worldwide in science and engineering. The U.S. dominance began to wane in the 1980s as PhD programs in Europe and Asia grew. Growth in the number of PhDs awarded has been particularly strong in Europe and Asia since the early 1990s. Both continents now surpass the United States in terms of the number of PhDs awarded (National Science Board 2004, figure 2-38). See discussion in Chapter 8.

60. Bonetta 2009.
61. Approximately 70 percent of all postdocs are supported on funds received from the federal government. Slightly fewer than 10 percent of these are supported on a fellowship rather than by a research or training grant. Data are not readily available on the division of support between fellowships and research grants for the 30 percent supported on nonfederal funds (National Science Foundation 2008, table 50).
62. According to her webpage, "Postdoctoral fellows in the laboratory generally secure independent funding through grants and fellowships" (Lindquist 2011).
63. NIH guidelines in 2010 called for a minimum salary of $37,740 for postdocs with one or fewer years of experience, rising to $47,940 for postdocs in the fifth year. (See Stanford 2010a). Some institutions pay more than this. Stanford, for example, starts at $42,645 and Whitehead starts at $49,145 (See Whitehead 2010). However, some institutions pay less than this, especially for positions in fields other than the biomedical sciences that are not covered by the NIH guidelines.
64. Lindquist 2011.
65. Ibid.
66. American Institute of Physics 2010.
67. Measures of the strength of the job market are notoriously difficult to construct. For example, information on academic job vacancies is not readily available (Ma and Stephan 2005).
68. Ibid.
69. Mervis 2008a.
70. The best postdoc position, in terms of independence, is often the first. Thereafter, the postdoc is more likely to move into a supporting research role.
71. Stanford University 2010a. The NIH guidelines for 2010 state a minimum of $52,058 for postdocs with seven or more years of experience.
72. National Postdoctoral Association 2010.
73. Benderly 2008. The union represents approximately 6,500 postdocs. Some of these work in hospitals and are not included in the counts of postdocs presented in this and other chapters.
74. Minogue 2010.
75. U.S. Bureau of Labor Statistics 2011b.
76. American Institute of Physics 2010.
77. Geoff Davis (2005) reports that 1,110 of the 2,770 respondents indicated that they were looking for a job. Among these, 72.7 percent were "very interested" in a job at a research university and 23.0 percent were "somewhat interested."
78. Puljak and Sharif 2009.
79. Fox and Stephan 2001. The National S&E PhD & Postdoc Survey (SEPPS) conducted in 2010 found that 50 percent or more of individuals in biological/life sciences, physics, and computer sciences PhD programs reported that, putting job availability aside, their most preferred career was as a faculty member doing research. Sauermann, 2011.
80. Ehrenberg and Zhang 2005.

81. Competing claims for state revenue have also come from primary and secondary education as well as from welfare programs.
82. National Association of State Universities and Land-Grant Colleges (NASULGC) Discussion Paper, 2009.
83. Data for University of Washington and Pennsylvania State University come from Ghose (2009).
84. Computed from University of Michigan 2010, which shows the total budget to be $5.067 billion and the amount received from the state to be $320 million.
85. Bunton and Mallon 2007. At 12 other institutions the financial guarantee is not clearly defined and at 3 it is "other."
86. Stephan 2008.
87. See Rilevazione Nuclei 2007 for information regarding faculty positions in Italy.
88. Schulze 2008.
89. The ratios are for the period 1996 to 2004 (ibid.).
90. The typical academic career path in Germany involves preparing the Habilitation. After completion, and pending availability of a position, one is hired into a C3 position, which must be at an institution other than where the Habilitation was prepared.
91. Ibid.
92. Kim 2007.
93. Stephan and Levin 2002.
94. This is not to say that one gets job security upon graduating. One may have to wait many years to get such a position, serving as a postdoc or teaching assistant, or even providing "free service" in an academic department. But once an individual does land such a job, it comes with job security.
95. Cruz-Castro and Sanz-Menéndez 2009.
96. University of California Newsroom 2009. The furlough policy effectively cuts salaries from 4 to 10 percent, with those in the higher salary brackets taking the largest cut. This means that most tenure-track faculty took a 10 percent cut.
97. Faculty are also civil servants in Norway and Spain. However, in Norway, some negotiation over salary occurs at the time one is hired; although faculty in the same wage class get the same raise, there is also some room for adjustment based on performance. In Spain, a review process has been in place for more than eighteen years that evaluates tenured individuals based on their performance for a "sexenio," which is accompanied by a 3 percent raise (Franzoni, Scellato, and Stephan 2011).
98. Lissoni et al. 2010.
99. The process is different for medicine, law, and engineering.
100. Lissoni et al. 2010.
101. Recent reform in France gives university presidents the power to appoint ad hoc recruitment committees with 50 percent external members (Brézin and Triller 2008).
102. Pezzoni, Sterzi, and Lissoni 2009.

103. There is another reason, not discussed here, why cohort may matter; this relates to what is occurring in scientific theory and practice when the scientist is being trained. The key to what is often referred to as the "vintage" hypothesis is that some scientists are particularly lucky in that they learn theories or techniques while in graduate school that remain relevant for an extended period of time. But other scientists are not so lucky, receiving their training in theories and techniques that rapidly fade from importance. Particularly fortunate are scientists and engineers who receive their training at the time the change is actually occurring and thus get in on the ground floor of a new approach or school of thought. Stephan and Levin (1992).

104. Black and Stephan 2004.

105. Borjas and Doran (2011, 33) show that mathematicians who wrote dissertations "on topics similar to those that interested the newly hired Soviets experienced a substantial decline in the quality of their first academic placement after 1992." The authors define a Soviet émigré mathematician to be one employed at a U.S. institution and who also published one or more papers after coming to the United States. Using this definition, there were 272 Soviet émigré mathematicians; they represented approximately 13 percent of the Soviet mathematics population and were drawn from the elite-tail of the Soviet distribution. See also note 2.

106. Carpenter 2009.

107. Ibid. The discussion focuses on individuals seeking positions in academe. But cohort effects can also be felt by those seeking positions in industry or government. Because job market conditions are tied to the overall performance of the economy, cohort effects are strongly correlated across sectors.

108. For early work on the relationship of productivity to place, see Blackburn, Behymer, and Hall 1978; Blau 1973; Long 1978; Long and McGinnis 1981; Pelz and Andrews 1976.

109. Oyer 2006. To quote Borjas and Doran (2011, 28) "It is very difficult for academics to reenter the publications market once they have taken some years off from successful active research. In academia, the short run is the long run."

110. Cohort effects have been studied by others. For example, Oreopoulos, Von Wachter, and Heisz (2008) have studied the effect of graduating during a recession. The results "point to an important role for initial job placement in determining long-term labor market success." New entrants hired during a typical recession usually start out taking a 9 percent loss compared with those whose careers start in nonrecession times. This halves within about five years and disappears after ten.

111. Merton 1968, 58. See Chapter 2.

112. National Research Council 1998.

113. Data in the report allow one to differentiate between general life sciences and the biomedical sciences. The data provided here are for the biomedical sciences.

114. All data are taken from National Research Council 1998.

115. Ibid., 8.

116. Garrison and McGuire 2008, slide 18. Training grants were established in 1974 when Congress established the National Research Service Awards

(NRSA). In the early years, the program provided over two-thirds of the support for graduate and postdoctoral training. Today it funds about 15 percent of the total number of trainees. See Committee to Study the Changing Needs for Biomedical, Behavioral, and Clinical Research Personnel (2008).

117. National Research Council 1998, 91.
118. These programs are only now beginning to be evaluated.
119. Calculations are based on the Survey of Doctorate Recipients. See National Science Foundation 2011b and the Appendix.
120. Data are for PhDs five to six years after receiving their PhD in the biomedical sciences and are calculated from the 2006 Survey of Doctorate Recipients. See National Science Foundation 2011b and the Appendix.
121. Stephan 2007a.
122. Elias Zerhuni and the NIH leadership put special emphasis on the young; the number awarded to new investigators has grown recently. See Chapter 6.
123. Data come from the NIH Office of Extramural Research (OER), and were prepared for The Association of American Medical Colleges's GREAT group (Graduate Research Education and Training) (Stephan 2007a).
124. *Nature* Editors 2007. The editorial was based on data released by the Federation of American Societies for Experimental Biology (FASEB) summarizing the career trajectories of young life scientists.
125. National Research Council 2005. The report also made several other recommendations, including a small grants program for individuals who do not have principal investigator status.
126. Twenty-three of the thirty entering-class members could be located in the fall of 2008 when Jeffrey Mervis set about interviewing them (Mervis 2008a).
127. Ibid., 1624.
128. By way of example, fewer U.S. citizens now opt to take a postdoc position, a necessary step to becoming a faculty member.
129. Levitt 2010.
130. National Research Council 2011, 3. The committee was charged with evaluating National Research Service Awards, administered by the NIH.
131. National Research Council. 2011, viii.
132. Ibid.
133. Ibid., 5.

8. The Foreign Born

1. Data are for 2008. See Figures 8.1 and 8.2.
2. The five highest, in terms of country of origin, are China (7.5 percent), India (4.9 percent), the United Kingdom (2.3 percent), the former Soviet Union (2.0 percent), and Canada (1.5 percent). Data are as of 2003; National Science Board 2010, appendix, table 3-10.
3. Graduate students generally enter on an F-1 visa, although if they come on a certain type of fellowship, usually foreign-funded (such as a Fulbright), they hold a J-1 visa (Hunt 2009, 7).
4. Postdocs are generally here on a J-1 visa, although since a policy change in 2001 universities have increasingly applied for H-1B visas for postdoctoral fellows.

5. Occasionally, students or postdocs win one of the 50,000 green cards in the U.S. Government Diversity Visa Lottery program.

6. Naturalization is the process whereby U.S. citizenship is granted to a foreign citizen or national. Usually an applicant for naturalization has already established permanent residency. See U.S. Citizenship and Immigration Services 2011. Data often differentiate between citizens who are naturalized and those who are born citizens.

7. Some databases also include information regarding the place of birth.

8. Data were calculated by Patrick Gaulé (e-mail to Paula Stephan, 2010); they are computed for professors listed in the 2007 Directory of Graduate Research of the American Chemical Society as belonging to a chemistry department in a U.S. research-intensive university, using the Carnegie classification. Of the 6,008 faculty in these departments, the country of undergraduate education can be determined for all but 626.

9. By contrast, the thirteen physicists at Stanford who received their bachelor's training abroad attended college in a wide variety of countries. Germany headed the list with three, two faculty trained in Russia and another two in the United Kingdom. The others trained in Canada, Australia, Israel, Italy, Taiwan, and China.

10. See Ding and Li 2008.

11. Ben-David 2008.

12. The H-1B is a nonimmigrant visa that allows U.S. employers to hire noncitizens on a temporary basis in occupations requiring specialized knowledge.

13. To be more specific, *foreign born* refers to permanent and temporary residents and those who indicated they had applied for citizenship by the time the doctorate was received.

14. Association of American Medical Colleges 2003. The data for medical schools are for 2000.

15. Patrick Gaulé, e-mail to Paula Stephan, 2010.

16. Data are for 2003 and come from the National Survey of College Graduates (National Science Foundation 2011d and the Appendix). The analysis includes those working at four-year colleges and universities, medical schools, and research institutes. It is restricted to those in the labor force who are age 70 or younger. It excludes the social and behavioral sciences.

17. The methodology assumes that approximately 20 percent of new faculty hired in the last ten years received their PhDs outside the United States. See Stephan 2010b.

18. This is not to say that foreign students were not a presence in U.S. graduate programs prior to the 1960s. Between 1936 and 1956, foreign students made up 19 percent of PhDs awarded by U.S. universities in engineering, 10 percent in the physical sciences, and 12 percent in the life sciences (National Academy of Sciences 1958).

19. The Act was designed to prevent political persecution of Chinese students in the aftermath of the 1989 Tiananmen Square protests. It granted permanent residency to all Chinese student nationals who arrived in the United States on or before April 11, 1990.

20. National Science Board 2010, appendix, table 2–18. For 2007, see Burns, Einaudi, and Green 2009, table 3.
21. All data come from WebCASPER (National Science Foundation, 2010c). *Foreign* is defined to include temporary as well as permanent residents. If the analysis is restricted to temporary residents, the percentages (for 2008) are as follows: engineering, 57.1 percent; math and computer science, 52.1 percent; physical sciences, 40.8 percent; and life sciences, 29.3 percent.
22. The data come from National Science Board (2008, appendix, table 2-11). They are calculated for individuals who received their PhD in 2005, and S&E includes health fields. Alternative modes of primary support not mentioned above are "personal," "teaching assistantships," "other assistantships," "traineeships," and "other." Note that although there are a large number of training grants (the NIH alone supports over 3,200 students on training grants each year) only 276 of all new PhDs reported this as their primary means of support. This reflects the fact that the duration of most training grants is one to two years and thus is not the primary means of support while in graduate school.
23. Falkenheim 2007, table 10.
24. The calculations are for degrees awarded between 2004 and 2006. A few years earlier, Berkeley had been the number-one undergraduate source institution (Mervis 2008c, 185).
25. In the early part of the twentieth century, many of China's leading scientists trained in the United States (Bound, Turner, and Walsh 2009, 81).
26. Data from National Science Foundation 2006, table S-2.
27. A large number of Iranians left Iran as a result of the fall of the Shah. A number of those who left eventually ended up in PhD programs in the United States. As a result, the percentage of U.S. PhDs awarded to Iranians increased in the 1980s to 4.8 percent, but the number of new PhD students coming from Iran declined.
28. Kim 2010. A similar phenomenon is occurring among Japanese, but in this instance among Japanese postdoctoral students. Although in the past many young Japanese used to come to the United States and Europe for postdoctoral training, today, facing a challenging job market, they stay close to home, fearing that they may not find a job upon their return. See Arai 2010, 1207.
29. The cohort that entered college in China in 1978 contributed 46.6 percent of the 11,197 PhDs awarded to Chinese students in the United States between 1985 and 1994 (Blanchard, Bound, and Turner 2008, 239).
30. This number comes from National Science Foundation 2009b, table 12.
31. Blanchard, Bound, and Turner 2008, 241.
32. Ibid., table 16.1. The percentage attending top-five programs is considerably lower: 5.3 percent of Chinese students in chemistry received a PhD from a top-five department, 8.3 percent of those in physics received a degree from a top physics department, and 6.3 percent of those in biochemistry received a degree from a top biochemistry department (Bound, Turner, and Walsh 2009, table 2.2). The data are for Chinese students who received a Ph.D. between 1991 and 2003. Only one of the PhD students from Yale's program in

molecular biophysics and biochemistry (discussed in Chapter 7) was Chinese, indicative of the strong interest among U.S. students in biochemistry programs (Mervis 2008c).

33. Blanchard, Bound, and Turner 2008, table 16.1.

34. Students from foreign baccalaureate institutions also cluster at certain U.S. institutions. According to the National Science Foundation (2009b) tabulations, Texas A&M produces the second largest number of PhDs with temporary visas in the United States. Texas A&M reports that Seoul National University is second only to Texas A&M itself in supplying doctoral candidates to A&M. Moreover, among the top fourteen institutions, only five are outside of Texas: Seoul National, National Taiwan University, Tsinghua University, University of Mombai, and Oklahoma State University. See Texas A&M University 2009, figure 17.

35. Tanyildiz 2008. Tanyildiz estimates a random utility model of the choice of PhD institution by temporary residents from the four countries. Tanyildiz finds no support that Indian and Turkish students are more likely to attend institutions with heavier concentrations of Indian and Turkish faculty.

36. Gaulé and Piacentini 2010a.

37. Faculty were determined to be native on the basis of their last name and the undergraduate institution they attended. Nationalities of faculty were determined on the basis of name, as was the nationality of students (Tanyildiz 2008).

38. The methodology for computing the rates matches social security numbers to earnings records for groups of doctoral recipients and was developed by Michael G. Finn at Oak Ridge. The latest report was published in 2010 and uses 2007 data.

39. Michael G. Finn, personal correspondence with Paula Stephan, 2010.

40. Finn 2010, table 14.

41. According to the Korean Research Foundation, 52.8 percent of recipients of PhDs from foreign countries who registered their degrees during the period from 2000 through August 2007 received their training in the United States. At prestigious South Korean universities, U.S. PhDs dominate. For example, at Seoul National University, 52.6 percent of the professors with PhDs received their training in the United States. The two other premier science and engineering universities in South Korea—Korea Advanced Institute of Science and Technology, and Pohang School of Technology—also have high proportions of U.S. PhDs. At the former, 84 percent of science professors received their doctorates in the United States, and almost three-quarters of the engineering faculty were trained in the United States. At the latter, seven-eighths of the science professorate were trained in the United States, and five-sixths of the engineering professorate were trained in the United States. See Stephan 2010b.

42. Blanchard, Bound, and Turner 2008, table 16.1.

43. It is not possible to get estimates of the number of postdoctoral fellows who have permanent visa status. See Chapter 5, note 55 for data limitations.

44. Regets 2005. The estimate that five out of ten earned their PhDs abroad is based on a comparison of counts from the NSF Survey of Doctorate Recipients

and the NSF Survey of Graduate Students and Postdoctorates in 2001 (National Science Foundation 2011b and 2011d and the Appendix). For example, in 2001, 17,900 academic postdocs with temporary visas were reported through the Survey of Graduate Students and Postdoctorates, while only 3,500 postdocs with temporary visas were reported in the Survey of Earned Doctorates, which only collects data on doctorates earned in the United States. The difference in these counts is attributed to postdocs with PhDs earned outside the United States.

45. The number of temporary-resident postdoctoral fellows in the life sciences went from 3,341 in 1985 to 11,958 in 2008. This does not include medical or "other life sciences." Data come from National Science Foundation Survey of Graduate Students and Postdoctorates (National Science Foundation 2010c and the Appendix. Also available on WebCASPAR).

46. Davis 2005, http://postdoc.sigmaxi org/results/tables/table8.

47. Stephan and Ma 2005.

48. We know little about how the support mechanism for postdocs on temporary visas differs from that of U.S. citizens. This is because the National Science Foundation Graduate Students and Postdoctorates in Science and Engineering Survey that collects data on postdocs does not collect source of support by visa status (National Science Foundation 2011d and the Appendix). What we do know, however, is that the number of postdocs supported in science and engineering on federal funds exceeds the number on temporary visas. See National Science Foundation 2008, table 50.

49. Phillips 1996.

50. Zhang 2008. The analysis controls for field fixed effects, year fixed effects, and other covariates, including the number of college graduates for the cohort. Note that a variety of crowd-out effects could occur, but Zhang only tests one type of effect. For example, the number of both foreign and domestic doctoral students could increase, but the number of U.S. doctorate recipients might increase more if there had not been an increase in the foreign born.

51. Attiyeh and Attiyeh (1997) find that graduate schools, in four out of the five fields they studied, gave preferential treatment to native applicants over foreign applicants. The four fields are biochemistry, mechanical engineering, mathematics, and economics. The one field that did not show preferential treatment was English.

52. George Borjas's research is consistent with a crowd-out effect for whites, especially white men, for programs whose institutions rank in the top half. He finds no evidence of a crowd-out effect at other institutions. His sample, however, includes all graduate programs, not just those in science and engineering. Thus, it is difficult to draw conclusions for science and engineering PhD programs from this work (Borjas 2007). An alternative explanation for the effect Borjas finds is that universities increase their enrollment of foreign graduate students because white men are pulled into other careers. This is consistent with the work of Attiyeh and Attiyeh (1997), which finds that graduate schools, in four out of the five fields they studied, gave preferential treatment to native applicants over foreign applicants.

53. Bound, Turner, and Walsh 2009, 89.

54. The estimate holds field, time, and cohort constant (Borjas 2009).
55. Ibid., 134.
56. Details are spelled out in Stephan and Levin (2007) and Levin et al. (2004). The analysis is for the period 1979 to 1997. The analysis adapts a technique originally developed in the regional science literature known as "shift-share."
57. Crowd-out effects can be significant if there is a sudden increase in the supply of foreign-born scientists but resources (both in terms of academic positions and journal space) do not increase. Borjas and Doran (2011) document that there was considerable downward mobility of U.S. mathematicians whose research closely overlapped that of the highly productive Soviet mathematicians hired by U.S. institutions after the breakup of the Soviet Union. Downward mobility is measured in terms of the rank of the academic department in which the mathematician worked.
58. The methodology uses both first and last names and thus minimizes ambiguity in assigning names with multiple ethnicities, such as Lee and Park. The methodology identifies eight ethnicities: Chinese, Indian/Hindi, Japanese, Korean, Russian, English, European, and Hispanic.
59. See the discussion in Black and Stephan 2010.
60. The sample of papers is restricted to those with fewer than ten authors that have a last author with a U.S. academic address. The sample is discussed in Chapter 4.
61. The study is for individuals scoring 700 or above on the quantitative test (*Science* Editors 2000). Foreign-born students may also cost more to educate in terms of faculty time. To the extent that this is true, one would expect departments only to admit foreign students who can contribute relatively more in terms of productivity.
62. Gaulé and Piacentini 2010b.
63. Levin and Stephan 1999.
64. Citation classics are journal articles that, according to the Institute of Scientific Information (ISI), have a "lasting effect on the whole of science." The study examined the 138 papers declared classics by the ISI during the period June 1992 to June 1993. The ISI discontinued the practice of declaring citation classics in the late 1990s. Each issue of *Science Watch,* published by the ISI in the 1980s and 1990s, contained a list of the ten most cited or "hot papers" in chemistry and physics or medicine and biology. The Levin and Stephan study chose the 251 papers declared "hot" between January 1991 and April 1993. From the list of 250 most cited authors, the study examined the 183 authors who were based in the United States.
65. In the case of the life sciences, the proportion of foreign-born authors of hot papers is not significantly different from the proportion of foreign-born life scientists in the United States in 1990.
66. The methodology determines nationality by country of undergraduate degree. The study uses listings in the Directory of Graduate Research of the American Chemical Society and examines the careers of those who appear at least once in the directory between 1993 and 2007 and were born after 1944. The study searches on Google and LinkedIn to determine location of those who cease to be listed in the directory (Gaulé and Piacentini 2010a).

67. Stephan 2010b.
68. Visiting positions, known as *jiangzuo,* or lecture chairs, were created with an eye to attracting top researchers to universities and institutes (Xin and Normile 2006).
69. National Science Foundation 2007a, figure 5.
70. The number of science bachelor degrees awarded in China doubled between 1990 and 2002; the number awarded in engineering nearly tripled. By way of contrast, during the same period the number of science bachelor degrees grew in the United States by 25 percent; the number of bachelors degrees in engineering declined by about 6 percent (National Science Foundation 2007a, table 2). Data on size of population come from table 1 of the same report.
71. See "Transcript: Obama's State of the Union Address," January 25, 2011, http://www.npr.org/2011/01/26/133224933/transcript-obamas-state-of-union-address for the text.
72. Adams et al. 2005. The top-twelve research countries are Australia, Canada, France, Germany, Israel, Italy, Japan, the Netherlands, New Zealand, Sweden, Switzerland, and the United Kingdom.
73. The United States could also make careers in science and engineering more attractive by increasing demand. One way to do this is to increase funding for public research. Another way is to stimulate demand for industrial R&D by implementing an R&D tax credit.
74. The recommendation calls for providing funds to universities in response to proposals submitted to support graduate students in key areas; 20 percent of the total fellowship funding could be used to support international students (Wendler et al. 2010).
75. Teitelbaum 2003, 52.

9. The Relationship of Science to Economic Growth

1. Measured in terms of real-world per capita GDP (DeLong 2000).
2. Mokyr 2010. Another change during the period was the emergence of a number of large industrial towns.
3. DeLong 2000.
4. U.S. Department of Labor 2009, 12, 13. The time it takes for a quantity to double can be computed by dividing 72 by the rate of growth. Thus, an economy that grows at 7.2 percent doubles every ten years.
5. World Bank website, search term "per capita income growth," http://search.worldbank.org.
6. See, for example, Jorgenson, Ho, and Stiroh 2008.
7. Rosenberg and Birdzell 1986.
8. Kuznets 1965, 9. The full Nobel citation read "for his empirically founded interpretation of economic growth which has led to new and deepened insight into the economic and social structure and process of development" (Nobel Foundation 2011).
9. Varian 2004, 805. Information sharing also expanded considerably during the period due to improved postal services, the spread of libraries and encyclopedias, and the publication of papers by learned societies (Mokyr 2010).

10. Mokyr 2010, 28.
11. Romer 2002.
12. Ibid.
13. Growth rates are the average for the ten years 2000–2009. See World Bank website, search term "economic growth," http://search.worldbank.org.
14. Firms do engage in basic research if the payoff, after accounting for competitive effects and lags, is sufficiently large. Large industrial laboratories have historically supported some basic research, although the amount of basic research performed in industry has declined in recent years. Some fundamental discoveries have been made by scientists conducting basic research at industrial laboratories—often with the goal of solving a practical problem.

    Bell Labs created a group of physicists led by William Shockley in the 1940s to conduct basic research with an eye to "solving" the vacuum tube problem. The result was the transistor—an invention that transformed the world and earned its three Bell Lab inventors—John Bardeen, Walter Brattain, and William Shockley—a Nobel Prize in 1956. Earlier, Karl Jansky had laid the foundation for radio astronomy while working at Bell Labs when he discovered that radio waves were emitted from the galaxy. The impetus for his research was Bell's interest in discovering the origins of static on long-distance communications. Bell Labs, which was financed by a "tax" on Bell's operating companies, faded when the monopoly was broken up in 1982.

    Four researchers who have worked at IBM were awarded the Nobel Prize in physics for work they did while at IBM; another IBM researcher was awarded the Nobel Prize in physics for work he did while employed by Sony. See IBM 2010.
15. Cox 2008.
16. Stokes 1997.
17. Gordon cites five great inventions that transformed the late nineteenth and early twentieth centuries: electricity; the internal combustion engine; inventions focused around the rearranging of molecules; inventions that focus on entertainment, communication, and information; and innovations in running water, indoor plumbing, and urban sanitation (Gordon 2000).
18. International Brotherhood of Boilermakers 2008. From 1891 to 1897 Purdue University kept a fully operational steam locomotive on campus for research purposes.
19. Rosenberg and Nelson 1994.
20. "Heterosis: Hybrid Corn," 2011, *Wikipedia,* http://en.wikipedia.org/wiki/Hybrid_corn.
21. The U.S. patent office denied Gordon Gould's application for a patent and awarded the patent to Bell Labs instead in 1960. It was not until 1987 that Gould won the first significant patent lawsuit related to the laser. See "Laser," 2011, *Wikipedia,* http://en.wikipedia.org/wiki/Laser. Charles Townes, the Columbia faculty member, earlier in the decade had developed the maser while working with two graduate students at Columbia.
22. Fishman 2001.
23. Resistance causes the electricity grid to lose approximately 10 percent of all electricity generated.

24. Cockburn and Henderson 1998. The enabling discovery for fourteen of the twenty-one drugs judged to have the "highest therapeutic impact" occurred in the public sector. The origins of the enabling research cannot be determined for two of the drugs.
25. Kneller 2010.
26. Edwards, Murray, and Yu 2003. Note that in many cases the biotechnology company licenses the intellectual property from the university and then within a period of months sublicenses it to a pharmaceutical company.
27. Other biotech drugs that have had a significant impact on public health include epoetin alfa (Procrit, Epogen) for the treatment of anemia, filgrastim (Neupogen), which treats a lack of white blood cells caused by cancer, and infliximab (Remicade) for the treatment of rheumatoid arthritis (ibid.).
28. Stevens et al. 2011.
29. Reduced mortality from cardiovascular disease accounts for over five of the almost nine-year increase in life expectancy in the past half century. Reduction in infant mortality is second in importance, accounting for more than an additional year. Cutler 2004a, 7–8.
30. Lichtenberg 2002.
31. Cutler 2004a, 10. Cutler estimates that the decline in smoking explains at least 10 percent of the decline. Cutler 2004b, 53.
32. Cole (2010) provides a 150-page inventory of contributions American universities have made to new products and processes in the past fifty or so years.
33. Townes 2003.
34. Griliches 1960.
35. The point is articulated well in David, Mowery, and Steinmueller 1992.
36. Foray and Lizionni 2010.
37. David, Mowery, and Steinmueller 1992, 73.
38. Rosenberg and Nelson 1994. The same year (1882) that Thomas Edison opened the Pearl Street Station in New York City, MIT introduced its first course in electrical engineering. Cornell introduced a course the next year and awarded the first doctorate in the field in 1885.
39. Jong 2006.
40. Rosenberg 2004. Jobs for students benefit universities in at least two ways. First, the growth of jobs outside academe in industry has allowed academic departments to expand their research programs through the use of graduate research assistants and postdocs. Second, the placement of students in industry enhances the relationships between universities and firms.
41. There is also a literature that examines the relationship between what can be thought of as the stock of R&D and a measure of output. This research builds on the work of Zvi Griliches. Almost invariably such research finds a positive and significant relationship between the stock of publicly funded R&D and output.
42. Adams 1990. Two measures of productivity are commonly used in industry growth studies. The first and simplest is real output per hours worked, which is called *labor productivity*. The second and more complicated measure is *total* or *multifactor productivity*, the real output per unit of input (based on an index of all inputs used).

43. Adams also investigates the impact of what he calls knowledge spillover stocks by seeing how the stock of research not directly relevant to a specific industry affects the industry. He finds spillover knowledge measured in this way to account for 25 percent of industry total factor productivity growth—but the lag is on the order of 30 years.
44. Adams, Clemmons, and Stephan 2006.
45. Branstetter and Yoshiaki 2005.
46. National Science Board 2004, appendix, figure 5-45.
47. The study surveyed 3,240 labs and received 1,478 responses. The discussion here is based on the 1,267 cases of firms whose focus was in manufacturing and were not foreign owned (Cohen, Nelson, and Walsh 2002).
48. The research of Fleming and Sorenson (2004) suggests that science is most helpful when inventors face the difficult task of trying to combine "tightly coupled components"; science plays a smaller role when inventors seek to combine independent components.
49. Mansfield 1991a, 1992. Mansfield did a follow-up study, gathering similar data for a sample of seventy-seven firms concerning the contribution of academic research to new products and processes introduced from 1986 to 1994. The evidence of the follow-up study is consistent with that of the earlier study: in the absence of recent academic research, approximately 10 percent of new products and processes could not have been developed without substantial delay (Mansfield 1998).
50. Mansfield 1995.
51. Agrawal and Henderson 2002, 58.
52. Foray and Lissoni 2010, 292.
53. Jaffe 1989b.
54. See, for example, Acs, Audretsch, and Feldman 1992; Black 2004; Autant-Bernard 2001.
55. The approach is not limited to examining the relationship between innovation and university research but often includes a measure of private R&D expenditures in the geographic area to determine the extent to which spillovers occur within the private sector as well.
56. Stanford University 2009b.
57. Stanford University 2009a.
58. MIT News 1997.
59. Zucker, Darby, and Brewer 1998; Zucker, Darby, and Armstrong 1999.
60. The research classifies a patent as important if the patent has been granted in at least two of three major world markets: Japan, the United States, and Europe (Cockburn and Henderson 1998).
61. Deng, Lev, and Narin 1999.
62. Cohen, Nelson, and Walsh 2002.
63. National Science Board 2010, appendix, table 5-46.
64. Association of University Technology Managers (AUTM) 2004 and 2007 data.
65. Mansfield 1995.
66. Adams 2001, table 5.

67. Ibid., 266.
68. Ibid., table 3.
69. Cohen and Leventhal 1989.
70. Sauermann and Stephan 2010.
71. Data are taken from appendix, table 5-42, National Science Board 2010. Fractional counts allocate articles with collaborators from multiple sectors on a proportional basis according to contribution. Statistics exclude articles in the social sciences and psychology.
72. Data are for 2005 and come from table 6-29 and 6-30 of National Science Board 2008.
73. The development of absorptive capacity and connectedness are not the only reasons why firms participate in open science, allowing and, in some instances, encouraging scientists and engineers to publish. Foremost among these other reasons is the recruitment of talent. Scientists and engineers working in industry value the ability to publish and are willing to pay for the privilege. Firms that allow new hires who recently completed a postdoctoral position in biology to participate in the norms of science by publishing pay on average 25 percent less than firms that do not allow new hires to publish (Stern 2004).

    It is not only an interest in priority: the ability to publish allows scientists to maintain the option to work outside the for-profit sector. The reputation of the laboratory, which is directly related to publication activity, also affects the ability of the company to hire scientists and engineers (Scherer 1967). It may also affect its ability to attract government contracts (Lichtenberg 1988).

    A number of other factors lead companies to opt for disclosure through publication. A critical element is the company's ability to screen the material that is published, thereby ensuring that its proprietary interests are maintained (Hicks 1995).
74. Data are for 2003 and come from the National Survey of College Graduates. The count excludes the social and behavioral sciences. Only those in the labor force who are age 70 or younger are counted. See National Science Foundation 2011a and the Appendix.
75. Ibid., with the same applicable restrictions.
76. Ibid. The category "life sciences" includes biological, agricultural, and environmental life sciences.
77. Despite the fact that industrial scientists report being less satisfied with the amount of independence that they have at work than do academic scientists, over 50 percent of industrial scientists indicate that they are "very satisfied" with their level of independence. Research scientists in industry earn on average about 30 percent more than their research-active colleagues in academe (Sauermann and Stephan 2010).
78. Lohr 2006, C1, C4. Expenditures on such functions represent an investment in what could be termed "intangible capital." Corrado and Hulten (2010) argue that these innovation-related expenditures on intangibles should be included in GDP as business investment.
79. Oppenheimer, as quoted in *Time* Staff (1948, 81).

80. Alberts 2008.
81. The survey is officially known as the Survey of Earned Doctorates. See National Science Foundation 2011c and the Appendix.
82. Stephan 2007c. The data have not been coded for other years.
83. Approximately 29 percent of postdoctoral fellows supported by the National Institute of General Medical Sciences in 1992–1994 were working in industry in 2010 (Levitt 2010).
84. The heavy representation of midwestern institutions reflects in part the fact that the Midwest produces a large number of engineering PhDs who, compared with those in the biomedical sciences, rarely take postdoctoral positions in academe before working in industry.
85. The percentage would be somewhat larger if the data permitted following the employment patterns of those who first take a postdoctoral position and eventually end up working for a large pharmaceutical company.
86. The percentage increases to 44 percent when those working in the United States at a top-200 foreign-owned R&D firm (or subsidiary) are included. It increases by an additional 5 percent when those working at firms ranked 201–500 in terms of R&D expenditures are added. The study concludes that a large number of newly minted PhDs—indeed just over 50 percent—work in firms that spend a relatively small amount on R&D activities. The relatively low percentage going to work at high-intensive R&D firms may also reflect that some PhDs, who become frustrated with research while in graduate school, seek alternative types of employment.
87. Taken together, the five cities employ approximately 18.4 percent of those going directly to work in industry upon graduation.
88. Thompson 2003,9.
89. Fitzgerald 2008, 563; FDA 2010.
90. Blume-Kohout 2009, 29.
91. Harris 2011, A21.
92. The Howard Hughes Medical Institute has funded a few programs in recent years that seek to bridge basic science PhD programs with course work on human physiology. One, at Stanford University, provides PhD students with a year's worth of classes in medicine for which they receive a Master's degree.
93. To return to the terminology of the economic historian Joel Mokyr (2010), prescriptive knowledge (technology) informs propositional knowledge (science), and propositional knowledge informs prescriptive knowledge.
94. Although it is impossible to get an exact accounting, estimates find the private return to R&D to be positive and somewhat higher than the returns to ordinary capital (Hall, Mairesse, and Mohnen 2010, esp. 1034).
95. Saxenian 1995. Tom Wolfe (1983) described the Wagon Wheel in a 1983 article in *Esquire* Magazine: "Every year there was some place, the Wagon Wheel, Chez Yvonne, Rickey's, the Roundhouse, where members of this esoteric fraternity, the young men and women of the semiconductor industry, would head after work to have a drink and gossip and brag and trade war stories about contacts, burst modes, bubble memories, pulse trains, bounceless modes, slow-death episodes, RAMs, NAKs, MOSes, PCMs, PROMs, PROM

blowers, PROM blasters, and teramagnitudes, meaning multiples of a million millions."

96. The mobility of researchers between firms is a major mechanism by which knowledge spills over among Italian firms (Breschi and Lissoni 2003). Almeida and Kogut (1999) find high interfirm mobility among patent holders in the semiconductor industry.

97. Reid 1985, 65.

98. Jaffe 1989b; Acs, Audretsch, and Feldman 1992; Black 2004; Autant-Bernard 2001. In an earlier work, Jaffe constructed a "spillover pool," defined as the sum of all other firms' R&D weighted by a measure of relatedness; he found that the size of the pool had a strong positive effect on a firm's patents, R&D, and total factor productivity (Jaffe 1986, 1989a).

99. Patent citations convey information about the source and location of knowledge embodied in the patent (Jaffe, Trajtenberg, and Henderson 1993).

100. Ibid.

101. Increasing returns to scale means that if all inputs were to increase by a factor of x, output would increase by a factor of more than x.

102. Romer 1990 and 1994.

103. National Science Board 2010, appendix, table 4-3.

10. Can We Do Better?

1. Alberts 2010, 1257.

2. Lee 2007. Kornburg continued by saying "And of course, the kind of work that we would most like to see take place, which is groundbreaking and innovative, lies at the other extreme."

3. Quake 2009.

4. Arrow 1959.

5. Carmichael and Begley 2010.

6. Cummings and Kiesler (2005) find multi-university research projects to be less successful than projects that take place entirely within one university.

7. National Research Council 2011.

8. Ibid.

9. When Sauermann and Roach asked graduate students and postdocs "How would you rate your research ability relative to your peers in your specific field of study?" the average rating on a scale of 0 to 10 was 6.48. Although this could reflect the composition of the sample, drawn from 39 institutions with large programs, it is highly likely that it also reflects the tendency of individuals in graduate school to think that they are better than average (Sauermann and Roach, 2011).

10. Vance, 2011, 44.

11. Ibid.

12. One member was from industry, and one member was from the American Medical Information Association. All other members came from academe. One of these members was a postdoc at the time he was appointed to the committee. He became the manager of the postdoctoral program and ethics

program coordinator at New York University's School of Medicine during the time the report was being written.

13. Alberts 2010, 1257.

14. An alternative possibility, also proposed by Alberts, is to place a maximum on the amount of money that the NIH will contribute to the salary of a faculty member. Alberts also suggested introducing an overhead cost penalty in proportion to the number of soft-money positions an institution has (Alberts 2010, 1257).

15. A joint University-Fermilab PhD program does exist. The program was started in 1985. But PhD production is minimal: To date, 36 individuals have graduated with a joint PhD. See http://apc.fnal.gov/programs2/joint_university.shtml.

16. There are other ways to accomplish this goal. The NSF IGERT (Integrative Graduate Education and Research Traineeship) program, for example, which supports interdisciplinary training, is designed precisely to decouple students from advisors in particular disciplines. See http://www.igert.org/public/about for a description.

17. Several eminent scientists support such an idea. For example, Roald Hoffmann, a 1981 Nobel laureate in chemistry, proposed that the government stop supporting graduate students on research grants and use the money for competitive fellowships that students could use at the university of their choice. The proposal was made in a May 8, 2009 editorial in the *Chronicle of Higher Education;* it was elaborated upon in an interview with Jeffrey Mervis (2009b). Shirley Tilghman, president of Princeton University and a highly respected geneticist, also supports the idea. The 1998 National Academy of Sciences study that she chaired also recommended a substitution of training grants for graduate research assistantships (see Chapter 7). Thomas Cech, a Nobel laureate in chemistry and former president of the Howard Hughes Medical Institute, is supportive of such a move as well.

18. Mervis 2009b, 529.

19. Policies can also increase stratification in science, making research for those at the margin more difficult. A case in point was the U.S. Administration human embryonic stem cell (hESC) research policy implemented in 2001 under Present Bush, which restricted publicly funded research to stem cell lines already in existence. Using a methodology similar to that employed for studying the impact of lifted restrictions on the use of certain mice as well as that used to study BRC's, researchers have studied how hESC affected research practices in the United States. Not surprisingly, the policy was found to have a significantly more chilling impact on researchers working at non-top-25 research institutions. Furman and Murray 2011.

20. Ben Jones deserves the priority for the idea. See Jones 2010b.

21. Funding First 2000. The organization was an initiative of the Mary Woodward Lasker Charitable Trust.

22. For beer see http://www.wallstats.com/blog/50-billion-bottles-of-beer-on-the-wall/. The calculation assumes that one-third of beer is drunk in restaurants or bars; that two-thirds is drunk at home and that the average price of a pint

consumed in the United States is $1.88. For defense see http://comptroller.
defense.gov/defbudget/fy2012/FY2012_Budget_Request_Overview_Book.
pdf.

23. Arrow 1955.
24. Acemoglu 2009. The John Bates Clark Medal is awarded by the American
    Economics Association to the "American economist under the age of forty
    who is adjudged to have made a significant contribution to economic thought
    and knowledge." It was awarded biennially until 2007. Since then it has been
    awarded annually. "John Bates Clark Medal," 2010, *Wikipedia,* http://en.
    wikipedia.org/wiki/John_Bates_Clark_Medal.
25. Azoulay, Zivin, and Manso 2009.
26. Berg 2010.
27. Sacks 2007.
28. Cummings and Kiesler 2005.
29. National Science Foundation 2011. http://www.nsf.gov/funding/pgm_summ
    .jsp?pims_id=501084.
30. The financing for the institute comes from Stephen and Connie Lieber, who
    have a daughter with schizophrenia.
31. Collins 2010b, 37
32. Kaiser 2011.
33. Collins 2010b.

References

Abdo, Aous A., M. Ackermann, M. Arimoto, K. Asano, W. B. Atwood, M. Axelsson, L. Baldini, et al. 2009. "Fermi Observations of High-Energy Gamma-Ray Emissions from GRB 080916C." *Science* 323:1688–93.

Acemoglu, Daron. 2009. "A Note on Diversity and Technological Progress." Unpublished manuscript, Massachusetts Institute of Technology, July 2009. http://www.idei.fr/tnit/papers/acemoglu1.pdf.Kaiser.

Acs, Zoltan, David Audretsch, and Maryann Feldman. 1992. "Real Effects of Academic Research: Comment." *American Economic Review* 83:363–67.

Adams, James D. 1990. "Fundamental Stocks of Knowledge and Productivity Growth." *Journal of Political Economy* 98:673–702.

———. 2001. "Comparative Localization of Academic and Industrial Spillovers." *Journal of Economic Geography* 2:253–78.

Adams, James D., Grant Black, Roger Clemmons, and Paula Stephan. 2005. "Scientific Teams and Institutional Collaborations: Evidence from U.S. Universities, 1981–1999." *Research Policy* 34:259–85.

Adams, James D., J. Roger Clemmons, and Paula E. Stephan. 2006. "How Rapidly Does Science Leak Out?" NBER Working Paper 11997. National Bureau of Economic Research, Cambridge, MA.

Agin, Dan. 2007. *Junk Science: An Overdue Indictment of Government, Industry, and Faith Groups That Twist Science for Their Own Gain.* New York: Macmillan.

Agrawal, Ajay, and Avi Goldfarb. 2008. "Restructuring Research: Communication Costs and the Democratization of University Innovation." *American Economic Review* 98:1578–90.

Agrawal, Ajay, and Rebecca Henderson. 2002. "Putting Patents in Context: Exploring Knowledge Transfer from MIT." *Management Science* 48:44–60.

Agre, Peter. 2003. "Autobiography." Nobelprize.org (website). http://nobelprize.org/nobel_prizes/chemistry/laureates/2003/agre-autobio.html.

Ainsworth, Claire. 2008. "Stretching the Imagination." *Nature* 456:696–99.

Alberts, Bruce. 2008. "Hybrid Vigor in Science." *Science* 320:155.

———. 2009. "On Incentives for Innovation." *Science* 326:1163.

———. 2010. "Overbuilding Research Capacity." *Science* 329:1257.

Allison, Paul, and J. Scott Long. 1990. "Departmental Effects on Scientific Productivity." *American Sociological Review* 55:469–78.

Allison, Paul, Scott Long, and Tad Krauze. 1982. "Cumulative Advantage and Inequality in Science." *American Sociological Review* 47:615—25.

Allison, Paul, and John Stewart. 1974. "Productivity Differences among Scientists: Evidence for Accumulative Advantage." *American Sociological Review* 39:596–606.

ALLWHOIS. http://www.allwhois.com/.

Almeida, Paul, and Bruce Kogut. 1999. "Localization of Knowledge and the Mobility of Engineers in Regional Networks." *Management Science* 45:905–17.

Alonso, S., F. J. Cabrerizo, E. Herrera-Viedma, and F. Herrera. 2009. "h-Index: A Review Focused in Its Variants, Computation and Standardization for Different Scientific Fields." *Journal of Informetrics* 3:273–89.

Alston, Julian M., Matthew Andersen, Jennifer S. James, and Philip G. Pardey. 2009. *Persistence Pays: U.S. Agricultural Productivity Growth and the Benefits from Public R&D Spending.* New York: Springer.

American Academy of Arts and Sciences. 2008. *ARISE: Advancing Research in Science and Engineering: Investing in Early-Career Scientists and High-Risk, High-Reward Research.* Cambridge, MA: American Academy of Arts and Sciences. http://www.amacad.org/AriseFolder/default.aspx.

American Association of University Professors. 2009. *Facts and Figures: AAUP Faculty Salary Survey 2008–2009.* http://chronicle.com/stats/aaup/.

———. 2010. *No Refuge: The Annual Report on the Economic Status of the Profession 2009–2010.* Washington, DC: American Association of University Professors.

American Institute of Physics. 2010. "Table 6. Long-term Career Goals of Physics PhDs, classes of 2005 & 2006." Initial Employment Report, AIP Statistical Research Center. http://www.aip.org/statistics/trends/highlite/emp3/table6.htm.

Anft, Michael. 2008. "Of Mice and Medicine." *Johns Hopkins Magazine* 60:31–7.

"Anton (Computer)." 2009. *Wikipedia.* http://en.wikipedia.org/wiki/Anton_(computer).

Arai, K. 2010. "Japanese Science in a Global World." *Science* 328:1207.

"Arecibo Observatory." 2011. *Wikipedia.* http://en.wikipedia.org/wiki/Arecibo_Observatory.

Argyres, Nicholas, and Julia Liebeskind. 1998. "Privatizing the Intellectual Commons: Universities and the Commercialization of Biotechnology." *Journal of Economic Behavior and Organization* 35:427–54.

Arrow, Kenneth. 1955. *Economic Aspects of Military Research and Development.* Santa Monica, CA: RAND Corporation.

———. 1959. *Economic Welfare and the Allocation of Resources for Invention.* P1856-RC. Santa Monica, CA: RAND Corporation. Also published in *The Rate and Direction of Inventive Activity: Economic and Social Factors,* 609–26. National Bureau of Economic Research. New York: Arno Press, 1975 (repr. 1962).

———. 1987. "Reflections on the Essays." In *Arrow and the Ascent of Modern Economic Theory,* 685–9. Edited by George R. Feiwel. New York: New York University Press.

Arrow, Kenneth J., and W. M. Capron. 1959. "Dynamic Shortages and Price Rises: The Engineering-Scientist Case." *Quarterly Journal of Economics* 73:292–308.

Association of American Medical Colleges. 2003. "Trends among Foreign-Graduate Faculty at U.S. Medical Schools, 1981–2000." http://www.aamc.org/data/aib/aibissues/aibvol3_no1.pdf.

———. 2011. "Sponsored Program Salary Support to Medical School Faculty in 2009." In Brief. https://www.aamc.org/download/170836/data/aibvol11_no1.pdf.

Association of University Technology Managers. 1996. *FY 1996 Licensing Activity Survey.* Deerfield, IL: AUTM.

"Atomic Clock." 2010. *Wikipedia.* http://en.wikipedia.org/wiki/Atomic_clock.

Attiyeh, Gregory, and Richard Attiyeh. 1997. "Testing for Bias in Graduate School Admission." *Journal of Human Resources* 32:29–97.

Austin, James. 2010. "NIH Impact Scores: Which Criteria Matter Most?" *Science Careers Blog,* July 22. http://blogs.sciencemag.org/sciencecareers/2010/07/nih-impact-scor.html.

Autant-Bernard, Corinne. 2001. "Science and Knowledge Flows: The French Case." *Research Policy* 30:1069–78.

Azoulay, Pierre, Waverly Ding, and Toby Stuart. 2009. "The Impact of Academic Patenting on the Rate, Quality, and Direction of (Public) Research Output." *Journal of Industrial Economics* 57:637–76.

Azoulay, Pierre, Joshua Graff Zivin, and Gustavo Manso. 2009. "Incentives and Creativity: Evidence from the Academic Life Sciences." NBER Working Paper 15466. National Bureau of Economic Research, Cambridge, MA.

BankBoston Economics Department. 1997. *MIT: The Impact of Innovation.* Boston: BankBoston. http://web.mit.edu/newsoffice/founders/.

Basken, Paul. 2009. "NIH Is Deluged with 21,000 Grant Applications for Stimulus Funds." *Chronicle of Higher Education,* June 9.

———. 2010. "Lawmakers Renew Commitment to Science Spending, Despite Budget-Deficit Fears." *Chronicle of Higher Education,* April 29.

Ben-David, Dan. 2008. "Brain Drained: Soaring Minds." *Vox,* March 13.

Benderly, Beryl Lieff. 2008. "University of California Postdoc Union Wins Official Recognition." *Science Careers,* August 28.

"Benford's Law." 2010. *Wikipedia.* http://en.wikipedia.org/wiki/Benford's_law.

Bera, Rajendra K. 2009. "The Story of the Cohen–Boyer Patents." *Current Science* 96:760–3.

Berardelli, Phil. 2010. "'Impossible' Soccer Goal Explained by New Twist on Curveball Physics." *Science Now,* September 2. http://news.sciencemag.org/science now/2010/09/impossible-soccer-goal-explained.html.

Berg, Jeremy. 2010. "Another Look at Measuring the Scientific Output and Impact of NIGMS Grants." *NIGMS Feedback Loop,* November 22. https://loop.nigms .nih.gov/index.php/2010/11/22/another-look-at-measuring-the-scientific-out put-and-impact-of-nigms-grants/.

Berg, Jeremy, John L. Tymoczko, and Lubert Stryer. 2010. *Biochemistry.* 6th ed. New York: W. H. Freeman.

Berrill, Norman J. 1983. "The Pleasure and Practice of Biology." *Canadian Journal of Zoology* 61:947–51.

Bertrand, Marianne, Claudia Goldin, and Lawrence Katz. 2009. "Dynamics of the Gender Gap for Young Professionals in the Corporate and Financial Sectors." NBER Working Paper 14681. National Bureau of Economic Research, Cambridge, MA.

Bhattacharjee, Yudhijit. 2006. "U.S. Research Funding. Industry Shrinks Academic Support." *Science* 312:671a.

———. 2008a. "Combating Terrorism. New Efforts to Detect Explosives Require Advances on Many Fronts." *Science* 320:1416–7.

———. 2008b. "Scientific Honors. The Cost of a Genuine Collaboration." *Science* 320:959.

———. 2009. "Race for the Heavens." *Science* 326:512–15.

Bill and Melinda Gates Foundation. 2009. "Grant Search." Bill and Melinda Gates Foundation (website). http://www.gatesfoundation.org/grants/Pages/search.aspx.

Biophysical Society. 2003. "Biophysicist in Profile: Lila Gierasch." *Biophysical Society Newsletter,* January/February. http://www.biophysics.org/LinkClick. aspx?fileticket=fM0uqLnEvsw%3D&tabid=524.

Biotechnology Industry Organization. 2011. "Russ Prize Winner: Leroy Hood Revolutionized DNA Research." *BioTechNow,* January 24. http://biotech-now .org/section/bio-matters/2011/01/24/russ-prize-winner-leroy-hood-revolu tionized-dna-research.

Black, Grant. 2004. *The Geography of Small Firm Innovation.* New York: Kluwer.

Black, Grant, and Paula Stephan. 2004. *Bioinformatics: Recent Trends in Programs, Placements and Job Opportunities Final Report.* New York: Alfred P. Sloan Foundation.

———. 2010. "The Economics of University Science and the Role of Foreign Graduate Students and Postdoctoral Scholars." In *American Universities in a Global Market,* 129–61. Edited by Charles T. Clotfelter. Chicago: University of Chicago Press.

Blackburn, Robert T., Charles E. Behymer, and David E. Hall. 1978. "Research Note: Correlates of Faculty Publications." *Sociology of Education* 51:132–41.

Blanchard, Emily, John Bound, and Sarah Turner. 2008. "Opening (and Closing) Doors: Country Specific Shocks in U.S. Doctorate Education." In *Doctoral Education and the Faculty of the Future,* 224–8. Edited by Ronald G. Ehrenberg and Charlotte V. Kuh. Ithaca, NY: Cornell University Press.

Blank, David, and George J. Stigler. 1957. *The Demand and Supply of Scientific Personnel*. New York: National Bureau of Economic Research.

Blau, Judith R. 1973. "Sociometric Structure of a Scientific Discipline." In *Research in Sociology of Knowledge, Sciences and Art*, 91–206. Edited by Robert A. Jones. Greenwich, CT: JAI Press.

Blume-Kohout, Margaret E. 2009. "Drug Development and Public Research Funding: Evidence of Lagged Effects." Unpublished paper. University of Waterloo, Canada. http://sites.google.com/site/mblumekohout/documents/Blume-Kohout_Paper.pdf.

Blumenthal, David, Nancyanne Causino, Eric Campbell, and Karen Seashore Louis. 1996. "Relationships between Academic Institutions and Industry in the Life Sciences: An Industry Survey." *New England Journal of Medicine* 334:368–74.

Blumenthal, David, Michael Gluck, Karen Seashore Lewis, Michael Stotto, and David Wise. 1986. "University-Industry Research Relationships in Biotechnology: Implications for the University." *Science* 232:1361–66.

Bohannon, John. 2011. "National Science Foundation. Meeting for Peer Review at a Resort That's Virtually Free." *Science* 331:27.

Bok, Derek C. 1982. *Beyond the Ivory Tower: Social Responsibilities of the Modern University*. Cambridge, MA: Harvard University Press.

Bole, Kristen. 2010. "UCSF Receives $15 Million to Advance Personalized Medicine." *UCSF News Center*. University of San Francisco, CA (website). http://www.ucsf.edu/news/2010/09/4451/ucsf-receives-15-million-advance-personalized-medicine.

Bolon, Brad, Stephen W. Barthold, Kelli L. Boyd, Cory Brayton, Robert D. Cardiff, Linda C. Cork, Kathryn A. Easton, Trenton R. Schoeb, John P. Sundberg, and Jerrold M. Ward. 2010. "Letter to the Editor. Male Mice Not Alone in Research." *Science* 328:1103.

Bonetta, Laura. 2009. "Advice for Beginning Faculty: How to Find the Best Postdoc" *Science Careers*, February 6.

Borjas, George. 2007. "Do Foreign Students Crowd Out Native Students from Graduate Programs?" In *Science and the University*, 134–49. Edited by Paula Stephan and Ronald Ehrenberg. Madison: University of Wisconsin Press.

———. 2009. "Immigration in High Skilled Labor Markets: The Impact of Foreign Students on the Earnings of Doctorates." In *Science and Engineering Careers in the United States: An Analysis of Markets and Employment*, 131–62. Edited by Richard Freeman and Daniel Goroff. Chicago: University of Chicago Press.

Borjas, George, and Kirk Doran. 2011. "The Collapse of the Soviet Union and the Productivity of American Mathematicians." Unpublished paper, Harvard University.

Bound, John, Sarah Turner, and Patrick Walsh. 2009. "Internationalization of U.S. Doctorate Education." In *Science and Engineering Careers in the United States: An Analysis of Markets and Employment*, 59–97. Edited by Richard Freeman and Daniel Goroff. Chicago: University of Chicago Press.

Bowen, William G., and Julie Ann Sosa. 1989. *Prospects for Faculty in the Arts and Sciences: A Study of Factors Affecting Demand and Supply, 1987–2012.* Princeton, NJ: Princeton University Press.

Bowen, William G., Sarah Turner, and Marcia Witte. 1992. "The BA-PhD Nexus." *Journal of Higher Education* 63:65–86.

Bowers, Keith. 2009. "Biotech Firm Complete Genomics Takes the Lead in Genome Sequencing." *Silicon Valley/San Jose Business Journal,* December 6. http://www.bizjournals.com/sanjose/stories/2009/12/07/focus5.html.

Branstetter, Lee, and Ogura Yoshiaki. 2005. "Is Academic Science Driving a Surge in Industrial Innovation? Evidence from Patent Citations." NBER Working Paper 11561. National Bureau of Economic Research, Cambridge, MA.

Breschi, Stefano, and Francesco Lissoni. 2003. "Mobility and Social Networks: Localized Knowledge Spillovers Revisited." Working Papers 142. Centre for Research on Innovation and Internationalisation (CESPRI), Luigi Bocconi University, Milan, Italy.

Breschi, Stefano, Francesco Lissoni, and Fabio Montobbio. 2007. "The Scientific Productivity of Academic Inventors: New Evidence from Italian Data." *Economics of Innovation and New Technology* 16:101–18.

Brezin, Edouard, and Antoine Triller. 2008. "Long Road to Reform in France." *Science* 320:1695.

Brinster, Ralph L., Howard Y. Chen, Myrna Trumbauer, Allen W. Senear, Raphael Warren, and Richard D. Palmiter. 1981. "Somatic Expression of Herpes Thymidine Kinase in Mice Following Injection of a Fusion Gene into Eggs." *Cell* 27:223–31.

Britt, Ronda. 2009. "Federal Government Is Largest Source of University R&D Funding in S&E; Share Drops in FY 2008." NSF 09-318. Arlington, VA: Division of Science Resources Statistics, National Science Foundation. http://www.nsf.gov/statistics/infbrief/nsf09318.

Brown, Jeffrey R., Stephen G. Dimmock, Jun-Koo Kang, and Scott J. Weisbenner. 2010. "Why I Lost My Secretary: The Effect of Endowment Shocks on University Operations." NBER Working Paper 15861. National Bureau of Economic Research, Cambridge, MA.

Buckman, Rebecca. 2008. "Scientist Gives VC an Edge." *Wall Street Journal.* April 14.

Bunton, Sarah, and William Mallon. 2007. "The Continued Evolution of Faculty Appointment and Tenure Policies at U.S. Medical Schools." *Academic Medicine* 82:281–9.

Burns, Laura, Peter Einaudi, and Patricia Green. 2009. "S&E Graduate Enrollments Accelerate in 2007; Enrollments of Foreign Students Reach New High." NSF 09-314, June. Arlington, VA: National Center for Science and Engineering Statistics (NCSES), National Science Foundation. http://www.nsf.gov/statistics/infbrief/nsf09314/.

Burrelli, Joan, Alan Rapoport, and Rolf Lehming. 2008. "Baccalaureate Origins of S&E Doctorate Recipients." NSF 08-311, July. Arlington, VA: National Center for Science and Engineering Statistics (NCSES), National Science Foundation. http://www.nsf.gov/statistics/infbrief/nsf08311/.

Butkus, Ben. 2007a. "NYU Sells Portion of Royalty Interest in Remicade to Royalty Pharma for $650m." *Biotech Transfer Week,* May 14.

———. 2007b. "Texas A&M's Use of Tech Commercialization as Basis for Awarding Tenure Gains Traction." *Biotech Transfer Week,* August 6. http://www.ge nomeweb.com/biotechtransferweek/texas-am%E2%80%99s-use-tech-commercialization-basis-awarding-tenure-gains-traction.

Butler, Linda. 2004. "What Happens When Funding Is Linked to Publication Counts?" In *Handbook of Quantitative Science and Technology Research: The Use of Publication and Patent Statistics in Studies of S&T Systems,* 389–406. Edited by Henk F. Moed, Wolfgang Glänzel, and Ulrich Schmoch. Dordrecht, the Netherlands: Kluwer Academic.

Byrne, Richard. 2008. "Gap Persists between Faculty Salaries at Public and Private Institutions." *Chronicle of Higher Education* 54:32.

Cameron, David. 2010. "Mining the 'Wisdom of Crowds' to Attack Disease." *Harvard Medical School News Alert,* September 29. http://hms.harvard.edu/public/ news/2010/092910_innocentive/index.html.

Campbell, Kenneth D. 1997. "Merck, MIT Announces Collaboration." *MIT Tech Talk,* March 19. http://web.mit.edu/newsoffice/1997/merck-0319.html.

Campus Grotto. 2009. "Average Starting Salary by Degree for 2009." Campus Grotto website. July 15. http://www.campusgrotto.com/average-starting-salary-by-degree-for-2009.html.

Carayol, Nicholas. 2007. "Academic Incentives, Research Organization and Patenting at a Large French University." *Economics of Innovation and New Technology* 16:71–99.

Carely, Flanigan. 1998. "Prevalence of Articles with Honorary Authors and Ghost Authors in Peer-Reviewed Medical Journals." *Journal of the American Medical Association* 280:222–24.

Carmichael, Mary and Sharon Begley. 2010. "Desperately Seeking Cures." *Newsweek,* May 15. http://www.newsweek.com/2010/05/15/desperately-seeking -cures.html.

Carpenter, Siri. 2009. "Discouraging Days for Jobseekers." *Science Careers,* February 13. http://sciencecareers.sciencemag.org/career_magazine/previous_issues/ articles/2009_02_13/caredit.a0900022.

Ceci, Stephen, and Wendy Williams. 2009. *The Mathematics of Sex: How Biology and Society Limit Talented Women.* Oxford: Oxford University Press.

Center for High Angular Resolution Astronomy. 2009. "The CHARA Array." Georgia State University, Atlanta. http://www.chara.gsu.edu/CHARA/array.php.

Center on Congress at Indiana University. 2008. "Members of Congress Questions and Answers." Center on Congress (website). http://www.centeroncongress. org/members-congress-questions-and-answers.

Children's Memorial Research Center. 2009. "Why Use Zebrafish as a Model?" Children's Memorial Research Center (Chicago) website. http://www.childrensmrc.org/topczewski/why_zebrafish/.

Chiswick, Barry R., Nicholas Larsen, and Paul J. Pieper. 2010. "The Production of PhDs in the United States and Canada." IZA Discussion Paper No. 5367.

Institute for the Study of Labor (IZA), Bonn, Germany. http://ftp.iza.org/dp5367.pdf.

Cho, Adrian. 2006. "Embracing Small Science in a Big Way." *Science* 313:1872–75.

———. 2008. "The Hot Question: How New Are the New Superconductors?" *Science* 320:870–71.

Cho, Adrian, and Daniel Clery. 2009. "International Year of Astronomy. Astronomy Hits the Big Time." *Science* 323:332–5.

Chronicle of Higher Education. 2009. *Stipends for Graduate Assistants, 2008–9.* Survey online database. http://chronicle.com/stats/stipends/?inst=1172.

Church, George M. 2005. "Can a Sequencing Method Be 100 Times Faster Than ABI but More Expensive?" *Polny Technology FAQ.* Harvard Molecular Technology Group, Cambridge, MA. http://arep.med.harvard.edu/Polonator/speed.html.

Clery, Daniel. 2009a. "Exotic Telescopes Prepare to Probe Era of First Stars and Galaxies." *Science* 325:1617–9.

———. 2009b. "Herschel Will Open a New Vista on Infant Stars and Galaxies." *Science* 324:584–6.

———. 2009c. "ITER Blueprints near Completion, but Financial Hurdles Lie Ahead." *Science* 326:932–3.

———. 2009d. "Research Funding. England Spreads Its Funds Widely, Sparking Debate." *Science* 323:1413.

———. 2010a. "Budget Red Tape in Europe Brings New Delay to ITER." *Science* 327:1434.

———. 2010b. "ITER Cost Estimates Leave Europe Struggling to Find Ways to Pay." *Science* 328:798.

Coase, Robert. 1974. "The Lighthouse in Economics." *Journal of Law and Economics* 17:357–76.

Cockburn, Iain M., and Rebecca Henderson. 1998. "Absorptive Capacity, Coauthoring Behavior, and the Organization of Research in Drug Discovery." *Journal of Industrial Economics* 46:157–82.

Coelho, Sarah. 2009. "Profile: Jorge Cham. Piled Higher and Deeper: The Everyday Life of a Grad Student." *Science* 323:1668–9.

Cohen, Jon. 2007. "Gene Sequencing in a Flash: New Machines Are Opening up Novel Areas of Research." *Technology Review* 110:72–7.

Cohen, Wesley, Richard Nelson, and John P Walsh. 2002. "Links and Impacts: The Influence of Public Research on Industrial R&D." *Management Science* 48:1–23.

Cohen, Wesley M., and Daniel A. Leventhal. 1989. "Innovation and Learning: The Two Faces of R&D." *Economic Journal* 99:569–96.

Cole, Jonathan R. 2010. *The Great American University: Its Rise to Preeminence, Its Indispensable National Role, Why It Must Be Protected.* New York: Public Affairs.

Cole, Jonathan R., and Stephen Cole. 1973. *Social Stratification in Science.* Chicago: University of Chicago Press.

Collins, Francis S. 2010a. "A Genome Story: 10th Anniversary Commentary." *Scientific American* Guest Blog, June 25. http://www.scientificamerican.com/blog/post.cfm?id=a-genome-story-10th-anniversary-com-2010-06-25.

———. 2010b. "Opportunities for Research and NIH." *Science* 327:36–7.

Collins, Francis S., Michael Morgan, and Aristides Patrinos. 2003. "The Human Genome Project: Lessons from Large-Scale Biology." *Science* 300:286–90.

Commission of the European Communities. 2003. "Investing in Research: An Action Plan for Europe." Brussels, 4.6.2003, COM(2003) 226 final/2. July 30. http://ec.europa.eu/invest-in-research/pdf/226/en.pdf.

Committee to Study the Changing Needs for Biomedical, Behavioral, and Clinical Research Personnel. 2008. Paper presented at the National Institute of General Medical Sciences. Bethesda, Maryland.

Congressional Quarterly. 2007. *Guide to Congress.* 6th ed., 2 vols. Washington, DC: GQ Press.

Corrado, Carol A., and Charles Hulten. 2010. "Measuring Intangible Capital: How Do You Measure a 'Technological Revolution'?" *American Economic Review: Papers and Proceedings* 100:99–104.

Costantini, Franklin, and Elizabeth Lacy. 1981. "Introduction of a Rabbit-Globin Gene into the Mouse Germ Line." *Nature* 294:92–94.

Council of Graduate Schools. 2009. "Findings from the 2009 CGS International Graduate Admissions Survey. Phase II: Applications and Initial Offers of Admission." August 2009. Washington, DC: CGS. http://www.cgsnet.org/portals/0/pdf/R_IntlAdm09_II.pdf.

Couzin, Jennifer. 2006. "Scientific Misconduct: Truth and Consequences." *Science* 313:1222–6.

———. 2008. "Science and Commerce: Gene Tests for Psychiatric Risk Polarize Researchers." *Science* 319:274–7.

———. 2009. "Research Funding. For Many Scientists, the Madoff Scandal Suddenly Hits Home." *Science* 323:25.

Couzin-Frankel, Jennifer. 2009. "Genetics. The Promise of a Cure: 20 Years and Counting." *Science* 324:1504–7.

Cox, Brian. 2008. "Gravity: The 'Holy Grail' of Physics." *BBC Online,* January 29. http://news.bbc.co.uk/2/hi/science/nature/7215972.stm.

Coyle, Daniel. 2009. *The Talent Code: Unlocking the Secret of Skill in Sports, Art, Music, Math, and Just about Anything.* New York: Bantam.

Coyne, Jerry A. 2010. "Harvard Dean: Hauser Guilty of Scientific Misconduct." *Why Evolution Is True* (blog), August 20. http://whyevolutionistrue.wordpress.com/2010/08/20/harvard-dean-hauser-guilty-of-scientific-misconduct/.

Critser, Greg. 2007. "Of Men and Mice: How a Twenty-Gram Rodent Conquered the World of Science." *Harper's Magazine* 315 (December): 65–76.

Cruz-Castro, Laura, and Luis Sanz-Menéndez. 2009. "Mobility versus Job Stability: Assessing Tenure and Productivity Outcomes." *Research Policy* 39:27–38.

Cummings, Jonathan N., and Sara Kiesler. 2005. "Collaborative Research across Disciplinary and Organizational Boundaries." *Social Studies of Science* 35(5): 703–22.

Cutler, David. 2004a. "Are the Benefits of Medicine Worth What We Pay for It?" Policy Brief, 15th Annual Herbert Lourie Memorial Lecture on Health Policy, Maxwell School, Syracuse University.

————.2004b. *Your Money or Your Life: Strong Medicine for America's Health Care System,* Oxford University Press, New York.

Cutler, David, and Srikanth Kadiyala. 2003. "The Return to Biomedical Research: Treatment and Behavioral Effects," in *Measuring the Gains from Medical Research: An Economic Approach,* edited by Kevin Murphy and Robert Topel, Chicago, University of Chicago Press, 2003.

Czarnitzki, Dirk, Christoph Grimpe, and Andrew A. Toole. 2011. "Delay and Secrecy: Does Industry Sponsorship Jeopardize Disclosure of Academic Research?" Zentrum für Europäische Wirtschaftsforschung GimbH (ZEW) Discussion Paper No. 11-009.

Czarnitzki, Dirk, Katrin Hussinger, and Cedric Schneider. 2009. "The Nexus between Science and Industry: Evidence from Faculty Inventions." ZEW Discussion Paper No. 09-028. Zentrum für Europäische Wirtschaftsforschung/ Center for European Economic Research, Mannheim, Germany.

Danielson, Amy, ed. 2009. Research News Online, May 8. Office of the Vice President, University of Minnesota. http://www.research.umn.edu/communications /publications/rno/5-8-09.html.

Darwin, Charles. 1945. *The Voyage of the Beagle.* Raleigh, NC: Hayes Barton Press. First published in 1839.

Dasgupta, Partha, and Paul David. 1987. "Information Disclosure and the Economics of Science and Technology." In *Arrow and the Ascent of Modern Economic Theory,* 519–42. Edited by George Feiwel. New York: New York University Press.

————. 1994. "Toward a new economics of science." *Research Policy* 23, 487–521.

David, Paul. 1994. "Positive Feedbacks and Research Productivity in Science: Reopening Another Black Box." In *The Economics of Technology,* 65–89. Edited by O. Granstrand. Amsterdam: Elsevier Science.

David, Paul A., David Mowery, and W. Edward Steinmueller. 1992. "Analyzing the Economic Payoffs from Basic Research." *Economics of Innovation and New Technology* 2:73–90.

David, Paul, and Andrea Pozzi. 2010. "Scientific Misconduct in Theory and Practice: Quantitative Realities of Falsification, Fabrication and Plagiary in U.S. Publicly Funded Biomedical Research." Paper presented at the International Conference in Honor of Jacques Mairesse, "R&D, Science, Innovation and Intellectual Property," ENSAE. Paris, September 16–17.

"David Quéré." 2010. *Wikipédia.* http://fr.wikipedia.org/wiki/David_Quéré.

Davis, Geoff. 1997. "Mathematicians and the Market." Online preprint. Mathematics Department, Dartmouth College, Hanover, NH. http://www.geoffdavis .net/dartmouth/policy/papers.html.

————. 2005. "Doctors without Orders: Highlights of the Sigma Xi Postdoc Survey." *American Scientist* 93 (3): special supplement, May–June. http://postdoc .sigmaxi.org.

————. 2007. "NIH Budget Doubling: Side Effects and Solutions." Presentation at a seminar, Cambridge, MA: Harvard University, March 12.

————. 2010. *Find the Graduate School That's Right for You.* http://graduate -school.phds.org.

De Figueiredo, John M., and Brian S. Silverman. 2007. "How Does the Government (Want to) Fund Science? Politics, Lobbying, and Academic Earmarks." In *Science and the University*, 36–54. Edited by Paula Stephan and Ronald Ehrenberg. Madison: University of Wisconsin Press.

DeLong, J. Bradford. 2000. "Cornucopia: The Pace of Economic Growth in the Twentieth Century." NBER Working Paper 7602. National Bureau of Economic Research, Cambridge, MA.

Deng, Zhen, Baruch Lev, and Francis Narin. 1999. "Science and Technology as Predictors of Stock Performance." *Financial Analysts Journal* 55:20–32.

de Solla Price, Derek J. 1986. *Little Science, Big Science . . . And Beyond.* New York: Columbia University Press.

Diamond, A. M., Jr. 1986. "The Life-Cycle Research Productivity of Mathematicians and Scientists." *Journal of Gerontology* 41:520–5.

Dimsdale, John. 2009. "Inventor, 89, Has His Eye on Diamonds." Zalman Shapiro, interviewed by Kai Ryssdal. *American Public Media*, June 16. http://marketplace.publicradio.org/display/web/2009/06/16/pm_serial_inventor/.

Ding, Lan, and Haizheng Li. 2008. "Social Network and Study Abroad: The Case of Chinese Students in the U.S." Paper presented at Chinese Economists Society 2008 North America Conference. University of Regina, Saskatchewan, Canada, August 20–22.

Ding, Waverly, Sharon Levin, Paula Stephan, and Anne E. Winkler. 2010. "The Impact of Information Technology on Scientists' Productivity, Quality and Collaboration Patterns." *Management Science* 56:1439–61.

Ding, Waverly, Fiona Murray, and Toby Stuart. 2009. "Commercial Science: A New Arena for Gender Differences in Scientific Careers?" Unpublished paper.

"DNA Sequencing." 2011. *Wikipedia.* http://en.wikipedia.org/wiki/DNA_sequencing.

Dolan DNA Learning Center. 2010. "Making Sequencing Automated, Michael Hunkapiller." ID 15098. Cold Spring Harbor Laboratory, Harlem DNA Lab and DNA Learning Center West (website). http://www.dnalc.org/view/15098-Making-sequencing-automated-Michael-Hunkpiller.html.

Drmanac, Radoje, Andrew B. Sparks, Matthew J. Callow, Aaron L. Halpern, Norman L. Burns, Bahram G. Kermani, Paolo Carnevali, Igor Nazarenko, Geoffrey B. Nilsen, and George Yeung. 2010. "Human Genome Sequencing Using Unchained Base Reads on Self-Assembling DNA Nanoarrays." *Science* 327:78–81.

Ducor, Phillipe. 2000. "Intellectual Property: Coauthorship and Coinventorship." *Science* 289:873–75.

Edelman, Benjamin, and Ian Larkin. 2009. "Demographics, Career Concerns or Social Comparison: Who Games SSRN Download Counts?" Harvard Business School Working Paper 09–0906. Harvard University, Cambridge, MA.

Edwards, Mark, Fiona Murray, and Robert Yu. 2003. "Value Creation and Sharing among Universities, Biotechnology and Pharma." *Nature Biotechnology* 21:618–24.

———. 2006. "Gold in the Ivory Tower: Equity Rewards of Outlicensing." *Nature Biotechnology* 24:509–16.

Egghe, Leo. 2006. "Theory and Practice of the g-Index." *Scientometrics* 69:131–52.

Ehrenberg, Ronald G., Marquise McGraw, and Jesenka Mrdjenovic. 2006. "Why Do Field Differentials in Average Faculty Salaries Vary across Universities?" *Economics of Education Review* 25:241–8.

Ehrenberg, Ronald G., Paul J. Pieper, and Rachel A. Willis. 1998. "Do Economics Departments with Lower Tenure Probabilities Pay Higher Faculty Salaries?" *Review of Economics and Statistics* 80:503–12.

Ehrenberg, Ronald G., Michael J. Rizzo , and George Jakubson. 2007. "Who Bears the Growing Cost of Science at Universities?" In *Science and the University,* 19–35. Edited by Paula Stephan and Ronald Ehrenberg. Madison: University of Wisconsin.

Ehrenberg, Ronald G., and Liang Zhang. 2005. "The Changing Nature of Faculty Employment." In *Recruitment, Retention and Retirement in Higher Education: Building and Managing the Faculty of the Future,* 32–52. Edited by Robert Clark and Jennifer Ma. Northampton, MA: Edward Elgar.

Eisenberg, Rebecca. 1987. "Proprietary Rights and the Norms of Science in Biotechnology Research." *Yale Law Journal* 97:177–231.

Eisenstein, Ronald I., and David S. Resnick. 2001. "Going for the Big One." *Nature Biotechnology* 19:881–82.

Ellard, David. 2002. "The History of MRI." Clinical Radiology Department, University of Manchester website. http://www.isbe.man.ac.uk/personal/dellard/dje/history_mri/history%20of%20mri.htm.

Enserink, Martin. 2006. "Stem Cell Research: A Season of Generosity . . . and Jeremiads." *Science* 314:1525a.

———. 2008a. "Valérie Pécresse interview. After Initial Reforms, French Minister Promises More Changes." *Science* 319:152.

———. 2008b. "Will French Science Swallow Zerhouni's Strong Medicine?" *Science* 322:1312.

European Commission. 2007a. *China, EU and the World: Growing Harmony?* Brussels: Bureau of European Policy Advisers. http://ec.europa.eu/dgs/policy_advisers/publications/docs/china_report_27_july_06_en.pdf.

———. 2007b. *Sixth Framework Programme, 2002–2006.* Research and Innovation. http://ec.europa.eu/research/fp6/index_en.cfm.

———. 2010. "Participate in FP7," *Seventh Framework Programme (FP7).* Community Research and Development Information Service for Science, Research and Development (CORDIS). http://cordis.europa.eu/fp7/who_en.html.

"European Extremely Large Telescope." 2010. *Wikipedia.* http://en.wikipedia.org/wiki/European_Extremely_Large_Telescope.

European Southern Observatory. 2010. The European Extremely Large Telescope. http://www.eso.org/public/teles-instr/e-elt.html.

European University Institute. 2010. Academic Careers Observatory: Salary Comparisons. http://www.eui.eu/ProgrammesAndFellowships/AcademicCareersObservatory/CareerComparisons/SalaryComparisons.aspx.

Everdell, William R. 2003. Review of *Einstein's Clocks, Poincaré's Maps: Empires of Time* by Peter Galison. *New York Times Book Review,* August 17.

Fabrizio, Kira R., and Alberto Di Minin. 2008. "Commercializing the Laboratory: Faculty Patenting and the Open Science Environment." *Research Policy* 37:914–31.

"Fact and Fiction." *Science* 320:857.

FDA. 2010. "NMEs Approved by CDER." http://www.fda.gov/downloads/Drugs/DevelopmentApprovalProcess/HowDrugsareDevelopedandApproved/DrugandBiologicApprovalReports/UCM242695.pdf

Falkenheim, Jaquelina C. 2007. "U.S. Doctoral Awards in Science and Engineering Continue Upward Trend in 2006." NSF 08-301, November. Arlington, VA: National Center for Science and Engineering Statistics (NCSES), National Science Foundation. http://www.nsf.gov/statistics/infbrief/nsf08301/.

Feldman, Maryann P., Alessandra Colaianni, and Connie Kang Liu. 2007. "Lessons from the Commercialization of the Cohen-Boyer Patents: The Stanford University Licensing Program." In *Intellectual Property Management in Health and Agricultural Innovation: A Handbook of Best Practices,* Chapter 17.22. Edited by Anatole Krattiger, Richard Mahoney, Lita Nelsen, Jennifer Thomson, Alan Bennett, Kanikaram Satyanarayana, Gregory Graff, Carlos Fernandez, and Stanley Kowalski. Davis, CA: PIPRA. http://www.iphandbook.org/handbook/ch17/p22/index.html.

Feynman, Richard. 1985. *Surely You're Joking, Mr. Feynman.* New York: Bantam Books.

⏤⏤⏤. 1999. *The Pleasure of Finding Things Out: The Best Short Works of Richard P. Feynman.* Edited by Jeffrey Robbins. Cambridge, MA: Helix Books/Perseus.

Finn, Michael G. 2010. "Stay Rates of Foreign Doctorate Recipients from U.S. Universities, 2007." *Oak Ridge Institute for Science and Education.* November. http://orise.orau.gov/files/sep/stay-rates-foreign-doctorate-recipients-2007.pdf.

Fishman, Charles. 2001. "The Killer App—Bar None." *American Way* Magazine, August 1. http://www.americanwaymag.com/so-woodland-bar-code-bernard-silver-drexel-university.

Fitzgerald, Garrett. 2008. "Drugs, Industry and Academia." *Science* 320:1563.

Fleming, Lee, and Olav Sorenson. 2004. "Science as a Map in Technological Search." *Strategic Management Journal* 25:909–28.

Florida State University, Office of Research. 2010. Office of IP Development and Commercialization (website), Tallahassee. http://www.research.fsu.edu/techtransfer/.

Foray, Dominique, and Francesco Lissoni. 2010. "University Research and Public-Private Interaction." In *Handbook of the Economics of Innovation,* Vol. 1, Chapter 6. Edited by Bronwyn Hall and Nathan Rosenberg. London: Elsevier Press.

"454 Life Sciences." 2011. *Wikipedia.* http://en.wikipedia.org/wiki/454_Life_Sciences.

Fox, Mary Frank. 1983. "Publication Productivity among Scientists: A Critical Review." *Social Studies of Science* 13:285–305.

⏤⏤⏤. 1994. "Scientific Misconduct and Editorial and Peer Review Processes." *Journal of Higher Education* 65:298–309.

⏤⏤⏤. 2010. Book review of *How Institutions Affect Academic Careers* by Joseph C. Hermanowicz, University of Chicago Press, 2009. *American Journal of Sociology* 116:663–5.

Fox, Mary Frank, and Sushanta Mohapatra. 2007. "Social-Organizational Characteristics of Work and Publication Productivity among Academic Scientists in Doctoral-Granting Departments." *Journal of Higher Education* 78:542–71.

Fox, Mary Frank, and Paula Stephan. 2001. "Careers of Young Scientists: Preferences, Prospects and Realities by Gender and Field." *Social Studies of Science* 31:109–22.

Frank, Robert, and Philip Cook. 1992. *Winner-Take-All Markets*. Ithaca, NY: Cornell University Press.

Frankson, Christine. 2010. "Faculty Spotlight—Dr. John Criscione." *CNVE Newsletter* 6.3, September. http://cnve.tamu.edu/newsletter/sept2010b/.

Franzoni, Chiara. 2009. "Do Scientists Get Fundamental Research Ideas by Solving Practical Problems?" *Industrial and Corporate Change* 18:671–99.

Franzoni, Chiara, Giuseppe Scellato, and Paula Stephan. 2011. "Changing Incentives to Publish." *Science* 333: 702–703.

Freeman, Richard. 1989. *Labor Markets in Action*. Cambridge, MA: Harvard University Press.

Freeman, Richard, Tanwin Chang, and Hanley Chiang. 2009. "Supporting 'the Best and Brightest' in Science and Engineering: NSF Graduate Research Fellowships." In *Science and Engineering Careers in the United States: An Analysis of Markets and Employment,* 19–57. Edited by Richard Freeman and Daniel Goroff. Chicago: University of Chicago Press.

Freeman, Richard, and Daniel Goroff. 2009. "Introduction." In *Science and Engineering Careers in the United States: An Analysis of Markets and Employment,* 1–26. Edited by Richard Freeman and Daniel Goroff. Chicago: University of Chicago Press.

Freeman, Richard, Emily Jin, and Chia-Yu Shen. 2007. "Where Do New U.S.-Trained Science-Engineering PhDs Come From?" In *Science and the University,* 197–220. Edited by Paula Stephan and Ron Ehrenberg. Ithaca, NY: Cornell University Press.

Freeman, Richard, and John Van Reenen. 2008. "Be Careful What You Wish For: A Cautionary Tale about Budget Doubling." *Issues in Science and Technology,* Fall.

———. 2009. "What If Congress Doubled R&D Spending on the Physical Sciences?" In *Innovation Policy and the Economy,* Vol. 9, Chapter 1. Edited by Josh Lerner and Scott Stern. Cambridge, MA: National Bureau of Economic Research.

Freeman, Richard, Eric Weinstein, Elizabeth Marincola, Janet Rosenbaum, and Frank Solomon. 2001a. "Careers and Rewards in Bio Sciences: The Disconnect between Scientific Progress and Career Progression." American Society for Cell Biology. http://www.ascb.org/newsfiles/careers_rewards.pdf.

———. 2001b. "Competition and Careers in Biosciences." *Science* 294:2293–4.

Funding First. 2000. *Exceptional Returns: The Economic Value of America's Investment in Medical Research*. Monograph. New York: Mary Woodard Lasker Charitable Trust. http://www.laskerfoundation.org/media/pdf/exceptional.pdf.

Furman, Jeffrey L., and Fiona Murray. 2011. "Does Open Access Democratize Innovation? Examining the Impact of Open Institutions on the Inner and Outer Circles of Science." Working paper, MIT.

Furman, Jeffrey L., Fiona Murray, and Scott Stern. 2010. "More for the Research Dollar." *Nature* 468:757–58.

Furman, Jeffrey L., and Scott Stern. 2011. "Climbing atop the Shoulders of Giants: The Impact of Institutions on Cumulative Research." *American Economic Review* 101:1933–63.

Gaglani, Shiv. 2009. "Investing in our Future: Ways to Attract and Keep Young People in Science and Technology." Presented at "Toward an R&D Agenda for the New Administration and Congress: Perspectives from Scientists and Economists," Science and Engineering Workforce Project Workshop, National Bureau for Economic Research Conference (NBER). Cambridge, MA.

Galison, Peter. 2004. *Einstein's Clocks, Poincaré's Maps: Empires of Time*. New York: W. W. Norton.

Gans, Joshua S., and Fiona Murray. 2010. "Funding Conditions, the Public-Private Portfolio and the Disclosure of Scientific Knowledge." Paper presented at NBER Conference Celebrating the Fiftieth Anniversary of the Publication of *The Rate and Direction of Inventive Activity*. Aerlie Conference Center, Warrenton, VA, September 30–October 2.

Gardner, Martin. 1977. "A New Kind of Cipher That Would Take Millions of Years to Break [RSA Challenge]." *Scientific American* 237:120-4.

Garrison, Howard, and Kimberly McGuire. 2008. "Education and Employment of Biological and Medical Scientists: Data from National Surveys." Paper presented at the Federation of American Societies for Experimental Biology (FASEB). Bethesda, MD. http://www.faseb.org/Policy-and-Government-Affairs/Data-Compilations/Education-and-Employment-of-Scientists.aspx.

Garrison, Howard, and Kim Ngo. 2010. "NIH Funding and Grants to Investigators." FASEB PowerPoint Slides. Presentation made by Garrison, at conference "How Can We Maintain Biomedical Research and Development at the End of ARRA?" Cold Spring Harbor, NY, April 25–27, 2010.

Gaulé, Patrick, and Mario Piacentini. 2010a. "Chinese Graduate Students and U.S. Scientific Productivity: Evidence from Chemistry." Unpublished draft manuscript. Sloan School of Management, Massachusetts Institute of Technology, Cambridge; Department of Economics, University of Geneva. http://www.uclouvain.be/cps/ucl/doc/econ/documents/IRS_Piacentini.pdf.

———. 2010b. "Return Migration of the Very High Skilled: Evidence from U.S.-Based Faculty." Massachusetts Institute of Technology Working Paper, Cambridge, MA.

Geisler, Iris, and Ronald L. Oaxaca. 2005. "Faculty Salary Determination at a Research I University." Unpublished manuscript. http://www.nber.org/~sewp/events/2005.01.14/Bios%2BLinks/Oaxaca-rec1-Academic-Salary05.pdf.

"Gemini Observatory." 2011. *Wikipedia*. http://en.wikipedia.org/wiki/Gemini_Observatory.

Geuna, Aldo. 2001. "The Changing Rationale for European University Research Funding: Are There Negative Unintended Consequences?" *Journal of Economic Issues* 35:607–32.

Geuna, Aldo, and Lionel J. J. Nesta. 2006. "University Patenting and Its Effects on Academic Research: The Emerging European Evidence." *Research Policy* 35:790–807.

Ghose, Tia. 2009. "State Schools Feeling the Pinch." *The Scientist,* February 16. http://www.the-scientist.com/blog/display/55426/.

Giacomini, Kathleen. 2011. Giacomini Lab, University of California, San Francisco. Department of Bioengineering and Therapeutic Sciences. http://bts.ucsf.edu/giacomini/.

Gieryn, Thomas, and Richard Hirsh. 1983. "Marginality and Innovation in Science." *Social Studies of Science* 13:87–106.

"Gini Coefficient," 2010, *Wikipedia,* http://en.wikipedia.org/wiki/Gini_coefficient.

Ginther, Donna, and Shulamit Kahn. 2009. "Does Science Promote Women? Evidence from Academia 1973–2001." In *Science and Engineering Careers in the United States: An Analysis of Markets and Employment,* 163–194. Edited by Richard Freeman and Daniel Goroff. Chicago: University of Chicago Press.

Gittelman, Michelle. 2006. "National Institutions, Public–Private Knowledge Flows, and Innovation Performance: A Comparative Study of the Biotechnology Industry in the U.S. and France." *Research Policy* 35:1052–68.

Goldfarb, Brent, and Magnus Henrekson. 2003. "Bottom-up versus Top-down Policies towards the Commercialization of the University Intellectual Property." *Research Policy* 32:639–58.

Goldin, Claudia, and Lawrence F. Katz. 1998. "The Origins of State-Level Differences in the Public Provision of Higher Education: 1890–1940." *American Economic Review* 88:303–08.

———. 1999. "The Shaping of Higher Education: The Formative Years in the United States, 1890 to 1940." *Journal of Economic Perspectives* 13:37–62.

Goldman, Charles, Traci Williams, David Adamson, and Kathy Rosenblat. 2000. *Paying for University Research Facilities and Administration.* Santa Monica, CA: RAND Corporation.

Gomez-Mejia, Luis, and David Balkin. 1992. "Determinants of Faculty Pay: An Agency Theory Perspective." *Academy of Management Journal* 35:921–55.

Goodman, Laurie. 2004. "Clearing a Roadmap." *Journal of Clinical Investigation* 113:1512–3. doi:10.1172/JCI22106.

Goodwin, Margarette, Ann Bonham, Anthony Mazzaschi, Hershel Alexander, and Jack Krakower. 2011. "Sponsored Program Salary Support to Medical School Faculty in 2009." *Analysis in Brief* (Association of American Medical Colleges) 11 (1), January. https://www.aamc.org/download/170836/data/aibvol11_no1.pdf.

Gordon, J. W., G. A. Scangos, D. J. Plotkin, J. A. Barbosa, and F. H. Ruddle. 1980. "Genetic Transformation of Mouse Embryos by Microinjection of Purified DNA." *Proceedings of the National Academy of Sciences of the United States of America* 77:7380–84.

Gordon, Robert R. 2000. "Does the 'New Economy' Measure up to the Great Innovations of the Past?" *Journal of Economic Perspectives* 14:49–74.

Graves, Philip, Dwight Lee, and Robert Sexton. 1987. "A Note on Interfirm Implications of Wages and Status." *Journal of Labor Research* 8:209–12.

Griliches, Zvi. 1960. "Hybrid Corn and the Economics of Innovation." *Science* 132:275–80.

———. 1979. "Issues in Assessing the Contribution of Research and Development to Productivity Growth." *The Bell Journal of Economics*, 10(1):92-116.

Grimm, David. 2006. "Spending Itself out of Existence, Whitaker Brings a Field to Life." *Science* 311:600–1.

Groen, Jeffrey, and Michael Rizzo. 2007. "The Changing Composition of U.S. Citizen PhDs." In *Science and the University*, 177–96. Edited by Paula Stephan and Ronald Ehrenberg. Madison: University of Wisconsin Press.

Groll, Elias J., and William White. 2010. "Allston Construction Pause Imposes Space Constraints on Harvard Science Schools." *Harvard Crimson,* March 31.

Grueber, Martin, and Tim Studt. 2010. "2011 Global R&D Funding Forecast: China's R&D Growth Engine." *R&D Daily,* December 15.

Hagstrom, Warren O. 1965. *The Scientific Community*. New York: Basic Books.

Halford, Bethany. 2011. "Is Chemistry Facing a Glut of PhDs?" *Science and Technology* 89:46–52.

Hall, Bronwyn, Jacques Mairesse, and Pierre Mohnen. 2010. "Returns to R&D and Productivity." In *Handbook of the Economics of Innovation,* Vol. 2, Chapter 24. Edited by Bronwyn Hall, and Nathan Rosenberg. London: Elsevier.

Halzin, Francis. 2010. "Icecube Neutrino Observatory." Conference at Hitosubashi University, Tokyo, Japan, March 25, 2010.

Hamermesh, Daniel, George Johnson, and Burton Weisbrod. 1982. "Scholarship, Citations and Salaries: Economic Rewards in Economics." *Southern Economic Journal* 49:472–81.

Harhoff, Dietmar, Frederic Scherer, and Katrin Vopel. 2005. "Exploring the Tail of Patented Invention Value Distributions." In *Patents: Economics, Policy, and Measurement*, 251–81. Edited by Frederic Scherer. Northampton, MA: Edward Elgar.

Harmon, Lindsey. 1961. "High School Backgrounds of Science Doctorates." *Science* 133:679–81.

Harré, Rom. 1979. *Social Being*. Oxford: Basil Blackwell.

Harris, Gardiner. 2011. "New Federal Research Center Will Help Develop Medicines." *New York Times,* January 22, A1, A21.

Harzing, Anne-Wil. 2010. *Publish or Perish* (software). Harzing.com. http://www.harzing.com/pop.htm.

Hegde, Deepak, and David C. Mowery. 2008. "Politics and Funding in the U.S. Public Biomedical R&D System." *Science* 322:1797–8.

Heinig, Stephen J., Jack Y. Krakower, Howard B. Dickler, and David Korn. 2007. "Sustaining the Engine of U.S. Biomedical Discovery." *New England Journal of Medicine* 357:1042–7.

Heller, Michael, and Rebecca Eisenberg. 1998. "Can Patents Deter Innovation? The Anticommons in Biomedical Research." *Science* 280:698–701.

Hendrick, Bill. 2009. "Lifesaving Science." *Delta Sky Magazine*. May.

Hermanowicz, Joseph C. 2006. "What Does It Take to Be Successful?" *Science, Technology and Human Values* 31:135–52.

Herper, Matthew. 2011. "Gene Machine." *Forbes,* January 17.

"Heterosis." 2010. Wikip*edia.* http://en.wikipedia.org/wiki/Heterosis.

"Heterosis: Hybrid Corn." 2011. *Wikipedia.* http://en.wikipedia.org/wiki/Hybrid_corn.

Hicks, Diana. 1995. "Published Papers, Tacit Competencies and Corporate Management of the Public/Private Character of Knowledge." *Industrial and Corporate Change* 4:401–24.

———. 2009. "Evolving Regimes of Multi-University Research Evaluation." *Higher Education* 57:393–404.

"High Temperature Conductivity." 2010. *Wikipedia.* http://en.wikipedia.org/wiki/High-temperature_superconductivity.

Hill, Susan, and Einaudi, Peter. 2010. "Jump in Fall 2008 Enrollments of First-Time, Full-Time S&E Graduate Students." NSF 10-320, June. Arlington, VA: National Center for Science and Engineering Statistics (NCSES), National Science Foundation. http://www.nsf.gov/statistics/infbrief/nsf10320/.

Hirsch, Jorge. 2005. "An Index to Quantify an Individual's Scientific Research Output." *Proceedings of the National Academy of Sciences of the United States of America* 102:16569–72.

Hirschler, Ben. 2010. "Small Study of Glaxo 'Red Wine' Drug Suspended." *Reuters,* May 4. http://www.reuters.com/article/idUSTRE6435A620100504.

Hoffer, Thomas B., Carolina Milesi, Lance Selfa, Karen Grigorian, Daniel J. Foley, Lynn M. Milan, Steven L. Proudfoot, and Emilda B. Rivers. 2011. "Unemployment among Doctoral Scientists and Engineers Remained below the National Average in 2008." NSF 11-308. Arlington, VA: National Center for Science and Engineering Statistics (NCSES), National Science Foundation. http://www.nsf.gov/statistics/infbrief/nsf11308/.

Howard Hughes Medical Institute. 2009a. "Financials: Endowment." Howard Hughes Medical Institute (website). http://www.hhmi.org/about/financials/endowment.html.

———. 2009b. "Financials: Scientific Research." Howard Hughes Medical Institute (website). http://www.hhmi.org/about/financials/scientific.html.

———. 2009c. "Growth: 1984–1992." Howard Hughes Medical Institute (website). http://www.hhmi.org/about/growth.html.

———. 2009d. "HHMI Investigators: Frequently Asked Questions about the HHMI Investigator Program." Howard Hughes Medical Institute (website). http://www.hhmi.org/research/investigators/investigator_faq.html.

———. 2009e. "HHMI Scientists & Research." Howard Hughes Medical Institute (website). http://www.hhmi.org/research/.

Hsu, Stephen D. H. 2010. Curriculum vitae. http://duende.uoregon.edu/~hsu/MyCV1.pdf.

Hull, David L. 1988. *Science as a Process.* Chicago: University of Chicago Press.

"The Human Genome: Unsung Heroes." 2007. *Science* 291:1207.

Hunt, Jennifer. 2009. "Which Immigrants Are Most Innovative and Entrepreneurial? Distinctions by Entry Visa." NBER Working Paper 14920. National Bureau of Economic Research, Cambridge, MA.

Hunter, Rosalind S., Andrew J. Oswald and Bruce Charlton. 2009. "The Elite Brain Drain." *Economic Journal* 119:231–251.

IBM. 2010. "Awards & Achievements." IBM Research (website). http://www.re-search.ibm.com/resources/awards.shtml.

"IceCube Neutrino Observatory." 2010. *Wikipedia.* http://en.wikipedia.org/wiki/IceCube_Neutrino_Observatory.

Ignatius, David. 2007. "The Ideas Engine Needs a Tuneup." *Washington Post,* June 3, B07.

Illumina. 2009. "Genome Analyzer IIx." Illumina, Inc. (website). http://www.illumina.com/pages.ilmn?ID=204.

Imperial College London, Faculty of Medicine. 2008. "Research Excellence Framework—Briefing Document: Faculty of Medicine." http://www1.imperial.ac.uk/resources/4BF62CE0-0147-4E30-9126-002531583473/.

"Income Inequality in the United States." *Wikipedia,* http://en.wikipedia.org/wiki/Income_incquality_in_the_United_States.

Information Please Database. 2007. "United States, U.S. Statistics, Mortality: Life Expectancy at Birth by Race and Sex, 1930–2005." Infoplease.com (website). http://www.infoplease.com/ipa/A0005148.html.

"Inktomi Corporation." 2010. *Wikipedia.* http://en.wikipedia.org/wiki/Inktomi_Corporation.

Institute for Systems Biology. 2010. "Hood Group." Institute for Systems Biology (website). http://www.systemsbiology.org/Scientists_and_Research/Faculty_Groups/Hood_Group.

Interfaces & Co. 2011. Physique et Mécanique des Milieux Hétérogènes (ESPCI) and Laboratoire d'Hydrodynamique (École Polytechnique). Centre National de la Recherche Scientifique, Paris. http://www.pmmh.cspci.fr/fr/gouttes/AccueilUS.html.

International Brotherhood of Boilermakers, Iron Ship Builders, Blacksmiths, Forgers, and Helpers, AFL-CIO. 2008. "Why Are Purdue Students and Alumni Called Boilermakers?" International Brotherhood of Boilermakers (website). http://www.boilermakers.org/resources/what_is_a_boilermaker/purdue_boilermakers.

International Committee of Medical Journal Editors. 2010. "Uniform Requirements for Manuscripts Submitted to Biomedical Journals: Ethical Considerations in the Conduct and Reporting of Research, Authorship and Contributorship." ICMJE website. http://www.icmje.org/ethical_1author.html.

J. Craig Venter Institute. 2008. "J. Craig Venter Institute Consolidates Sequencing Center and Reduces 29 Sequencing Staff Positions." December 9. J. Craig Venter Institute (website). http://www.jcvi.org/cms/press/press-releases/full-text/article/j-craig-venter-institute-consolidates-sequencing-center-and-reduces-29-sequencing-staff-positions/.

Jacobsen, Jennifer. 2003. "Who's Hiring in Physics?" *Chronicle of Higher Education.* June 19.

Jaffe, Adam. 1986. "Technological Opportunity and Spillovers of R&D." *American Economic Review* 76:984–1000.

———. 1989a. "Characterizing the 'Technological Position' of Firms, with Applications to Quantifying Technological Opportunity and Research Spillovers." *Research Policy* 18:87–97.

————. 1989b. "Real Effects of Academic Research." *American Economic Review* 79:957–70.

Jaffe, Adam, Manuel Trajtenberg, and Rebecca Henderson. 1993. "Geographic Localization of Knowledge Sources as Evidenced by Patent Citations." *Quarterly Journal of Economics* 108:576–98.

Jefferson, Thomas. 1967. *The Jefferson Cyclopedia*, Vol. 1. Edited by John P. Foley. New York: Russell and Russell.

Jenk, Daniel. 2007. "NIH Funds Next Generation of DNA Sequencing Projects at ASU." *ASU Biodesign Institute News,* January 30. http://biodesign.asu.edu/news/nih-funds-next-generation-of-dna-sequencing-projects-at-asu.

Jensen, Richard, and Marie Thursby. 2001. "Proofs and Prototypes for Sale: The Licensing of University Inventions." *American Economic Review* 91: 240–59.

"John Bates Clark Medal." 2010. *Wikipedia.* http://en.wikipedia.org/wiki/John_Bates_Clark_Medal.

Jones, Benjamin F. 2009. "The Burden of Knowledge and the 'Death of the Renaissance Man': Is Innovation Getting Harder?" *Review of Economic Studies* 76:283–317.

————. 2010a. "As Science Evolves, How Can Science Policy?" NBER Working Paper No. 16002. National Bureau of Economic Research, Cambridge, MA.

————. 2010b. "Why Science Needs a Nudge from Washington, D.C." *Newsweek,* June 21.

Jones, Benjamin, Stefan Wuchty, and Brian Uzzi. 2008. "Multi-university Research Teams. Shifting Impact, Geography, and Stratification in Science." *Science* 322:1259–62.

Jong, Simcha. 2006. "How Organizational Structures in Science Shape Spin-Off Firms: The Biochemistry Departments of Berkeley, Stanford, and UCSF and the Birth of the Biotech Industry." *Industrial and Corporate Change* 15:251–3.

Jorgenson, Dale W., Mun S. Ho, and Kevin J. Stiroh. 2008. "A Retrospective Look at the U.S. Productivity Resurgence." *Journal of Economic Perspectives* 22:2–24.

Kaiser, Jocelyn. 2008a. "Biochemist Robert Tjian Named President of Hughes Institute." *Science* 322:35.

————. 2008b. "The Graying of NIH Research." *Science* 322:848–9.

————. 2008c. "HHMI's Cech Signs Off on His Biggest Experiment." *Science* 320:164.

————. 2008d. "NIH Urged to Focus on New Ideas, New Applicants." *Science* 319:1169.

————. 2008e. "Two Teams Report Progress in Reversing Loss of Sight." *Science* 320:606–7.

————. 2008f. "Zerhouni's Parting Message: Make Room for Young Scientists." *Science* 322:834–5.

————. 2009a. "Grants 'Below Payline' Rise to Help New Investigators." *Science* 325:1607.

————. 2009b. "NIH Stimulus Plan Triggers Flood of Applications—and Anxiety." *Science* 324:318–9.

————. 2009c. "Wellcome Trust to Shift from Projects to People." *Science* 326:921.

————. 2011. "Despite Dire Budget Outlook, Panel Tells NIH to Train More Scientists." *ScienceInsider,* January 7. http://news.sciencemag.org/scienceinsider/2011/01/despite-dire-budget-outlook-pane.html.

Kaiser, Jocelyn, and Lila Guterman. 2008. "National Institutes of Health. Researchers Could Face More Scrutiny of Outside Income." *Science* 322:1622a.

Kaiser, Jocelyn, and Eli Kintisch. 2008. "Conflicts of Interest. Cardiologists Come under the Glare of a Senate Inquiry." *Science* 322:513.

Kalil, Tom, and Robynn Sturm. 2010. "Congress Grants Broad Prize Authority to All Federal Agencies." *The White House: Open Government Initiative* (blog), December 21. http://www.whitehouse.gov/blog/2010/12/21/congress-grants-broad-prize-authority-all-federal-agencies.

Katz, Sylvan, and Diana Hicks. 2008. "Excellence vs. Equity: Performance and Resource Allocation in Publicly Funded Research." Paper presented at the DIME-BRICK Workshop "The Economics and Policy of Academic Research." Collegio Carlo Alberto, Moncalieri (Torino), Italy, July 14–15.

Kean, Sam. 2006. "Scientists Spend Nearly Half Their Time on Administrative Tasks, Survey Finds." *Chronicle of Higher Education,* July 14. http://chronicle.com/article/Scientists-Spend-Nearly-Half/23697.

Kelly, Janis. 2005. "The Chimera That Roared: Remicade Royalties to Fund $105 Million Biomedical Research, Education at NYU." *Medscape Today,* August 18.

Kenney, Martin. 1986. *Biotechnology: The University-Industrial Complex.* New Haven, CT: Yale University Press.

Kim, Sunwoong. 2007. "Brain Drain and/or Brain Gain: Education and International Migration of Highly Educated Koreans." University of Wisconsin-Milwaukee.

————. 2010. "From Brain Drain to Brain Competition: Changing Opportunities and the Career Patterns of US-Trained Korean Academics." In *American Universities in a Global Market,* 335–69. Edited by Charles T. Clotfelter. Chicago: University of Chicago Press.

Kneller, Robert. 2010. "The Importance of New Companies for Drug Discovery: Origins of a Decade of New Drugs." *Nature Reviews* 9:867–82.

Koenig, Robert. 2006. "Candidate Sites for World's Largest Telescope Face First Big Hurdle." *Science* 313:910–12.

Kohn, Alexander. 1986. *False Profits.* Oxford: Basil Blackwell.

Kolbert, Elizabeth. 2007. "Crash Course: The World's Largest Particle Accelerator." *New Yorker,* May 14, 68–78.

Kong, Wuyi, Shaowei Li, Michael T. Longaker, and H. Peter Lorenz. 2008. "Blood-Derived Small Dot Cells Reduce Scar in Wound Healing." *Experimental Cell Research* 314:1529–39.

Krimsky, Sheldon, L. S. Rothenberg, P. Stott, and G. Kyle. 1996. "Financial Interests of Authors in Scientific Journals: A Pilot Study of 14 Publications." *Science and Engineering Ethics* 2:395–410.

Kuhn, Thomas S. 1962. *The Structure of Scientific Revolutions.* Chicago: University of Chicago Press.

Kuznets, Simon. 1965. *Modern Economic Growth.* New Haven, CT: Yale University Press.

Lacetera, Nicola, and Lorenzo Zirulia. 2009. "The Economics of Scientific Misconduct." *Journal of Law, Economics, and Organization*, October 20. doi: 10.1093/jleo/ewp031.

Lach, Saul, and Mark Schankerman. 2008. "Incentives and Invention in Universities." *RAND Journal of Economics* 39:403–33.

La Jolla Institute for Allergy and Immunology. 2009. "La Jolla Institute Scientist Hilde Cheroutre Earns the 2009 NIH Director's Pioneer Award." *News Medical,* September 24. http://www.news-medical.net/news/20090924/La-Jolla-Institute-scientist-Hilde-Cheroutre-earns-the-2009-NIH-Directors-Pioneer-Award.aspx.

"Large Hadron Collider." 2011. *Wikipedia.* http://en.wikipedia.org/wiki/Large_Hadron_Collider#Cost.

"Laser." 2011. *Wikipedia.* http://en.wikipedia.org/wiki/Laser.

Latour, Bruno. 1987. *Science in Action: How to Follow Scientists and Engineers through Society.* Cambridge, MA: Harvard University Press.

Lavelle, Louis. 2008. "Higher Salaries for 2008 MBA Graduates." *Business Week,* November 13. http://www.businessweek.com/bschools/blogs/mba_admissions/archives/2008/11/higher_salaries.html.

Lawler, Andrew. 2008. "University Research. Steering Harvard toward Collaborative Science." *Science* 321:190–2.

Lazear, Edward P., and Sherwin Rosen. 1981. "Rank-Order Tournaments as Optimum Labor Contracts." *Journal of Political Economy* 89:841–64.

Lee, Christopher. 2007. "Slump in NIH Funding Is Taking Toll on Research." *Washington Post,* May 28, A06.

Lefevre, Christiane. 2008. *Destination Universe: The Incredible Journey of a Proton in the Large Hadron Collider.* Geneva: CERN.

Lehrer, Tom. [1993]. "Lobachevsky." In Tom Lehrer Revisited LP. *Demented Music Database* (website). http://dmdb.org/lyrics/lehrer.revisited.html#6.

Lemelson–MIT Program. 2003. "$500,000 Lemelson-MIT Prize awarded to Leroy Hood, M.D., Ph.D." April 24. Massachusetts Institute of Technology (website). http://web.mit.edu/Invent/n-pressreleases/n-press-03LMP.html.

———. [2007]. "Leroy Hood: 2003 Lemelson-MIT Prize Winner." Massachusetts Institute of Technology (website). http://web.mit.edu/invent/a-winners/a-hood.html.

Lerner, Josh, Antoinette Schoar, and Jialan Wang. 2008. "Secrets of the Academy: The Drivers of University Endowment Success." *Journal of Economic Perspectives* 22:207–22.

Leslie, Stuart W. 1993. *The Cold War and American Science: The Military-Industrial-Academic Complex at MIT and Stanford.* New York: Columbia University Press.

Levi-Montalcini, Rita. 1988. *In Praise of Imperfection: My Life and Work.* New York: Basic Books.

Levin, Sharon, Grant Black, Anne Winkler, and Paula Stephan. 2004. "Differential Employment Patterns for Citizens and Non-Citizens in Science and Engineering in the United States: Minting and Competitive Effects." *Growth and Change* 35:456–75.

Levin, Sharon, and Paula Stephan. 1997. "Gender Differences in the Rewards to Publishing in Academia: Science in the 1970's." *Sex Roles* 38:1049–604.

———. 1999. "Are the Foreign Born a Source of Strength for U.S. Science?" *Science* 285:1213–14.

Levitt, David G. 2010. "Careers of an Elite Cohort of U.S. Basic Life Science Postdoctoral Fellows and the Influence of Their Mentor's Citation Record." *BMC Medical Education* 10:80, November 15. doi: 10.1186/1472-6920-10-80.

Levy, Dawn. 2000. "Hennessy: Engineering Solutions." *Stanford Report,* October 18. http://news.stanford.edu/news/2000/october18/hensci-1018.html.

Lichtenberg, Frank R. 1988. "The Private R&D Investment Response to Federal Design and Technical Competitions." *American Economic Review* 78:550–59.

———. 2002. "New Drugs: Health and Economic Impacts." *NBER Reporter,* Winter, 5–7. http://www.nber.org/reporter/winter03/healthandeconomicimpacts.html.

Lindquist, Susan. 2011. Lindquist Lab (website). Whitehead Institute for Biomedical Research, Massachusetts Institute of Technology, Cambridge. http://web.wi.mit.edu/lindquist/pub/.

Lipowicz, Alice. 2010. "Apps for Healthy Kids Contest Winners Announced." *Federal Computer Week,* September 29. http://fcw.com/articles/2010/09/29/apps-for-healthy kids-winners-announced.aspx.

Lissoni, Francesco, Patrick Llerena, Maureen McKelvey, and Bulat Sanditov. 2008. "Academic Patenting in Europe: New Evidence from the KEINS Database." *Research Evaluation* 17:87–102.

———. 2010. "Scientific Productivity and Academic Promotion: A Study on French and Italian Physicists." NBER Working Paper No. 16341. National Bureau of Economic Research, Cambridge, MA.

Lissoni, Francesco, and Fabio Montobbio. 2010. "Inventorship and Authorship as Attribution Rights: An Enquiry into the Economics of Scientific Credit." Seminar presented at Entreprise, Économie et Société, École Doctorale de Sciences Économiques, Gestion et Démographie, Université Montesquieu - Bordeaux IV, Bordeaux, France, April 16. http://hp.gredeg.cnrs.fr/maurizio_iacopetta/LissoniMontobbio_11_1_2011.pdf.

Litan, Robert, Lesa Mitchell, and E. J. Reedy. 2008. "Commercializing University Innovations: Alternative Approaches." In *Innovation Policy and the Economy,* Vol. 8, 31–58. Edited by Adam B. Jaffe, Josh Lerner, and Scott Stern. Cambridge, MA: National Bureau of Economic Research.

———. 2009. "Crème de la Career." *New York Times,* April 12, 1, 6.

Lohr, Steve. 2006. "Academia Dissects the Service Sector, but Is it a Science?" *New York Times,* April 8, C1.

Long, J. Scott. 1978. "Productivity and Academic Position in the Scientific Career." *American Sociological Review* 43:889–908.

Long, J. Scott, and Robert McGinnis. 1981. "Organizational Context and Scientific Productivity." *American Sociological Review* 46:422–42.

Lotka, Alfred J. 1926. "The Frequency Distribution of Scientific Productivity." *Journal of the Washington Academy of Sciences* 16:317–23.

Ma, Jennifer, and Paula Stephan. 2005. "The Growing Postdoctorate Population at U.S. Research Universities." In *Recruitment, Retention and Retirement in*

Higher Education: Building and Managing the Faculty of the Future, 53–79. Edited by Robert Clark, and Jennifer Ma. Northampton: Edward Elgar.

Macintosh, Zoe. 2010. "Giant New Telescopy Gets $50 Million in Funding." *SPACE.com,* July 21. http://www.space.com/8791-giant-telescope-50-million -funding.html.

Malakoff, David. 2000. "The Rise of the Mouse, Biomedicine's Model Mammal." *Science* 288:248–53.

Mallon, William, and David Korn. 2004. "Bonus Pay for Research Faculty." *Science* 303:476–77.

Mansfield, Edwin. 1991a. "Academic Research and Industrial Innovation." *Research Policy* 20:1–12.

———. 1991b. "Social Returns from R&D: Findings, Methods and Limitations." *Research Technology Management,* 34:6, 24–27

———. 1992. "Academic Research and Industrial Innovation: A Further Note." *Research Policy* 21:295–6.

———. 1995. "Academic Research Underlying Industrial Innovations: Sources, Characteristics, and Financing." *Review of Economics and Statistics* 77:55–65.

———. 1998. "Academic Research and Industrial Innovation: An Update of Empirical Findings." *Research Policy* 26:773–6.

Markman, Gideon, Peter Gianiodis, and Phillip Phan. 2008. "Full-Time Faculty or Part-Time Entrepreneurs." *IEEE Transactions on Engineering Management* 55:29–36.

Marshall, Eliot. 2008. "Science Policy. Biosummit Seeks to Draw Obama's Attention to the Life Sciences." *Science* 322:1623.

———. 2009. "Recession Fallout. Harvard's Financial Crunch Raises Tensions among Biology Programs." *Science* 324:157–8.

Martin, Douglas. 2010. "W. E. Gordon, Creator of Link to Deep Space, Dies at 92." *New York Times,* February 27, 24.

Marty, Bernard, Russell L. Palma, Robert O. Pepin, Laurent Zimmermann, Dennis J. Schlutter, Peter G. Burnard, Andrew J. Westphal, Christopher J. Snead, Saša Bajt, Richard H. Becker, and Jacob E. Simones. 2008. "Helium and Neon Abundances and Compositions in Cometary Matter." *Science* 319:75–8.

Marx, Jean. 2007. "Molecular Biology. Trafficking Protein Suspected in Alzheimer's Disease." *Science* 315:314.

McCook, Alison. 2009. "Cuts in Funding at Wellcome." *The Scientist: Newsblog,* February 12. http://www.the-scientist.com/blog/print/55417/.

McCray, W. Patrick. 2000. "Large Telescopes and the Moral Economy of Recent Astronomy." *Social Studies of Science* 30:685–711.

McGraw-Herdeg, Michael. 2009. "24 Broad Institute DNA Scientists Were Laid Off on Tuesday." *The Tech* 128:65.

McKinsey & Company. 2009. *And the Winner Is . . . : Capturing the Promise of Philanthropic Prizes.* New York: McKinsey. http://www.mckinsey.com/App_ Media/Reports/SSO/And_the_winner_is.pdf.

McKnight, Steve. 2009. "Why Do We Choose to Be Scientists?" *Cell* 138:817–19.

Menard, Henry. 1971. *Science, Growth and Change.* Cambridge, MA: Harvard University Press.

Merton, Robert K. 1957. "Priorities in Scientific Discovery: A Chapter in the Sociology of Science." *American Sociological Review* 22:635–59.

———. 1961. "Singletons and Multiples in Scientific Discovery: A Chapter in the Sociology of Science." *Proceedings of the American Philosophical Society* 105:470–86.

———. 1968. "The Matthew Effect in Science: The Reward and Communication Systems of Science Are Considered." *Science* 159:56–63.

———. 1969. "Behavior Patterns of Scientists." *American Scientist* 57:1–23.

———. 1988. "The Matthew Effect in Science, II: Cumulative Advantage and the Symbolism of Intellectual Property." *Isis* 79:606–23.

Mervis, Jeffrey. 1998. "The Biocomplex World of Rita Colwell." *Science* 281:1944–7.

———. 2007a. "Harvard Proposes One for the Team." *Science* 315:449.

———. 2008a. "And Then There Was One." *Science* 321:1622–8.

———. 2008b. "Building a Scientific Legacy on a Controversial Foundation." *Science* 321:480–83.

———. 2008c. "Top Ph.D. Feeder Schools Are Now Chinese." *Science* 321:185.

———. 2009a. "The Money to Meet the President's Priorities." *Science* 324:1128–29.

———. 2009b. "Reshuffling Graduate Training." *Science* 325:528–30.

———. 2009c. "Senate Majority Leader Hands NSF a Gift to Serve the Exceptionally Gifted." *Science* 323:1548.

———. 2010. "NSF Turns Math Earmark on Its Ear to Fund New Institute." *Science* 329:1006–7.

Meyers, Michelle. 2008. LHC Shut Down until Early Spring. *CNET News,* September 23. http://news.cnet.com/8301-11386_3-10049188 76.html.

Mill, John Stuart. 1921. *Principles of Political Economy.* 7th ed. Edited by William J. Ashley. London: Longmans, Green. First published in 1848.

Miller, Gref. 2010. "Scientific Misconduct. Misconduct by Postdocs Leads to Retraction of Papers." *Science* 329:1583.

Minogue, Kristen. 2009. "Fluorescent Zebrafish Shed Light on Human Birth Defects." *Medill Reports Chicago,* February 5. http://news.medill.northwestern.edu/chicago/news.aspx?id=114601.

———. 2010. "California Postdocs Embrace Union Contract." *ScienceInsider,* August 13. http://news.sciencemag.org/scienceinsider/2010/08/california-postdocs-embrace-union.html.

MIT Museum. 2011. "Lab Life, Sharpies, Photo Mural Documenting Members of Prof. Philip Sharp's Laboratory, 1974–2010." The MIT 150 Exhibition, Massachusetts Institute of Technology, Cambridge, MA. http://museum.mit.edu/150/69.

MIT News. 1997. "MIT Graduates Have Started 4,000 Companies with 1,100,000 Jobs, $232 Billion in Sales in '94." *MIT News,* March 5. http://web.mit.edu/newsoffice/1997/jobs.html.

Mlodinow, Leonard. 2003. *Feynman's Rainbow: A Search for Beauty in Physics and in Life.* New York: Warner Books.

Mokyr, Joel. 2010. "The Contribution of Economic History to the Study of Innovation and Technical Change: 1750–1914." In *Handbook of the Economics of Innovation,* Vol. 1, Chapter 2. Edited by Bronwyn Hall and Nathan Rosenberg. London: Elsevier Press.

Morgan, Thomas. 1901. *Regeneration*. New York: Macmillan.

Mowatt, Graham, Liz Shirran, Jeremy M. Grimshaw, Drummond Rennie, Annette Flanagin, Veronica Yank, Graeme MacLennan, Peter C. Gøtzsche, and Lisa A. Bero. 2002. "Prevalence of Honorary and Ghost Authorship in Cochrane Reviews." *Journal of the American Medical Association* 287:2769–71.

Mowery, David, Richard R. Nelson, Bhaven N. Sampat, and Arvids A. Ziedonis. 2004. *Ivory Tower and Industrial Innovation: University-Industry Technology Transfer before and after the Bayh-Dole Act in the United States*. Stanford, CA: Stanford University Press.

Mowery, David, and Nathan Rosenberg. 1989. *Technology and the Pursuit of Economic Growth*. Cambridge, UK: Cambridge University Press.

Mulvey, Patrick J., and Casey Langer Tesfaye. 2004. "Graduate Student Report: First-Year Physics and Astronomy Students." American Institute of Physics (website). http://www.aip.org/statistics/trends/highlite/grad/gradhigh.pdf.

———. 2010. "Findings from the Initial Employment Survey of Physics PhDs, Classes of 2005 & 2006." American Insitute of Physics (website). http://www.aip.org/statistics/trends/highlite/emp3/emphigh.htm.

Murphy, Kevin, and Robert Topel. 2006. "The Value of Health and Longevity." *Journal of Political Economy* 114:871–904.

Murray, Fiona. 2010. "The Oncomouse That Roared: Hybrid Exchange Strategies as a Source of Productive Tension at the Boundary of Overlapping Institutions." *American Journal of Sociology* 116:341–88.

Murray, Fiona, Phillipe Aghion, Mathias Dewatripont, Julian Kolev, and Scott Stern. 2010. "Of Mice and Academics: Examining the Effect of Openness on Innovation." *American Journal of Sociology* 116:341–88.

Murray, Fiona, and Scott Stern. 2007. "Do Formal Intellectual Property Rights Hinder the Free Flow of Scientific Knowledge? An Empirical Test of the Anti-Commons Hypothesis." *Journal of Economic Behavior and Organization* 63:648–87.

Nadiri, M. Ishaq, and Theofanis P. Mamuneas. 1991. "The Effects of Public Infrastructure and R&D Capital on the Cost Structure and Performance of U.S. Manufacturing Industries." NBER working paper no. 3887. National Bureau of Economic Research, Cambridge, MA.

NASULGC, 2009. "Competitiveness of Public Research Universities & Consequences for the Country: Recommendations for change." http://www.aplu.org/document.doc?id=1561.

National Academy of Sciences. 1958. *Doctorate Production in United States Universities 1936–1956 with Baccalaureate Origins of Doctorates in Sciences, Arts and Humanities*. Washington, DC: National Research Council.

———. 2007. *Rising above the Gathering Storm: Energizing and Employing America for a Brighter Economic Future*. Washington, DC: National Academy of Sciences.

National Institute of General Medical Sciences. 2007a. *Report of the Protein Structure Initiative Assessment Panel*. National Advisory General Medical Sciences Council Working Group Panel for the Assessment of the Protein Structure Initiative. Bethesda, MD: NIGMS. http://www.nigms.nih.gov/News/Reports/PSIAssessmentPanel2007.htm.

———. [2007b]. "Update on NIH Peer Review." PowerPoint distributed to NIGMS Council. Bethesda, MD: NIGMS.

———. 2009a. *50 Years of Protein Structure Determination Timeline.* Bethesda, MD: NIGMS. http://publications.nigms.nih.gov/psi/timeline_text.html.

———. 2009b. *Glue Grants.* Bethesda, MD: NIGMS. http://www.nigms.nih.gov/Initiatives/Collaborative/GlueGrants.

———. 2009c. "NIGMS Invites Biologists to Join High-Throughput Structure Initiative." *NIH News,* February 12. http://www.nih.gov/news/health/feb2009/nigms-12.htm.

———. 2011. *Research Network.* (The NIH Pharmacogenomics Research Network [PGRN].) Bethesda, MD: NIGMS. http://www.nigms.nih.gov/Initiatives/PGRN/Network.

National Institutes of Health. 2008. "NIH Awards First EUREKA Grants for Exceptionally Innovative Research." *NIH News,* September 3. http://www.nih.gov/news/health/sep2008/nigms-03.htm.

———. 2009a. "Biographical Sketch Format Page," PHS 298/2590, April. Bethesda, MD: NIH. http://grants.nih.gov/grants/funding/phs398/biosketchsample.pdf.

———. 2009b. *Biomedical Research and Development Price Index.* Bethesda, MD: NIH. http://officeofbudget.od.nih.gov/pdfs/FY09/BRDPI%20Table%20of%20Annual%20Values_02_01_2009_2014.pdf.

———. 2009c. "NIH Announces 115 Awards to Encourage High-Risk Research and Innovation." *NIH News,* September 24. http://www.nih.gov/news/health/sep2009/od-24.htm.

———. 2009d. *NIH ARRA FY 2009 Funding.* Bethesda, MD: NIH. http://report.nih.gov/UploadDocs/Final_NIH_ARRA_FY2009_Funding.pdf.

———. 2009e. *National Institutes of Health (NIH) Extramural Data Book, Fiscal Year 2008.* Office of Extramural Research. Bethesda, MD: NIH. http://report.nih.gov/ndb/pdf/ndb_2008_Final.pdf.

———. 2009f. *Research Project Success Rates by NIH Institute for 2008.* Bethesda, MD: NIH. http://report.nih.gov/award/success/Success_ByIC.cfm.

———. 2009g. *Support of NIGMS Program Project Grants (P01).* Bethesda, MD: NIH. http://grants.nih.gov/grants/guide/pa-files/PA-07-030.html.

———. 2010. "Ruth L. Kirschstein National Research Service Award (NRSA) Stipends, Tuition/Fees and Other Budgetary Levels Effective for Fiscal Year 2010." Bethesda, MA: Office of Extramural Research. http://grants.nih.gov/grants/guide/notice-files/NOT-OD-10-047.html.

———. 2011. "Overview: NIH Director's Pioneer Award." NIH Common Fund, Division of Program Coordination, Planning and Strategic Initiatives. Bethesda, MA: NIH. http://commonfund.nih.gov/pioneer/.

National Opinion Research Center. 2008. *Doctorate Recipients from United States Universities, Selected Tables 2007.* Chicago: National Opinion Research Center.

National Postdoctoral Association. 2010. "About the NPA." National Postdoctoral Association website. http://www.nationalpostdoc.org/about-the-npa.

National Research Council. 1998. *Trends in the Early Careers of Life Scientists.* Committee on Dimensions, Causes and Implications of Recent Trends in the Careers of Life Scientists. Washington, DC: National Academies Press.

———. 2000. *Forecasting Demand and Supply of Doctoral Scientists and Engineers: Report of a Workshop on Methodology.* Washington, DC: National Academies Press.

———. 2005. *Bridges to Independence: Fostering the Independence of New Investigators in Biomedical Research.* Washington, DC: National Research Council.

———. 2011. *Research Training in the Biomedical, Behavioral, and Clinical Research Sciences.* Washington, DC: National Academies Press.

National Science Board. 2000. *Science and Engineering Indicators: 2000.* Arlington, VA: National Science Foundation. http://www.nsf.gov/statistics/seind00/.

———2002. *Science and Engineering Indicators* 2002. Artlinglton, VA., Nataional Science Foundation. http://www.nsf.gov/statistics/seind02/.

———. 2004. *Science and Engineering Indicators.* Arlington, VA: National Science Foundation. http://www.nsf.gov/statistics/seind04/.

———. 2006. Science and Engineering Indicators. Arlington, VA: National Science Foundation. http://www.nsf.gov/statistics/seind06/

———. 2007. "National Science Board Approves NSF Plan to Emphasize Transformative Research." Press release 07-097, August 9. Arlington, VA: National Science Foundation. http://www.nsf.gov/nsb/news/news_summ.jsp?cntn_id=109853&org=NSF.

———. 2008. *Science and Engineering Indicators.* Arlington, VA: National Science Foundation. http://www.nsf.gov/statistics/seind08/pdf/cov_v2.pdf.

———. 2010. *Science and Engineering Indicators: 2010.* Arlington, VA: National Science Foundation. http://www.nsf.gov/statistics/seind10/.

National Science Foundation. 1968. "Technology in Retrospect and Critical Events in Science." NSF C535. Unpublished manuscript prepared by IIT Research Institute, Chicago.

———. 1977. *Characteristics of Doctoral Scientists and Engineers in the United States 1975.* NSF-77-309.

———. 1989. *The State of Academic Science and Engineering.* Arlington, VA: National Science Foundation.

———. 1996. *Characteristics of Doctoral Scientists and Engineers in the United States 1993.* NSF-96-302.

———. 2004. *Federal Funds for Research and Development: Fiscal Years 1973–2003: Federal Obligations for Research to Universities and Colleges by Agency and Detailed Field of Science and Engineering.* NSF 04-332. National Center for Science and Engineering Statistics. Arlington, VA: National Science Foundation. http://www.nsf.gov/statistics/nsf04332/.

———. 2006. *Country of Citizenship of Non-U.S. Citizen Doctorate Recipients by Visa Status: 1960–1999.* U.S. Doctorates in the 20th Century. Arlington, VA: National Science Foundation.

———. 2007a. *Asia's Rising Science and Technology Strength: Comparative Indicators for Asia, the European Union, and the United States.* Arlington, VA: National Science Foundation.

———. 2007b. *Federal Funds for Research and Development: Fiscal Years 2004–2006. Detailed Statistical Tables.* NSF 07-323. Division of Science Resources

Statistics. Arlington, VA: National Science Foundation. http://www.nsf.gov/statistics/nsf07323/.

———. 2007c. *Impact of Proposal and Award Management Mechanisms, Final Report*. Arlington, VA: National Science Foundation. http://www.nsf.gov/pubs/2007/nsf0745/nsf0745.pdf.

———. 2007d. *Science and Engineering Research Facilities: Fiscal Year 2005*. NSF 07-325. National Center for Science and Engineering Statistics/Division of Science Resources Statistics. Arlington, VA: National Science Foundation. http://www.nsf.gov/statistics/nsf07325/.

———. 2008. *Graduate Students and Postdoctorates in Science and Engineering: Fall 2006*. Arlington, VA: National Science Foundation.

———. 2009a. *Characteristics of Doctoral Scientists and Engineers in the United States 2006*. National Center for Science and Engineering Statistics. Arlington, VA: National Science Foundation. http://www.nsf.gov/statistics/nsf09317/pdf/nsf09317.pdf.

———. 2009b. *Doctorate Recipients from U.S. Universities: Summary Report 2007–2008*. National Center for Science and Engineering Statistics. Arlington, VA: National Science Foundation. http://www.nsf.gov/statistics/nsf10309/pdf/nsf10309.pdf.

———. 2009c. *Report to the National Science Board on National Science Foundation's Merit Review Process, Fiscal Year 2008*. Arlington, VA: National Science Foundation. http://www.nsf.gov/nsb/publications/2009/nsb0943_merit_review_2008.pdf.

———. 2009d. *Survey of Research and Development Expenditures at Universities and Colleges*. National Center for Science and Engineering Statistics. Arlington, VA: National Science Foundation. http://www.nsf.gov/statistics/srvyrdexpenditures/.

———. 2010a. R&D Expenditures at Universities and Colleges by Source of Funds: FY 1953-2008. http://www.nsf.gov/statistics/nsf10311/pdf/tab1.pdf.

———. 2010b. Federal Funds for Research and Development Fiscal Years 2007-2009. NSF 10-305. Arlington, VA: National Science Foundation. http://www.sf.gov/statistics/nsf10305/.

———. 2010c. *WebCASPAR* (database). Arlington, VA: National Science Foundation. https://webcaspar.nsf.gov/;jsessionid=AC2E478221230456140B5016A9FF4292.

———. 2011a. National Survey of College Graduates. http://www.nsf.gov/statistics/showsrvy.cfm?srvy_CatID=3&srvy_Seri=7/.

———. 2011b. Survey of Doctorate Recipients. http://www.nsf.gov/statistics/srvydoctoratework/.

———. 2011c. Survey of Earned Doctorates. http://www.nsf.gov/statistics/srvydoctorates/.

———. 2011d. Survey of Graduate Students and Postdoctorates. http://www.nsf.gov/statistics/srvygradpostdoc/.

———. 2011e. Survey of Research and Development Expenditures at Universities. http://www.nsf.gov/statistics/srvyrdexpenditures/.

"Natural Experiments." 2011. *Wikipedia* http://en.wikipedia.org/wiki/Natural_experiment

Nature Editors. 2007. "Innovation versus Science?" *Nature* 448:839–40.

Nature Immunology Editor. 2006. "Mainstreaming the Alternative." *Nature Immunology* 7:535. doi:10.1038/ni0606-535.

Nelson, Richard R., Merton J. Peck, and Edward D. Kalachek. 1967. *Technology, Economic Growth, and Public Policy.* Washington, DC: Brookings Institution.

Nelson-Rees, Walter A. 2001. "Responsibility for Truth in Research." *Philosophical Transactions of the Royal Society B: Biological Sciences.* 356:849–51. doi 10.1098/rstb.2001.0873.

Newman, M. E. J. 2004. "Coauthorship Networks and Patterns of Scientific Collaboration." *Proceedings of the National Academy of Sciences of the United States of America* 101:5200–5.

New York Times Editors. 2010. "The Genome, 10 Years Later." *New York Times,* June 20, A28. http://www.nytimes.com/2010/06/21/opinion/21mon2.html.

Nikolai Lobachevsky. 2011. *Wilipedia.* http://en.wikipedia.org/wiki/Nikolai_Lobachevsky.

Nobel Foundation. 2011. "The Sveriges Riksbank Prize in Economic Sciences in Memory of Alfred Nobel 1971: Simon Kuznets." NobelPrize.org (website). http://nobelprize.org/nobel_prizes/economics/laureates/1971/.

Normile, Dennis. 2008. "Japan's Ocean Drilling Vessel Debuts to Rave Reviews." *Science* 319:1037.

———. 2009. "Science Windfall Stimulates High Hopes—and Political Maneuvering." *Science* 324:1375.

Northwestern University 2009, http://www.northwestern.edu/budget/documents/PDF5.pdf.

Norwegian Academy of Science and Letters. 2010. The Kavli Prize (website). http://www.kavliprize.no/.

Nyrén, Pal. 2007. "The History of Pyrosequencing." *Methods in Molecular Biology* 373:1–14.

Office of Research Integrity, U.S. Department of Health and Human Services. http://ori.hhs.gov/misconduct/cases/Goodwin_Elizabeth.shtml.

Office of the Executive Vice President. 2010. "Allston: Path Forward in Allston." Harvard University, Cambridge, MA. http://www.evp.harvard.edu/allston.

Oklahoma State University, 2009, *2008–2009 Faculty Salary Survey by Discipline.* Office of Institutional Research and Information Management.

Olson, Steve. 1986. *Biotechnology: An Industry Comes of Age.* Washington, DC: National Academy Press.

Oreopoulos, Philip, Till von Wachter, and Andew Heisz. 2008. "The Short- and Long-Term Career Effects of Graduating in a Recession: Hysteresis and Heterogeneity in the Market for Graduate Students." IZA Discussion Paper No. 3578. Institute for the Study of Labor (IZA), Bonn, Germany.

Organisation for Economic Co-operation and Development. 2008. *OECD Science, Technology, and Industry Outlook 2008.* Paris: Organisation for Economic Co-operation and Development. http://www.oecd.org/document/19/0,3746,en_2649_34273_46680723_1_1_1_1,00.html.

———. 2010. *Main Science and Technology Indicators.*

Overbye, Dennis. 2007. "A Giant Takes on Physics' Biggest Questions." *New York Times*, May 15, F1.

Oyer, Paul. 2006. "Initial Labor Market Conditions and Long-Term Outcomes for Economists." *Journal of Economic Perspectives* 20:143–60.

Pain, Elizabeth. 2008. "Science Careers. Playing Well with Industry." *Science* 319: 1548–51.

Paynter, Nina P., Daniel I. Chasman, Guillaume Paré, Julie E. Buring, Nancy R. Cook, Joseph P. Miletich, and Paul M Ridker. 2010. "Association between a Literature-Based Genetic Risk Score and Cardiovascular Events in Women." *Journal of the American Medical Association* 303:631–7.

Pelekanos, Adelle. 2008. "Money Management for Scientists: Lab Budgets and Funding Issues for Young PIs." *Science Alliance eBriefing* (New York Academy of Sciences), June 16.

Pelz, Donald C., and Frank M. Andrews. 1976. *Scientists in Organizations*. Ann Arbor: Institute for Social Research, University of Michigan.

Penning, Trevor. 1998. "The Postdoctoral Experience: An Associate Dean's Perspective." *The Scientist* 12:9.

Pennisi, Elizabeth. 2006. "Genomics. On Your Mark. Get Set. Sequence!" *Science* 314:232.

Peota, Carmen. 2007. "Biomedical Building Boom." *Minnesota Medicine* 90:18–9. http://www.minnesotamedicine.com/PastIssues/February2007/PulseBiomedicalFebruary2007/tabid/1705/Default.aspx.

Pezzoni, Michelle, Valerio Sterzi, and Francesco Lissoni. 2009. "Career Progress in Centralized Academic Systems: An Analysis of French and Italian Physicists." Knowledge, Internationalization, and Technology Studies (KITeS) Working Paper No. 26. Luigi Bocconi University, Milan, Italy.

Phillips, Michael. 1996. "Math PhDs Add to Anti-Foreigner Wave: Scholars Facing High Jobless Rate Seek Immigration Curbs." *Wall Street Journal*, September 4, A2.

Phipps, Polly, James W. Maxwell, and Colleen A. Rose. 2009. "2008 Annual Survey of the Mathematical Sciences in the United States (Second Report) (and Doctoral Degrees Conferred 2007–2008, Supplementary List)." *Notices of the American Mathematical Society* 56:828–43. http://www.ams.org/notices/200907/rtx090700828p.pdf.

Pines Lab. 2009. "The Pines Lab." Chemistry Department, University of California–Berkeley. http://waugh.cchem.berkeley.edu/.

Pollack, Andrew. 2011. "Taking DNA Sequencing to the Masses." *New York Times*, January 4. http://www.nytimes.com/2011/01/05/health/05gene.html.

"The Power of Serendipity." 2007. *CBS Sunday Morning* (website), October 7. http://www.cbsnews.com/stories/2007/10/05/sunday/main3336345.shtml.

"Protein Structure." 2009. *Wikipedia*. http://en.wikipedia.org/wiki/Protein_structure.

"PubChem." 2009. *Wikipedia*. http://en.wikipedia.org/wiki/PubChem.

Puljak, Livia, and Wallace D. Sharif. 2009. "Postdocs' Perceptions of Work Environment and Career Prospects at a US Academic Institution." *Research Evaluation* 18:411–5.

Quake, Stephen. 2009. "Letting Scientists Off the Leash." *New York Times Blog,* February 10.

Rabinow, Paul. 1997. *Making PCR: A Story of Biotechnology.* Chicago: University of Chicago Press.

RCSB Protein Data Bank. 2009. *A Resource for Studying Biological Macromolecules.* http://www.rcsb.org/pdb/.

Regets, Mark. 2005. "Foreign Students in the United States." Paper presented at Dialogue Meeting on Migration Governance: European and North American Perspectives. Brussels, Belgium, June 27.

Reid, T. R. 1985. *The Chip: How Two Americans Invented the Microchip and Launched a Revolution.* New York: Random House.

Research Assessment Exercise. 2008. "Quality Profile Will Provide Fuller and Fairer Assessment of Research." February 11. Higher Education Funding Council for England (HEFCE), the Scottish Funding Council (SFC), the Higher Education Funding Council for Wales (HEFCW), and the Department for Employment and Learning, Northern Ireland. http://www.rae.ac.uk/news/2004/fairer.htm.

"Richter Scale." 2010. *Wikipedia.* http://en.wikipedia.org/wiki/Richter_magnitude _scale.

Rilevazione Nuclei. 2007. "Ottavo Rapporto Sullo Stato Del Sistema Universitario." Comitato Nazionale per la Valutazione del Sistema Universitario (CNVSU), Ministero dell'Istruzione dell'Università e delle Ricerca, Italy. http://www .unisinforma.net/w2d3/v3/download/unisinforma/news/allegati/upload/sin-tesi%20del%20rapporto.pdf.

Rivest, Ron L., Adi Shamir, and Leonard Adleman. 1978. "A Method for Obtaining Digital Signatures and Public-Key Cryptosystems." *Communications of the ACM* 21:120–6.

Roberts, Richard J. 1993. "Autobiography." Nobelprize.org (website). http://nobel-prize.org/nobel_prizes/medicine/laureates/1993/roberts-autobio.html.

Robinson, Sara. 2003. "Still Guarding Secrets after Years of Attacks, RSA Earns Accolades for Its Founders." *SIAM News* 36 (5): 28.

Rockey, Sally. 2010. Presentation made at the 101st Advisory Committee to the Director, National Institutes of Health, December 9, 2010, Bethesda, Maryland.

Rockwell, Sara. 2009. "The FDP Faculty Burden Survey." *Research Management Review,* 61:29–44.

Roe, Anne. 1953. *The Making of a Scientist.* New York: Dodd, Mead.

Romer, Paul. 1990. "Endogenous Technological Change." *Journal of Political Economy* 98:S71-S102

———. 1994. "The Origins of Endogenous Growth." *Journal of Economic Perspectives* 8:3–22.

———. 2000. "Should the Government Subsidize Supply or Demand in the Market for Scientists and Engineers?" NBER Working Paper 7723. National Bureau of Economic Research, Cambridge, MA.

———. 2002. "Economic Growth." In *The Concise Encyclopedia of Economics.* Edited by David R. Henderson. Indianapolis, IN: Liberty Fund, Library of Economics and Liberty (website). http://www.econlib.org/library/Enc1/EconomicGrowth.html.

Rosenberg, Nathan. 2004. "Science and Technology: Which Way Does the Causation Run?" Paper presented at the opening of the Center for Interdisciplinary Studies of Science and Technology. Stanford, CA, November 1, 2004. http://www.crei.cat/activities/sc_conferences/23/papers/rosenberg.pdf.

———. 2007. "Endogenous Forces in Twentieth-Century America." In *Entrepreneurship, Innovation, and the Growth Mechanism of the Free-Enterprise Economies,* 80–99. Edited by Eytan Sheshinski, Robert J. Strom, and William J. Baumol. Princeton, NJ: Princeton University Press.

Rosenberg, Nathan, and L. E. Birdzell Jr. 1986. *How the West Grew Rich: The Economic Transformation of the Industrial World.* New York: Basic Books.

Rosenberg, Nathan, and Richard Nelson. 1994. "American Universities and Technical Advance in Industry." *Research Policy* 23:323–48.

Rosovsky, Henry. 1991. *The University: An Owner's Manual.* New York: W. W. Norton.

Ross, Joseph S., Kevin P. Hill, David S. Egilman, and Harlan M. Krumholz. 2008. "Guest Authorship and Ghostwriting in Publications Related to Rofecoxib: A Case Study of Industry Documents from Rofecoxib Litigation." *Journal of the American Medical Association* 299:1800–12.

Rothberg Institute for Childhood Diseases. 2009. "Board of Directors." http://www.childhooddiseases.org/scientists.html.

Roussel, Nicolas. 2011. *scHolar Index* (software). http://interaction.lille.inria.fr/~roussel/projects/scholarindex/index.cgi.

Ryoo, Jaewoo, and Sherwin Rosen. 2004. "The Engineering Labor Market." *Journal of Political Economy* 112:S110–38.

Sacks, Frederick. 2007. "Is the NIH Budget Saturated? Why Hasn't More Funding Meant More Publications?" *The Scientist,* November 19.

Sánchez Laboratory. 2010. "Thomas Hunt Morgan." Sánchez Laboratory Regeneration Research, Genetic Science Learning Center, University of Utah, Salt Lake City. http://planaria.neuro.utah.edu/research/Morgan.htm.

Sauermann, Henry. 2011. Presentation made April 19, at workshop "Measuring the Impacts of Federal Investments in Research." National Academies, Washington, DC.

Sauermann, Henry, Wesley Cohen, and Paula Stephan. 2010. "Complicating Merton: The Motives, Incentives and Innovative Activities of Academic Scientists and Engineers." Unpublished manuscript.

Sauermann, Henry, and Michael Roach. 2011. "The Price of Silence: Scientists' Trade Offs Between Publishing and Pay." Unpublished paper, Georgia Institute of Technology, Atlanta, GA.

Sauermann, Henry, and Paula Stephan. 2010. "Twins or Strangers: Differences and Similarities between Industrial and Academic Science." NBER Working Paper 16113. National Bureau of Economic Research, Cambridge, MA.

Saxenian, AnnaLee. 1995. "Creating a Twentieth Century Technical Community: Frederick Terman's Silicon Valley." Paper presented at the inaugural symposium on The Inventor and the Innovative Society, the Lemelson Center for the Study of Invention and Innovation, National Museum of American History, Smithsonian Institution. Washington, DC, November 10–11.

Scarpa, Toni. 2010. "Peer Review at NIH: A Conversation with CSR Director Toni Scarpa." *The Physiologist* 53:65, 67–9.

Scherer, Frederic M. 1967. Review of *Technology, Economic Growth and Public Policy,* by Richard R. Nelson, M. J. Peck, and E. D. Kalacheck. *Journal of Finance* 22:703–4.

———. 1998. "The Size Distribution of Profits from Innovation." *Annales d'Economie et de Statistique* 49/50:495–516.

Schulze, Günther. 2008. "Tertiary Education in a Federal System—the Case of Germany." In *Scientific Competition: Theory and Policy,* 35–66. Edited by Max Albert, Dieter Schmidtchen, and Stefan Voigt. Tübingen: Mohr Siebeck.

Science Editors. 2000. "Best and the Brightest Avoiding Science." *Science* 288:43.

Scientist Staff. 2010. "Top Ten Innovations 2010." *The Scientist* 24 (12): 47. http://www.the-scientist.com/2010/12/1/47/1/.

Service, Robert F. 2008. "Applied Physics. Tiny Transistor Gets a Good Sorting Out." *Science* 321:27.

Shapin, Steven. 2008. *The Scientific Life: A Moral History of a Late Modern Vocation.* Chicago: University of Chicago Press.

Shi, Yigong, and Yi Rao. 2010. "China's Research Culture." *Science* 328:1128.

Sigma Xi. 2003. *Postdoc Countries of Citizenship and Degree Earned.* http://postdoc.sigmaxi.org/results/tables/table8.

Simonton, Dean Keith. 2004. *Creativity in Science: Chance, Logic, Genius, and Zeitgeist.* Cambridge, United Kingdom: Cambridge University Press.

Simpson, John. 2007. "Share the Fruits of State Funded Research, Consumer Watchdog, August 11.

SKA 2011. http://www.skatelescope.org/the-location/.

SLAC National Accelerator Laboratory. 2010. *Linac Coherent Light Source News.* http://lcls.slac.stanford.edu/news.aspx.

Slaughter, Shelia, and Gary Rhodes. 2004. *Academic Capitalism and the New Economy: Markets, State and Higher Education.* Baltimore, MD: The Johns Hopkins University Press.

Sloan Digital Sky Survey. 2010. *Mapping the Universe: The Sloan Digital Sky Survey* (website). http://www.sdss.org.

Sobel, Dava. 1996. *Longitude: The True Story of a Lone Genius Who Solved the Greatest Scientific Problem of His Time.* London: Fourth Estate.

Sousa, Rui. 2008. "Research Funding: Less Should Be More." *Science* 322: 1324–25.

Stanford University. 2009a. "Economic Impact." Wellspring of Innovation (website). Palo Alto, CA. http://www.stanford.edu/group/wellspring/economic.html.

———. 2009b. *Wellspring of Innovation* (website). Palo Alto, CA. http://www.stanford.edu/group/wellspring/index.html.

———. 2009c. Stanford University Budget Plan. Palo Alto, CA. http://www.stanford.edu/dept/pres-provost/budget/plans/BudgetBookFY10.pdf.

———. 2010a. "Postdoctoral Scholars: Funding Guidelines." Palo Alto, CA. http://postdocs.stanford.edu/handbook/salary.html.

———. 2010b. "Stanford Graduate Fellowships in Science and Engineering." Vice Provost for Graduate Education. Palo Alto, CA. http://sgf.stanford.edu/.

———. 2010c. "Tuition and Fees." Palo Alton, CA. http://studentaffairs.stanford.edu/registrar/students/tuition-fees.

Stephan, Paula. 2004. "Robert K. Merton's Perspective on Priority and the Provision of the Public Good Knowledge." *Scientometrics* 60:81–87.

———. 2007a. "Early Careers for Biomedical Scientists: Doubling (and Troubling) Outcomes." Presentation at Harvard University for the Science and Engineering Workforce Project (SWEP), National Bureau of Economic Research (NBER). Cambridge, MA, February 27. http://www.nber.org/~sewp/Early%20Careers%20for%20Biomedical%20Scientists.pdf.

———. 2007b. "Social and Economic Perspective." Presentation at Modeling Scientific Workforce Diversity, National Institutes of General Medicine, National Institutes of Health. Bethesda, MD, October 3.

———. 2007c. "Wrapping It up in a Person: The Location Decision of New PhDs Going to Industry." In *Innovation Policy and the Economy*, Vol. 7, 71–98. Edited by Adam Jaffe, Josh Lerner, and Scott Stern. Cambridge, MA: MIT Press.

———. 2008. "Job Market Effects on Scientific Productivity." In *Scientific Competition: Theory and Policy*, 11–29. Edited by Max Albert, Dieter Schmidtchen, and Stefan Voigt. Tübingen: Mohr Siebeck.

———. 2009. "Tracking the Placement of Students as a Measure of Technology Transfer." In *Advances in the Study of Entrepreneurship, Innovation, and Economic Growth*, 113–40. Edited by Gary Libecap. London: Elsevier.

———. 2010a. "The Economics of Science." In *Handbook of the Economics of Innovation*, Vol. 1, Chapter 5. Edited by Bronwyn Hall and Nathan Rosenberg. London: Elseivier.

———. 2010b. "The 'I's' Have It: Immigration and Innovation, the Perspective from Academe." In *Innovation Policy and the Economy*, Vol. 10, 83–127. Edited by Josh Lerner and Scott Stern. Cambridge, MA: MIT University Press.

Stephan, Paula, Grant Black, and Tanwin Chang. 2007. "The Small Size of the Small Scale Market: The Early-Stage Labor Market for Highly Skilled Nanotechnology Workers." *Research Policy* 36:887–92.

Stephan, Paula, and Stephen Everhart. 1998. "The Changing Rewards to Science: The Case of Biotechnology." *Small Business Economics* 10:141–51.

Stephan, Paula, Shif Gurmu, A.J. Sumell, and Grant Black. 2007. "Who's Patenting in the University?" *Economics of Innovation and New Technology*, Vol 61(2): 71–99.

Stephan, Paula, and Sharon Levin. 1992. *Striking the Mother Lode in Science: The Importance of Age, Place, and Time*. New York: Oxford University Press.

———. 1993. "Age and the Nobel Prize Revisited." *Scientometrics* 28:387–99.

———. 2002. "The Importance of Implicit Contracts in Collaborative Scientific Research." In *Science Bought and Sold: Essays in the Economics of Science*, Edited by Philip Mirowski and Esther-Mirjam Sent. Chicago: University of Chicago Press.

———. 2007. "Foreign Scholars in U.S. Science: Contributions and Costs." In *Science and the University*, Edited by Paula Stephan and Ronald G. Ehrenberg. Madison, WI: University of Wisconsin Press.

Stephan, Paula, and Jennifer Ma. 2005. "The Increased Frequency and Duration of the Postdoctoral Career Stage." *American Economic Review Papers and Proceedings* 95:71–75.

Stephan, Paula, A. J. Sumell, Grant Black, and James D. Adams. 2004. "Doctoral Education and Economic Development: The Flow of New PhDs to Industry." *Economic Development Quarterly* 18:151–67.

Stern, Scott. 2004. "Do Scientists Pay to Be Scientists?" *Management Science* 50:835–53.

Stevens, Ashley, J. J. Jensen, K. Wyller, P. C. Kilgore, S. Chatterjee, and M. L. Rohrbaugh. 2011. "The Role of Public-Sector Research in the Discovery of Drugs and Vaccines." *The New England Journal of Medicine* 364, no. 6 (2011):535–41.

Stigler, Stephen. 1980. "Stigler's Law of Eponymy." *Transactions of the New York Academy of Sciences* 39:147–58.

Stokes, Donald. 1997. *Pasteur's Quadrant*. Washington, DC: Brookings Institution Press.

Stone, Richard, and Hao Xin. 2010. "Supercomputer Leaves Competition and Users in the Dust." *Science*, 330:746–747.

Subcommittee on Basic Research. 1995. *Reshaping the Graduate Education of Scientists and Engineers: NAS's Committee on Science, Engineering, and Public Policy Report*. (Hearing before the Subcommittee on Basic Research of the Committee on Science, U.S. House of Representatives, 104th Cong, 1st sess, July 13, 1995.) Washington, DC: U.S. Government Printing Office. http://www.archive.org/stream/reshapinggraduat1995unit/reshapinggraduat1995unit_djvu.txt.

Summers, Lawrence H. 2005. "Remarks at NBER Conference on Diversifying the Science & Engineering Workforce." January 14. Office of the President, Harvard University, Cambridge, MA. http://president.harvard.edu/speeches/summers_2005/nber.php.

"Supercomputer." 2009. *Wikipedia*. http://en.wikipedia.org/wiki/Supercomputer.

Tanyildiz, Esra. 2008. "The Effects of Networks on Institution Selection by Foreign Doctoral Students in the U.S." PhD diss., Georgia State University.

Teitelbaum, Michael S. 2003. "Do We Need More Scientists?" Alfred P. Sloan Foundation. *Public Interest*, No. 153, Fall. www.sloan.org/assets/files/teitelbaum/publicinterestteitelbaum2003.pdf.

Tenenbaum, David. 2003. "Nobel Prizefight." University of Wisconsin: *The Why? Files* (website), October 23. http://www.whyfiles.org/188nobel_mri/.

Texas A&M University. 2009. *Executive Summary. Survey of Earned Doctorates: 1958 through 2007*. Office of Institutional Studies and Planning. College Station: Texas A&M University. http://www.tamu.edu/customers/oisp/reports/survey-earned-doctorates-sed-1958–2007.pdf.

Thimann, Kenneth V., and Walton C. Galinat. 1991. "Paul Christoph Mangelsdorf (July 20, 1899–July 22, 1989)." *Proceedings of the American Philosophical Society*, 135:468–72.

Thompson, Tyler B. 2003, "An Industry Perspective on Intellectual Property from Sponsored Research." *Research Management Review*, 13:1-9.

Thursby, Jerry, Anne Fuller, and Marie Thursby. 2009. "U.S. Faculty Patenting: Inside and Outside the University." *Research Policy* 38:14–25.

Thursby, Jerry, and Marie Thursby. 2006. "Where Is the New Science in Corporate R&D?" *Science* 314:1547–48.

———. 2010a. "Has the Bayh-Dole Act Compromised Basic Research?" Unpublished manuscript. Georgia Institute of Technology, Atlanta.

———. 2010b. "University Licensing: Harnessing or Tarnishing Faculty Research?" In *Innovation, Policy and the Economy, Vol. 10*, Edited by Josh Lerner and Scott Stern. Cambridge, MA: MIT University Press.

Time Staff. 1948. "The Eternal Apprentice." Time Magazine 58, November 8. http://www.time.com/time/magazine/article/0,9171,853367,00.html.

Timmerman, Luke. 2010. "Illumina CEO Jay Flatley on How to Keep an Edge in the Fast-Paced World of Gene Squencing." *XConomy: San Diego*, April 6. http://www.xconomy.com/san-diego/2010/04/06/illumina-ceo-jay-flatley-on-how-to-keep-an-edge-in-the-fast-paced-world-of-gene-sequencing/.

TMT Project. 2009. "Thirty Meter Telescope Selects Mauna Kea." Thirty Meter Telescope Press Release, July 21. http://www.tmt.org/news/site-selection.htm.

TOP500. 2011. "Top500 2011: http://www.top500.org/."

Toutkoushian, Robert, and Valerie Conley. 2005. "Progress for Women in Academe, yet Inequities Persist: Evidence from NSOPF: 99." *Research in Higher Education* 46:1–28.

Townes, Charles H. 2003. "The First Laser." In *A Century of Nature: Twenty-One Discoveries That Changed Science and the World*, Edited by Laura Garwin and Tim Lincoln. Chicago: University of Chicago Press.

Trainer, Matthew. 2004. "The Patents of William Thomson (Lord Kelvin)." *World Patent Information* 26:311–17.

Tuition Remission Task Force. 2006. "Final Report: Tuition Remission Task Force." University of Wisconsin, Madison. February 17. http://www.secfac.wisc.edu/trtffinalreport.pdf.

Turkish Academic Network and Information Centre. 2008. Home Page. http://www.ulakbim.gov.tr/eng/.

United for Medical Research. 2011. *An Economic Engine: NIH Research, Employment, and the Future of the Medical Innovation Sector.*

U.S. Bureau of Labor Statistics. 2011a. "Consumer Price Index: All Urban Consumers." March 17. U.S. Department of Labor. ftp://ftp.bls.gov/pub/special.requests/cpi/cpiai.txt.

———. 2011b. "Table 1. Union Affiliation of Employed Wage and Salary Workers by Selected Characteristics." *Economic News Release*, January 21. U.S. Department of Labor, Division of Labor Force Statistics. http://www.bls.gov/news.release/union2.t01.htm.

U.S. Census Bureau. 2011. "Births, Deaths, Marriages, and Divorces: Life Expectancy." *The 2011 Statistical Abstract: The National Data Book*. U.S. Census Bureau (website). http://www.census.gov/compendia/statab/cats/births_deaths_marriages_divorces/life_expectancy.html.

U.S. Citizenship and Immigration Services. 2011. "Citizenship through Naturalization." April 08. U.S. Department of Homeland Security. http://www.uscis.gov/portal/site/uscis/menuitem.eb1d4c2a3e5b9ac89243c6a7543f6d1a/?vgnext

channel=d84d6811264a3210VgnVCM100000b92ca60aRCRD&vgnextoid=
d84d6811264a3210VgnVCM100000b92ca60aRCRD.

U.S. Department of Labor. 2009. *International Comparisons of GDP Per Capita and Per Employed Person: 17 Countries, 1960–2008.* Division of International Labor Comparisons. Washington, DC: U.S. Government Printing Office. http://www.bls.gov/fls/flsgdp.pdf.

U.S. Patent and Trademark Office. 2010. "U.S. Patent Statistics Chart, Calendar Years 1963–2010." Patent Technology Monitoring Team (PTMT). http://www.uspto.gov/web/offices/ac/ido/oeip/taf/us_stat.htm.

University of California Newsroom. 2009. "Regents Approve Fiscal Plan, Furloughs." July 16. University of California website. http://www.universityof-california.edu/news/article/21511.

University of Chicago, Office of Technology and Intellectual Property. [2007.] *Bringing Innovation to Life: Five-Year Report.* No. 4-07/8M/VPR07777. Chicago: University of Chicago Press. http://www.uchicago.edu/pdfs/UChicago-Tech_Bringing_Innovation_to_Life_5yrRpt.pdf.

University of Georgia. 2010. *Executive Summary: University of Georgia Proposal for Reuse of the Navy Supply Corps School Property.* Athens: University of Georgia. http://www.uga.edu/news/artman/publish/01–17_UGA_Navy_School_Proposal.shtml.

University of Michigan. 2010. "Budget Update: University Budget Information." http://www.vpcomm.umich.edu/budget/ubudget.html.

University of North Carolina at Chapel Hill. 2010. "Faculty Salaries at Research (Very High Research Activity) and AAU Institutions, 2009–2010." Office of Institutional Research and Assessment. http://oira.unc.edu/faculty-salaries-at-research-and-aau-universities.html.

University of Virginia. 2010. http://www.virginia.edu/budget/Docs/2010-2011%20Budget%20Summary%20All%20Divisions.pdf.

Uzzi, Brian, Luis Amaral, and Felix Reed-Tsochas. 2007. "Small-World Networks and Management Science Research: A Review." *European Management Review* 4:77–91.

Vance, Tracy. 2011. "Academia Faces PhD Overload," *Genome Technology,* March, pp. 38-44.

Varian, Hal R. 2004. "Review of Mokyr's *Gifts of Athena.*" *Journal of Economic Literature* 42:805–10.

Venkataram, Bina, 2011. "$1 Million to Inventor of Tracker for A.L.S., *New York Times,* February 3.

Veugelers, Reinhilde. 2011. "Higher Order Moments in Science." Presentation at the conference, "Economics of Science. Where Do We Stand?" Paris, *Observatoire des Sciences et Techniques,* April 4–5, 2011.

Vogel, Gretchen. 2000. "The Mouse House as a Recruiting Tool." *Science* 288:254–5.

———. 2006. "Basic Science Agency Gets a Tag-Team Leadership." *Science* 313:1371.

———. 2010. "To Scientists' Dismay, Mixed-up Cell Lines Strike Again." *Science* 329:104.

Von Hippel, Eric. 1994. "'Sticky Information' and the Locus of Problem Solving: Implications for Innovation." *Management Science* 40:429–43.

W. M. Keck Observatory. 2009. "About Keck: The Observatory." http://keckobservatory.org/about/the_observatory.

Wade, Nicholas. 2000. "Double Landmarks for Watson: Helix and Genome." *New York Times,* June 27.

———. 2009. "Cost of Decoding a Genome Is Lowered." *New York Times,* August 11.

Wagner, Erwin F., Timothy Stewart, and Beatrice Mintz. 1981. "The Human b-Globin Gene and a Functional Viral Thymidine Kinase Gene in Developing Mice." *Proceedings of the National Academy of Sciences of the United States of America* 78:5016–20.

Wagner, Thomas E., Peter Hoppe, Joseph Jollick, David Scholl, Richard Hodinka, and Janice Gault. 1981. "Microinjection of a Rabbit Beta Globin Gene into Zygotes and Its Subsequent Expression in Adult Mice and Their Offspring." *Proceedings of the National Academy of Sciences of the United States of America* 78:6376–80.

Wald, Chelsea, and Corinna Wu. 2010. "Of Mice and Women: The Bias in Animal Models." *Science* 327:1571–2.

Walsh, John P., Wesley M. Cohen, and Charlene Cho. 2007. "Where Excludability Matters: Material versus Intellectual Property in Academic Biomedical Research." *Research Policy* 36:1184–203.

Waltz, Emily. 2006. "Profile: Robert Tjian." *Biotechnology* 24:235.

Wang, Zhong L. 2011. Professor Zhong L. Wang's Nano Research Group (website). http://www.nanoscience.gatech.edu/zlwang/.

Weiss, Yoram, and Lee Lillard. 1982. "Output Variability, Academic Labor Contracts, and Waiting Times for Promotion." In *Research in Labor Economics,* Vol. 5, 157–88. Edited by Ronald G. Ehrenberg. Greenwich: JAI Press.

Wendler, Cathy, Brent Bridgeman, Fred Cline, Catherine Millett, JoAnn Rock, Nathan Bell, and Patricia McAllister. 2010. *The Path Forward: The Future of Graduate Education in the United States.* Princeton: Educational Testing Service.

Wenniger, Mary Dee. 2009. "Nancy Hopkins: 'The Exception' Relates Her Story at MIT." Women in Higher Education (website). http://wihe.com/printArticle.jsp?id=18218.

Wertheimer, Linda K. 2007. "Harvard Rethinks Allston." *Boston Globe,* December 12.

Wessel, David. 2010. "U.S. Keeps Foreign PhDs." *Wall Street Journal,* January 27.

Whitehead. 2010. http://www.wi.mit.edu/research/postdoc/home_ext.php?p=benes_ext.

White Research Group. 2011. White Lab: Synthesis-Diven Catalysis. (website). Department of Chemistry, University of Illinois, Urbana-Champaign. http://www.scs.illinois.edu/white/index.php.

Whitton, Michael. 2010. "Finding Your h-Index (Hirsch Index) in Google Scholar." University of Southhampton Library Factsheet no. 3 (April). http://www.soton.ac.uk/library/research/bibliometrics/factsheet03-hindex-gs.pdf.

Williams, Heidi. 2010. "Intellectual Property Rights and Innovation: Evidence from the Human Genome." NBER Working Paper 16213. National Bureau of Economic Research, Cambridge, MA.

Wilson, Robin. 2000. "They May Not Wear Armani to Class, but Some Professors Are Filthy Rich." *Chronicle of Higher Education.* March 3, p. A16–8.

———. 2008. "Wisconsin's Flagship Is Raided for Scholars." *Chronicle of Higher Education* 54:A19. http://chronicle.com/article/Wisconsin-s-Flagship-Is/33652.

Wines, Michale. 2011. "A U.S.-China Odyssey: Building a Better Mouse Map." *New York Times,* January 28. http://www.nytimes.com/2011/01/29/world/asia/29china.html.

Winkler, Anne, Sharon Levin, and Paula Stephan. 2010. "The Diffusion of IT in Higher Education: Publishing Productivity of Academic Life Scientists." *Economics of Innovation and New Technology* 19:475–97.

Winkler, Anne, Sharon Levin, Paula Stephan, and Wolfgang Glanzel. 2009. "The Diffusion of IT and the Increased Propensity of Teams to Transcend Institutional Boundaries." Unpublished paper. Georgia State University.

Wolfe, Tom. 1983. "The Tinkerings of Robert Noyce: How the Sun Rose on the Silicon Valley." *Esquire,* December, 346–74.

Wolpert, Lewis, and Alison Richards. 1988. *A Passion for Science: Renowned Scientists Offer Vivid Personal Portraits of Their Lives in Science.* Oxford: Oxford University Press.

Wuchty, Stefan, Benjamin Jones, and Brian Uzzi. 2007. "The Increasing Dominance of Teams in Production of Knowledge." *Science* 316:1036–9.

Xie, Yu, and Kimberlee A. Shauman. 2003. *Women in Science: Career Processes and Outcomes.* Cambridge, MA: Harvard University Press.

Xin, Hao, and Dennis Normile. 2006. "Frustrations Mount over China's High-priced Hunt for Trophy Professors." *Science* 313:1721–3.

X Prize Foundation. 2009a. "About the Google Lunar X Prize." Google Lunar X prize website. http://www.googlelunarxprize.org/lunar/about-the-prize.

———. 2009b. "The Teams: Astrobotic." Google Lunar X Prize website. http://www.googlelunarxprize.org/lunar/teams/astrobotic.

———. 2011. Archon Genomics X Prize (website). http://genomics.xprize.org.

"X-Ray Crystallography." 2011. *Wikipedia.* http://en.wikipedia.org/wiki/X-ray_crystallography.

Zhang, Liang. 2008. "Do Foreign Doctorate Recipients Displace U.S. Doctorate Recipients at U.S. Universities?" In *Doctoral Education and the Faculty of the Future,* 209–23. Edited by Ronald G. Ehrenberg and Charlotte V. Kuh. Ithaca, NY: Cornell University Press.

Ziman, John M. 1968. *Public Knowledge: An Essay Concerning the Social Dimension of Science.* Cambridge, United Kingdom: Cambridge University Press.

Zimmer, Carl. 2010. "The Search for Genes Leads to Unexpected Places." *New York Times,* April 26, 17.

Zucker, Lynne G., Michael R. Darby, and Jeff Armstrong. 1998. "Geographically Localized Knowledge: Spillovers or Markets?" *Economic Inquiry* 36: 65–86.

————. 1999. "Intellectual Capital and the Firm: The Technology of Geographically Localized Knowledge Spillovers." NBER Working Paper 4946. National Bureau of Economic Research, Cambridge, MA.

Zucker, Lynne G., Michael R. Darby, and Marilynn B. Brewer. 1998. "Intellectual Human Capital and the Birth of U.S. Biotechnology Enterprises." *American Economic Review* 88:290–306.

Zuckerman, Harriet. 1992. "The Proliferation of Prizes: Nobel Complements and Nobel Surrogates in the Reward System of Science." *Theoretical Medicine and Bioethics* 13:217–31.

Acknowledgments

In 1996, I published an article in the *Journal of Economic Literature* entitled "The Economics of Science." I thought that phase of my life was over after it was published, and I returned to doing research on more narrowly focused topics in the area. This changed when, at a World Bank Conference in 2005, Nathan Rosenberg and Bronwyn Hall approached me, in their most persuasive way, about revisiting the topic for a chapter in a handbook they were editing on the economics of innovation. With more than a bit of trepidation I agreed, knowing that the field had grown considerably in the ensuing ten years since I had completed the original essay. I started the chapter in 2007, while a Wertheim fellow at Harvard University. The next year, on a follow-up stay at Harvard, I had occasion to have lunch with Elizabeth Knoll of Harvard University Press. In her ever-so-polite manner, she inquired if I were working on anything that might be of interest to the press. In a foolish moment, I sent her a copy of the chapter that I had just completed, not knowing quite what I had gotten myself into. Now I do, three years later, as I finish a manuscript that has taken almost two years of my professional life to write. The moral of the story: the next time Elizabeth invites me to lunch, say "No."

Along the way I have had help and encouragement from friends, colleagues, and family. I have also had the support of two foundations. The Alfred P. Sloan Foundation provided the funding that allowed me to focus full time on the project for a period of six months. They also provided resources to support a graduate research assistant, Erin Coffman, who has been extraordinarily helpful with analyzing data, preparing figures and tables, and organizing the references. The International Center for Economic Research (ICER) granted me a fellowship, which provided me the opportunity to spend three months in Turin, Italy, at the University of Torino, during the fall of 2009, working full time on the manuscript. I have also had the good fortune that my home department of economics at Georgia State University has given me the freedom and institutional support to work on the project.

Before acknowledging the numerous individuals who have contributed to this project, let me also say that my research and perspective have benefited considerably from my participation in several government advisory boards and panels. My first such experience came in 1996 when I joined the National Research Council Committee on Trends in the Early Careers of Life Scientists. That experience provided me considerable insight into the workings of university labs and the way in which they are staffed. The strong leadership of Shirley Tilghman and her commitment to providing graduate students a fair chance at getting a reasonable job after years of training was a take-home I will not forget. Since then, I have served on several other National Research Council committees, including Policy Implications of International Graduate Students and Postdoctoral Scholars in the United States, and the Board on Higher Education and Workforce. I always leave these meetings with an increased understanding of science and an appreciation that I get more than I give as a member. Beginning in the early 2000s, I had the occasion to serve on the Social, Behavioral, and Economics Advisory Board of the National Science Foundation. This provided firsthand experience in looking at issues faced by federal agencies tasked with supporting research. In 2004, I served as a member of the European Commission's High Level Expert Group on "Maximizing the Wider Benefits of Competitive Basic Research Funding at the European Level," which helped lay the foundation for the European Research Council. Finally, but by no means last, during the years 2006 to 2009 I served on the National Advisory General Medical Sciences Council (NIGMS), National Institutes of Health. I learned much from fellow council members and the able personnel at NIGMS, as I participated in discussions concerning how the Institute would spend its annual $2 billion budget.

I have been fortunate in writing the book to have access to and use of data from the Survey of Doctorate Recipients and the Survey of Earned Doctorates, both from the National Center for Science and Engineering Statistics, National Science Foundation. I must point out that the use of NSF data does not imply NSF endorsement of the research methods or conclusions contained in this book.

Now to the people. First, and foremost, there are my coauthors, who have contributed to some of the work that I discuss in this book and, more importantly, have contributed to my understanding of science. Sharon Levin heads the list. Our pattern of collaboration began more than forty-two years ago in Ann Arbor, Michigan, when we were graduate students in economics and spent long hours studying together for comprehensive exams. In the 1980s, we collaborated on studying the degree to which science is a young person's game and the way in which one's cohort affects scientific productivity. That research was published in the *American Economic Review* and as a book from Oxford University Press. We have continued to collaborate long after that, studying the foreign born (with support from the Alfred P. Sloan Foundation) and, most recently, along with Anne Winkler, studying the relationship of the diffusion of information technology to the productivity of scientists. That research has been generously supported by the Andrew W. Mellon

Foundation. Other coauthors include James D. Adams, David Audretsch, Tanwin Chang, Roger Clemmons, Wesley Cohen, Waverly Ding, Ron Ehrenberg, Chiara Franzoni, Wolfgang Glänzel, Jennifer Ma, Fiona Murray, and Giuseppe Scellato.

My research has also benefitted from colleagues or former colleagues at Georgia State, most of whom have been coauthors. They include Grant Black, Asmaa El-Ganainy, the late Stephen Everhart, Shif Gurmu, Richard Hawkins, Barry Hirsch, Mary Kassis, Baoyun Qiao, Albert Sumell, and Mary Beth Walker. Grant also gave generously of his time in helping me analyze certain NSF data used in the book.

In recent years I have been fortunate to have colleagues who share similar interests at the Georgia Institute of Technology—only five miles from Georgia State. I have especially benefited from talks with Mary Frank Fox, Matthew Higgins, Henry Sauermann, Jerry Thursby, Marie Thursby, and John Walsh. All but Marie and John have been coauthors.

I also have learned a great deal from colleagues at the National Bureau of Economic Research (NBER). My participation in the "higher education" research group at the NBER, led by Charles Clotfelter for many years, gave me numerous opportunities to interact with others who study universities. Since 2000, I have had the good fortune to participate in the Science and Engineering Workforce Project at the NBER, led by Richard Freeman and, initially, by Daniel Goroff, with financial support from the Alfred P. Sloan Foundation. Michael Teitelbaum of the Foundation was particularly supportive of the project. Over the ensuing years, Richard Freeman has provided considerable input to and enthusiasm for the research I have undertaken regarding the economics of science. He has been particularly generous with his time regarding this book project. I have benefited considerably from his suggestions and comments.

A number of people at various foundations and companies have provided information and support. Nirmala Kannankutty at the National Science Foundation deserves special mention for her prompt replies to all my data inquiries. Harriett Zuckerman of the Andrew W. Mellon Foundation has been supportive of my research ever since we first met in the early 1980s, as has Michael Teitelbaum of the Alfred P. Sloan Foundation. Walter Schaffer, Office of Extramural Research, National Institutes of Health, has patiently answered the numerous questions I have posed. I also want to thank sales representatives from various equipment companies who good-naturedly got back to me with prices of equipment they knew I would never purchase. They are too numerous to name.

Several scientists proved particularly helpful. They include Fran Berman at Rennselaer Polytechnic Institute, Kathy Giacomini at the University of California, San Francisco, Francis Halzen of the University of Wisconsin—Madison, the late Bill Nelson at Georgia State University, David Quéré at École Superieure de Physique et Chimie Industrielles of France (ESCPI), Amy Rosenfeld at Stritch School of Medicine, Loyola University Chicago, and B. C. Wang at the University of Georgia. Chris Liu, a faculty member at the Rotman School of Management, University of Toronto, who has a PhD in biochemistry, also provided helpful suggestions.

The actual manuscript benefited considerably from careful reading by numerous colleagues. Four undertook the task of reading the entire manuscript. They are Richard Freeman (who read it twice), Francesco Lissoni, Henry Sauermann, and

Reinhilde Veugelers. The manuscript is infinitely better because of their insights and suggestions. Others gamely volunteered, or were recruited, to read certain chapters. They include Ron Ehrenberg, Mary Frank Fox, Chiara Franzoni, Howard Garrison, Aldo Geuna, Sharon Levin, Chris Liu, Amy Rosenfeld and Marie Thursby. Thank you! The manuscript also benefitted from two anonymous reviewers recruited by Harvard University Press. It goes without saying that I take full responsibility for all errors.

Throughout the process, I have been fortunate to work with Elizabeth Knoll at Harvard University Press. She sets the standard for editors: she has always been available with quick feedback, reading each chapter within a week of receiving it and dishing out both praise and caution when she thought it appropriate. Thank you, Elizabeth!

I have also been fortunate to have the support of friends, some of whom are neither economists nor particularly interested in the topic. They include Jim Gibbons, Françoise Palleau-Papin, J Stege, Laraine Tomassi, Dave Wolbert, and Kun (Quin) Zhang.

Last, but by no means least, there is my family. First, my son, David Amis, did an expert job of turning Erin Coffman's meticulous work into the final graphic product in this book. The manuscript also benefited from his careful reading and suggestions. Second, I am grateful to Jonathan DeLoach, David's partner, who good-naturedly suffered through countless family get-togethers when the book was the topic of discussion. Finally, there is my husband, Bill Amis, without whom I could never have begun or completed the book. Bill, professor emeritus of sociology, Georgia State University, has spent countless days reading the manuscript, making edits, and organizing the endnotes—always adjusting his schedule to meet that of the book's. Most importantly, he has provided the energy and support needed to complete the book. I will be forever thankful that our paths crossed forty years ago when I came to Georgia State University, just out of graduate school, and that he followed up on that less than auspicious first encounter. To Bill I dedicate this book.

Index

pharmaceuticals: contribution to increase in life expectancy, 7, 208; public research and, 207–208, 212, 215; and cardiovascular disease, 208; slowed pace of discovery, 224–225; and NIH, 224

PhD recipients, United States: by gender, 152–153; by race, 152; by number, 152–153, 187–188; by citizenship status, 152–153, 187–189; placement of, 158–161

PhDs: cohort effects and, 110, 174–176; information concerning jobs for, 151–163; number in science and engineering in United States, 152; scientists, number in United States, 152; reasons for getting degree, 152–158; market for obtaining, 152–158, 188, 195; earnings relative to bachelor degree holders, 153–155; time to degree, 156; present value of earnings of, 156–157; support for graduate study (*see* research assistantships and fellowships); draft deferment for study, 158; job availability for new, 158–161; part-time, out of labor force, unemployed, 160–161; placement in industry, 161, 222, 162; underrepresented minorities, 163; shortage, predictions of, 164–165; definite plans upon graduation, 168–169, 222; glut of 230; inefficiency and production of scientists and engineers, 230–232

PhDs, recent: placement in industry, 216, 218–221; placement with R&D intensive firms, 222; placement in industry (by state and city), 223; stay rate in state for those working in industry, 223

physical sciences: earnings of faculty in, 37–40; inequality of salaries in, 41; patenting, motives for, in, 50; hours worked by postdocs in, 69; postdocs working in, 70, 157; collaboration with authors outside United States, 73; expenditures for equipment in, 85; research space for, 106–107; share of federal research funds, 128–129, 237; and "Gathering Storm" report, 143; PhD earnings relative to BA earnings, 154–155; time to degree, 156; increase in men getting degrees in, 158; foreign born in PhD programs, 188; foreign born postdocs in, 193; support for during Cold War, 195; foreign born in faculty positions in, 196; displacement of citizens by foreign born, 199

physics: labs in, 68; coauthorship patterns in, 72–73; role of equipment in, 83–84; research space for, 107–108; number of doctorate granting programs in, 114–115; economic contributions of, 146, 205–206, 223, 236; market for new PhDs in 1970s, 158–159; job outcomes of PhDs trained in, 160–161; PhDs working in industry, 163, 218–220; postdoc position as indication of market softness, 168; aspirations of students in, 170; faculty at Stanford University in, 183; foreign born in faculty positions in, 186; hires in 2005, 186; funding for research in, 223, 236

PI. *See* principal investigator

Picower Foundation, 121

Piled Higher and Deeper (PHD), 163, 286n39

Pine, Alexander, 67

placement information for graduate students, 161–162, 233

plagiarism, 26, 27

planaria, 100, 283n87

policy issues: related to prizes and cumulative advantage, 33; related to the scientific commons, 57–59; related to intellectual property rights and the growth of collaboration in science, 80; related to the availability of equipment and materials, 108–109; related to amount and mix of funding for research, 145–147; related to size and structure of research grants, 149; related to the market for scientists and engineers, 180–181; related to the foreign born in the United States, 199–202; related to universities and economic growth, 223–225; possible solutions for a more efficient allocation of resources, 232–235

Pomerleau, Dean, 52

postdoctorates (postdocs): cost of, 68–69, 122; staffing of faculty labs by, 68–71; fellowships for, 69, 168; selection by PI, 69, 168; authors of papers in *Science*, 70; number of, 70, 166–167, 192–194; Sigma Xi survey of, 71; problems with enumeration of 166; market for training of, 166–169; and NIH doubling, 167; number by field, 167; by citizenship status, 167, 187, 192–194, 199–200; stipends for, 167–168, 288n61; propensity to take, factors affecting, 168–169; local associations of, 169; unionization of, 169; and H-1B visas, 291n4